# Processing Tropical Crops

*A Technical Approach*

**J.J. ASIEDU**

***Senior Lecturer in Processing Engineering, Physics and Mathematics, Bunda College of Agriculture, University of Malawi***

United Book Prints

*First reprint in India, 2017*

*Published by:*
United Book Prints
4806/24, Ansari Road,
Daryaganj,
New Delhi - 110 002
Tel.: 011-41511055
E-mail: unitedbookprints@gmail.com

*Print: 2017*

ISBN: 978-93-83692-75-0



Printed in India

# Contents

## Soya bean



## Yam



## Acknowledgements

This book, the product of three years of intensive research, would not exist in its present form without the active assistance and co-operation of many people and many organizations.

First of all I would like to record my appreciation for the courtesy shown by the staff of Macmillan Publishers Ltd in their co-operative and efficient handling of the project. Many thanks are also extended to the Bunda College Library staff for making available to me all the books and scientific papers relevant to the subject matter. I am also indebted to Dee, who on the one hand provided me with all the 'nutrients' necessary for this type of task and on the other hand patiently endured the many weekends and evenings that I had to be away during the gestation of this book.

In this book, despite the advice received from many experts, I have often in the final analysis opted for my own style and interpretation of facts and I remain solely responsible for the selection and mode of presentation of the material, including possible errors of facts, interpretation or emphasis that might remain.

For permission to reproduce illustrations and use copyright material I am grateful to the following:

Food and Agriculture Organization of the United Nations (FAO) for numerous figures from the following books:

—Coconut Palm Products by B. E. Grimwood.

—Coconut Oil Processing by J. G. Thieme.

—Coffee Processing by Rene Wilbaux.

Also, data on world production and hectarage of various crops were obtained mainly from FAO production yearbooks.

AVI Publishing Company, 250 Post Road East, P. O. Box 831, Westport, Conn. 06881 USA, for Figures 5.1, 5.2 and 5.3, Peanuts: production, processing, products 3rd Edition, Woodroof; for Figure 3.25, Coconuts: production, processing, products, Woodroof.

Longman Group Ltd for Figures 2.5, 2.6 and 2.7.

Cocoa 3rd Edition, by G.A.R. Wood: Longman Tropical Agriculture Series.

Springer Verlag Berlin/New York/Heidelberg for Figures 7.7, 7.8, 3.19 and 3.20.

While every effort has been made to acknowledge sources of information and any other copyright material, omissions may have been made inadvertently. For these I sincerely apologize.

# Preface

There is a considerable body of literature which would encourage us to believe that the food problems in the developing countries can be reduced, even eliminated, by intensifying agricultural production. Emphasis in agricultural research for many years has therefore been mainly on increased crop production. In fact, certain countries have experienced considerable increases in food production in the past years through the application of improved seeds, improved cultivation methods, fertilizers and irrigation. The term 'green revolution' was first used in 1968 to describe the increases in the production of wheat, rice and maize. However, the brilliant results of the green revolution did not alleviate the population – food imbalance problem. This has been so partly because the increase in food production is sometimes nullified by postharvest losses. Evidence from both scientific and empirical research suggests that about 25 per cent of food grown in the tropics is lost before it can be utilized. Indeed any postharvest losses do not only represent the waste of good arable land on which that food was grown but also represent the waste of all the human, energy and other inputs expended in the cultivation and harvesting of the crop.

Processing of crops is a widely acknowledged alternative approach which can be used as a significant means of reducing postharvest losses. In fact this acknowledgement is not new. In Africa and in Central and South America the preparation of storable maize or cassava products through fermentation is a widespread tradition. This is also true for Southeast Asia and China in the case of the preparation of soya sauce. This acknowledgement, however, has become more important in recent years with the realization that increased food production alone is not enough to challenge the population – food imbalance. The reduction of postharvest losses and production of convenient, safe and wholesome food through processing depend upon the proper use of existing traditional methods and recent developments derived from a broad spectrum of scientific disciplines. Biochemistry, physiology, physics and engineering all provide knowledge which has been useful and will continue to be useful in the future for improving traditional methods.

Reading through the literature on processing of tropical crops, it soon becomes apparent that a great amount of information has been scattered in a multitude of journals. Moreover, one rarely comes across a processing book that deals with many crops at a time in sufficient depth. It is against this background that this book has been written to compile and evaluate and to complement, where necessary, the available information on the processing and physico-chemical properties of ten selected crops.

This book describes the existing traditional methods of processing various tropical crops with detailed graphic flowcharts and the technology and equipment involved in the processes. The book also provides information on the chemical and physical properties of the respective crops. To a lesser extent advanced processing methods are included. The book offers food technologists, food engineers, biochemists, development aid workers, manufacturers of processing equipment and food scientists a useful source of information on the processing possibilities and potential of the selected crops. It should also serve as a useful text for a variety of science courses including Food Technology, Food Engineering and Chemistry at the undergraduate level.

Recent references have been cited at the end of each chapter to establish authority and to assist the reader who is interested in studying the related topics in more detail.

Given the importance of both the reduction of postharvest losses and the production of convenient wholesome and nutritious food, and the considerable contribution of these aspects to the achievement of major socio-economic objectives in the developing countries, it is hoped that this book will help to strike 'chords of awareness' amongst researchers, manufacturers, policy makers and others who are interested in improving the standard of living of the many less privileged people in the world.

Dr. J. J. Asiedu
Prof. Dr.-Ing. F. Wieneke

# 1

# Cassava

Cassava, *Manihot esculenta* Crantz (syn. *Manihot utilissima* Pohl), is a dicotyledonous perennial plant belonging to the botanical family Euphorbiaceae. It is a starchy root crop (Figure 1.1) that is grown almost entirely in the hotter lowland tropics. The crop is also known under a variety of names according to the region in which it is cultivated; cassava in the English-speaking countries of North America, Europe and Africa, manioc in French-speaking countries, tapioca in the English-speaking countries of Southeast Asia, mandioca in Brazil and yuka in the Spanish-speaking countries of South America. The total amount of cyanogenic glycosides in cassava roots is often used to place the numerous cassava cultivars into two major groups: the 'bitter' varieties, in which the cyanogenic glycosides are distributed throughout the tuber and are at a 'high' level, and the 'sweet' varieties, in which the glycosides are confined chiefly to the peel and are at a 'low' level.

## Main utilization

### *Human food*

Cassava is grown mainly in the equatorial belt bounded by latitudes 30°N and S and is restricted to zones less than 2000 m above sea level, receiving an annual rainfall of 200–2000 mm. Within this region, cassava is an important staple for about 800 million people (Truman, 1982). The global cassava production and consumption projections for 1985 and 1990 are 78 809 and 124 987 and 88 715 and 135 513 (figures in 100 tonnes) respectively (FAO, 1980). Annual per capita consumption of cassava is greatest in Africa, averaging 102 kg/year. In the Central African Republic, Congo, Gabon and Zaire, consumption exceeds 300 kg/person/year. In Latin America the average per capita consumption is about 35 kg/year (Phillips, 1982).

Fresh peeled cassava is eaten as a vegetable after boiling or roasting. In some West African countries, cassava is boiled and pounded with boiled plantain, to form an elastic dough called fufu, which is consumed with vegetable and meat soups. Peeled cassava roots are also often sliced, dried and ground into flour. Kokonte in Ghana is prepared in the same fashion except that the sliced, dried roots are not ground immediately after drying. The main form in which cassava is eaten in West Africa is a roasted granular product, prepared from peeled, grated and fermented cassava roots, known as gari. In South America and Latin America, a product called farinha de mandioca is very popular. It is similar to gari, except that considerably less fermentation is allowed to take place in its preparation.

Chickwangue is another African product, prepared by soaking the cassava

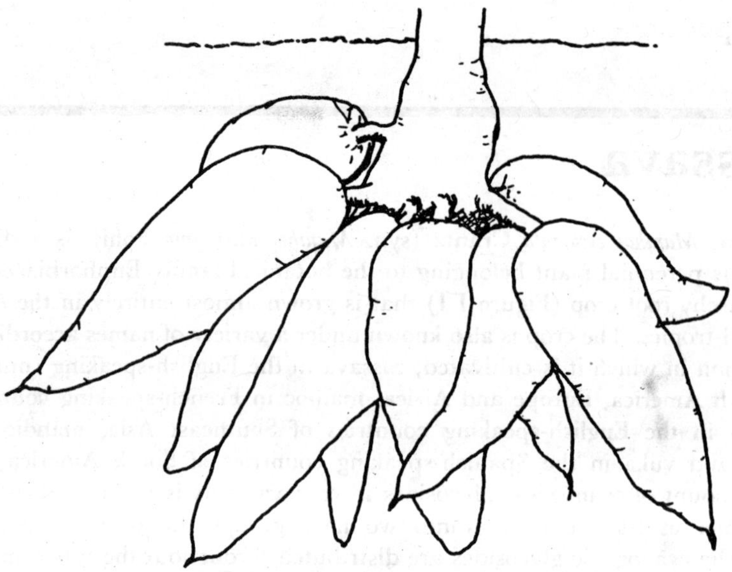

**Figure 1.1 Cassava tubers.**

in water for 2–7 days until it softens, after which the root is peeled and mashed. The fibres are removed, and the resulting paste of firm and elastic consistency is wrapped in palm or banana leaves. In the Philippines cassava is made into landang or cassava rice. Other products made from cassava include beer, biscuits and cakes.

### *Animal feeds*

Pellets and chips processed from cassava are a source of energy in animal feeds. For the production of chips, the fresh roots are washed, peeled and cut into slices 3–6 cm long. The slices are then dried on large concrete surfaces in the open air. Pellets are made from chips; dried chips are ground and hardened into cylindrical pellets about 2 cm long and up to 1 cm in diameter. The most prominent exporting countries of cassava chips and pellets are Thailand, Malaysia and Indonesia, exporting mostly to the European Community and the United States of America.

### *Industrial products*

Cassava is an important raw material for the non-food industries. The low amylose, high amylopectin content of cassava starch give it the necessary viscosity for high quality adhesives and for use in the paper and textile industries. Cassava starch is also used for the production of dextrins, which are utilized in glues. Another industrial product made from cassava is ethyl alcohol (ethanol).

## Composition

Depending on variety, age and maturity, the edible flesh of cassava makes up 80–90 per cent of the root. The chemical composition of peeled cassava varies with maturity, variety, cultural practices, storage environment and region. Cassava roots contain 30–40 per cent dry matter, of which starch and sugar are the predominant components (Table 1.1). The roots contain significant amounts of vitamin C, about 35 mg per 100 g fresh weight (Onwueme, 1978).

The amount of protein in cassava is low, and probably because of this kwashiorkor is believed to be prevalent in areas where cassava is the staple item of diet. The main amino-acids present in cassava proteins are arginine, histidine, isoleucine, leucine and lysine. Sulphur amino-acids are deficient. In addition to these constituents, cassava roots can contain a concentration of prussic acid in the range of 10–490 mg/kg root.

**Table 1.1 Composition of peeled cassava tubers**

| Constituent | Peeled tubers, total matter (%) | Dry matter (%) |
|---|---|---|
| Water | 66.2 | – |
| Starch | 27.5 | 81.5 |
| Sucrose | 1.0 | 3.0 |
| Glucose | 0.4 | 1.1 |
| Fructose | 0.3 | 0.8 |
| Protein | 0.4 | 1.3 |
| Fats | 0.2 | 0.6 |
| Minerals | 0.8 | 2.5 |
| Dietary fibre | 1.5 | 4.3 |
| Remainder | 1.7 | 4.9 |
| Cyanide (HCN) (ppm) | 150–300 | 300–900 |

Source: Meuser and Smolnik, 1980.

## Toxicity and detoxification

It is known that cassava contains two major cyanogenic glycosides: linamarin and lotaustralin. Both glycosides are hydrolysed to produce hydrocyanic or

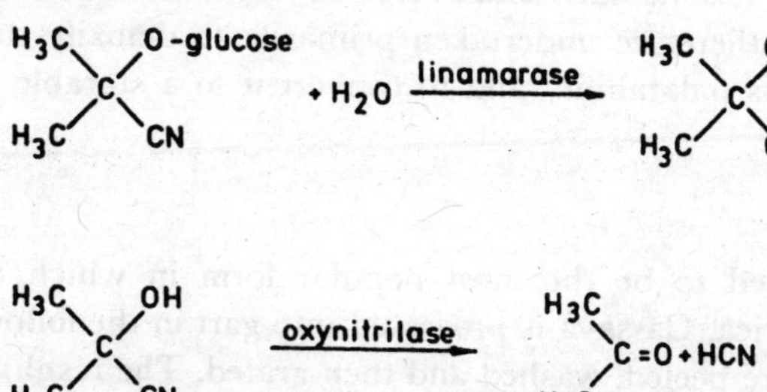


**Figure 1.2 Enzymatic degradation of linamarin (Conn, 1980).**

prussic acid (HCN), a poison, when they come in contact with the enzyme linamarase, which is released when the cells of cassava roots are ruptured. Figure 1.2 illustrates the degradation of linamarin and the subsequent production of cyanide. Cyanide toxicity in humans and animals on cassava diets is a well-recognized problem (Oke, 1968; Osuntokun *et al.*, 1969; Coursey, 1973; Erman *et al.*, 1980).

The presence of cyanogenic glycosides in most cassava cultivars necessitates a certain degree of detoxification before the roots can be consumed. The prussic acid in particular is lethal if more than about 0.1 g of it is contained in the food eaten by an individual at any one time (Onwueme, 1978). In general, three methods of detoxification are employed: (a) microbial detoxification through fermentation; (b) decomposition of the glycosides by heating them above 150°C; and (c) rupture of the roots to allow intimate interaction between linamarase and the glycoside, then expressing or volatilizing the resultant products of hydrolysis.

All forms of cassava processing therefore only decrease the levels of cyanogenic glycosides and prussic acid in the final product. For example, boiling is said to destroy the enzyme linamarase and eliminate the prussic acid. However, the linamarin itself is not destroyed by boiling, and its long-term ingestion may lead to cyanide toxicity in humans whose diets lack sufficient protein and iodine (Cock, 1985).

## Processing of cassava tubers

After harvesting, cassava roots are susceptible to spoilage, and without any preservation measures can only be stored for about 48 hours before they begin to deteriorate. Post-harvest deterioration of cassava is related to two separate processes: physiological changes and microbial changes. Physiological deterioration often sets in within 24 hours after harvest, while microbial deterioration usually begins within a week (Cock, 1985). Therefore the roots must be processed as soon as possible after harvest to arrest the physiological process and the subsequent deterioration.

Other factors favouring the processing of cassava are that the processed products are easier to store than raw cassava, they need less storage space and they can be stored for longer periods. For example gari, a dried cassava product, is less affected by biochemical changes during storage and in transit, and therefore does not lose its nutritional value as fast as fresh cassava does in storage. Processing is therefore undertaken primarily to detoxify the cassava product, to improve its palatability and to convert it to a storable form.

### *Gari*

Gari can be considered to be the most popular form in which cassava is consumed in West Africa. Cassava is processed into gari in the following way. The harvested roots are peeled, washed and then grated. The resulting pulp is put into a cloth bag and subjected to pressure by heaping stones on the bag. After pressing, the bag is left for 2–4 days, during which time the pulp

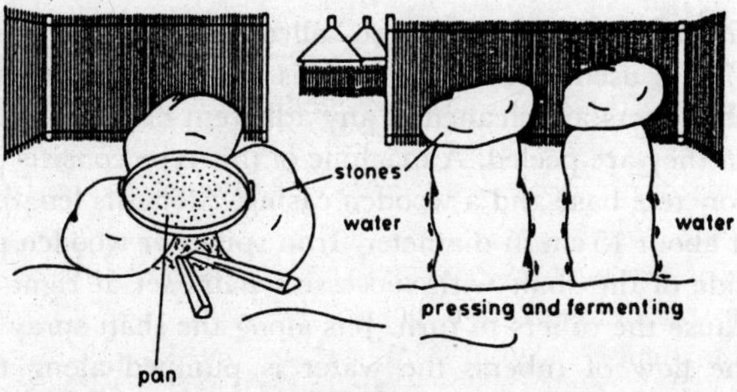


**Figure 1.3 Traditional processing of gari.**

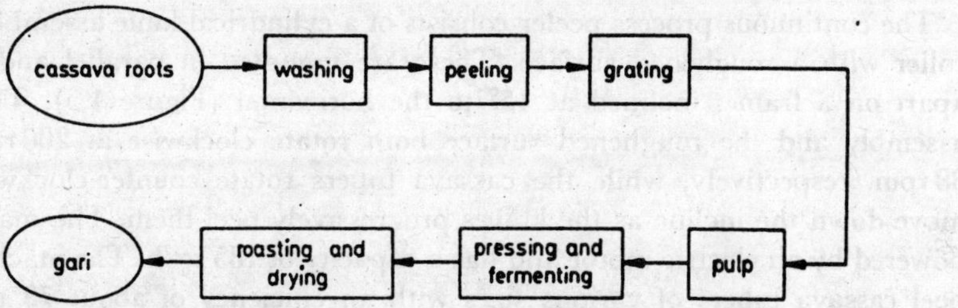


**Figure 1.4 Flowchart for gari processing in rural areas of West Africa.**

ferments. Most of the juice from the cassava pulp is expressed from the bag during this period. The fermented pulp is then removed from the bag, sieved and roasted or fried in wide, shallow metal pans, until dry (Figure 1.3).

The processing of cassava into gari involves several unit operations: peeling, washing, grating, pressing and fermenting, sieving, roasting and drying (Figure 1.4). Traditional gari production is laborious and time-consuming; on average it takes about 90 hours to process 100 kg of gari. About 65 per cent of the total time is spent on peeling and 25 per cent on roasting (Williams, 1979). Gari occupies an important place in the diets of the people of West Africa. It is customarily consumed in the form of meal, which is prepared by soaking the gari in water to swell the starch, and by making the swollen meal into a dough. The dough is then made into a ball with the fingers and is dipped into a stew containing ingredients such as palm oil, vegetables, meat or fish. Gari may also be eaten without a stew or soup by soaking it in cold water and adding sugar or milk.

## Peeling of cassava tubers

Peeling of cassava tubers presents a considerable problem in cassava processing. Traditionally, peeling is accomplished by hand; the roots are cut longitudinally and transversely to a depth corresponding to the thickness of the peel, which can then easily be removed. The structure of the root – the irregular shape and size – does not permit easy mechanical peeling.

However, in larger factories, the so-called washer-peeler, as described by Edwards (1974) is used. This equipment is divided into two sections; in the first section the tubers are cleaned of any adherent dirt and sand, while in the second section they are peeled. A machine of this type consists principally of a tank with a concrete base and a wooden casing. Along its length runs a hollow rotating shaft about 15 cm in diameter. Iron spikes or wooden paddles project from either side of the shaft with successive pairs set at right angles to each other; these cause the tubers to turn. Jets along the shaft spray water counter-current to the flow of tubers; the water is pumped along the shaft by a centrifugal pump. In the peeling section, the tubers are moved continuously along under the spray of water and are peeled by the friction and the action of the rotating spikes.

The continuous process peeler consists of a cylindrical knife assembly and a roller with a roughened surface. These are mounted in parallel and 20 mm apart on a frame, inclined at 15° to the horizontal (Figure 1.5). The knife assembly and the roughened surface both rotate clockwise at 200 rpm and 88 rpm, respectively, while the cassava tubers rotate counter-clockwise and move down the incline as the knives progressively peel them. The machine is powered by an electric motor and has a capacity of 185 kg/h. The machine can peel cassava tubers of various sizes with an efficiency of about 75 per cent (Odigboh, 1976a). The requirement that the roots must be trimmed, cut into straight lengths of approximately 100 mm and sized, is the major disadvantage of this peeler.

The batch abrasion peeler consists of a drum mounted eccentrically on a shaft on a frame (Figure 1.6). The drum is rotated at 40 rpm through V-belt drives from an electric motor. Unpeeled tubers are loaded into the drum

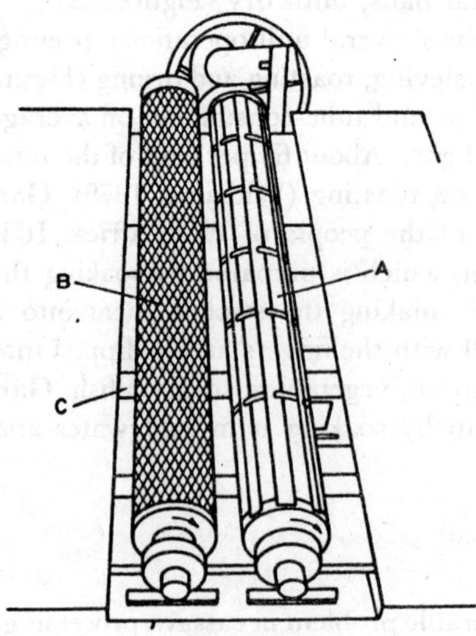


**Figure 1.5 Action zone of cassava peeler (Odigboh, 1976).**

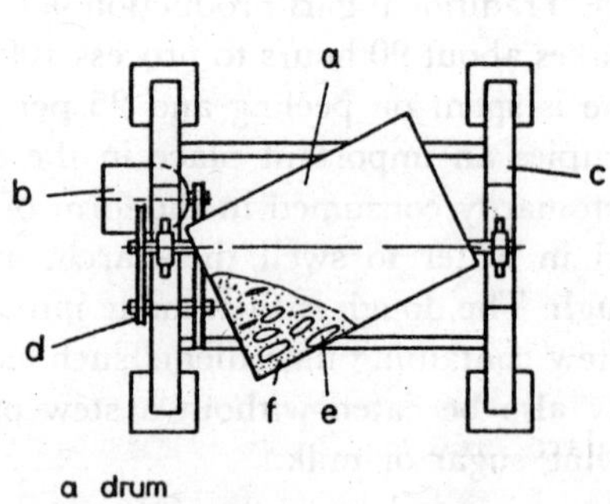


**Figure 1.6 Batch abrasion peeler (Odigboh, 1976).**

together with an inert abrasive material, for example, quarry sand. The drum, with its load of tubers, abrasive mixture and added water, is rotated until the cortex is completely removed from the tubers.

The maximum load of cassava roots per batch is reported to be 50 kg. The capacity of a peeler with a 210 litre drum is about 180 kg/h, excluding loading and unloading times (Odigboh, 1976b). The need for large quantities of water is a disadvantage.

## Rasping or pulping

The peeled tubers are ground to a homogeneous mixture by means of a mechanical action. This is carried out by slicing the tubers and then rasping, grating or crushing them into a fine pulp. At the village level the tubers are rasped by hand on bamboo mats. A simple but effective grater is obtained by perforating a sheet of galvanized iron with a nail and then clamping it around a wooden wheel with the sharp protruding rims of the nail openings turned outwards. The wheel may be driven by hand or by foot like a bicycle, while the worker presses the tubers from above onto the rasping surface. Otherwise the rasping surface can be attached to one side of a rotating disc equipped with a crank transmission driven by foot.

For medium and large-scale production, for example for the handling of 10 tonnes or more of fresh tubers a day, engine-driven raspers are used. The Jahn rasper, as described by Grace (1977) consists essentially of a rotor of hardwood or drawn steel tube, 50 cm in diameter, with a number of grooves milled longitudinally on which the rasping blades or saws are mounted. The number of saw teeth on the blades varies from 10–12/cm according to need. The blades are spaced 6–7 mm apart on the rotor.

## Fermentation and pressing of cassava pulp

The grating or rasping of the cassava root enhances contact between the enzyme linamarase and the cyanogenic glycosides, so that most of the glycoside can be hydrolysed to prussic acid. The pressing serves to remove the juice containing prussic acid, while the toasting–drying stage is aimed at vaporizing most of the remaining prussic acid.

The micro-organisms responsible for cassava fermentation are indigenous to the roots. Collard and Levi (1959) identified *Corynebacterium manihot* and *Geotricum candida* as the micro-organisms involved in a process similar to that of gari manufacture. More recent studies have questioned the validity of Levi and Collard's results. Ngaba and Lee (1977) isolated and identified *Lactobacillus* sp. and *Streptococcus* sp. as responsible for cassava fermentation and the subsequent acidity and flavour of gari. Ejiofor and Okafor (1981) observed, among others, high numbers of *Candida* and *Geotricum* spp. in pressed cassava pulp during the later stages of fermentation.

According to Meuser and Smolnik (1980) fermentation of cassava pulp at 37°C is complete within a maximum of 5 days. The water binding capacity of cassava mash is sufficiently changed only after about 3 days, after which the

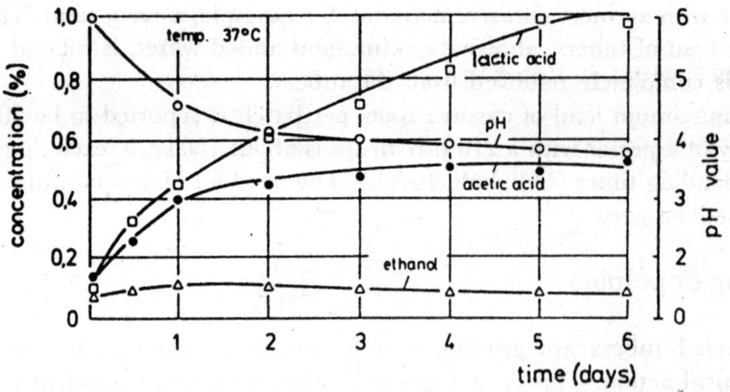


**Figure 1.7 Decomposition of soluble carbohydrates during fermentation of cassava pulp at 37°C (Meuser and Smolnik, 1980).**

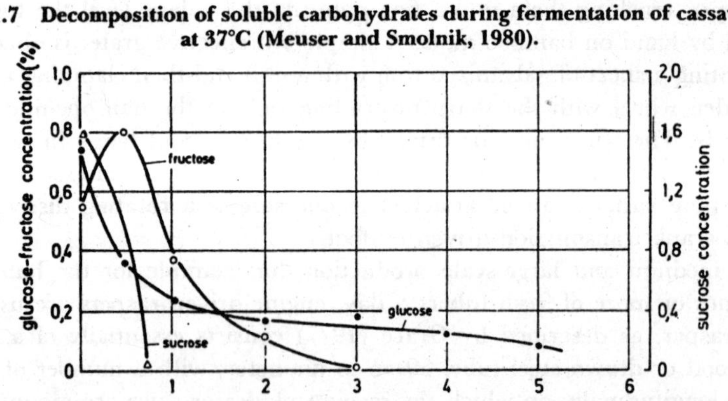


**Figure 1.8 Fructose, glucose and sucrose contents during fermentation of cassava pulp at 37°C (Meuser and Smolnik, 1980).**

fruit water can be pressed out to a reasonable degree. For this reason an accelerated fermentation would be of limited use. The decomposition of the soluble carbohydrates (sugars) to lactic and acetic acids and ethanol is illustrated in Figure 1.7.

The changes in concentration of fructose, sucrose and glucose during fermentation of cassava pulp are shown in Figure 1.8. The fermentation rate of sucrose is the highest. Within 18 hours of the start of fermentation the sucrose is totally hydrolysed. Fructose increases in the first 18 hours and falls steadily to a minimum after 72 hours of fermentation. Unlike sucrose and fructose, a certain amount of glucose still remains after a fermentation period of 5 days.

## Roasting and drying during gari production

The sieved cassava meal is roasted over a fire in the traditional method. Constant stirring is carried out during the roasting to circumvent burning and to prevent the meal from forming lumps. In the roasting process, some of the starch granules are gelatinized and aromatic compounds are formed from saccharides and soluble nitrogenous substances, so the fermented product has a characteristic taste and aroma.

**Figure 1.9 Drying cassava meal in the sun in Cameroon.**

Following roasting, gari is dried so that it can be stored for long periods and to further reduce the hydrocyanic acid content. Traditionally the drying of gari is done in the sun (Figure 1.9).

In a process developed by Meuser and Smolnik (1980) for mechanizing the production of gari, different drying systems were used to investigate the effect of the various systems on the hydrocyanic acid content of the product. The results of their investigation (Table 1.2) indicate that freeze drying or flash drying eliminates only the free acid, which in the study accounted for 50 per cent of the total cyanide present. In most of the methods used to process cassava into gari, there is usually some residual cyanide, the amount of which, among other factors, depends on the drying temperature. Drying cassava pulp in an air stream at 60°C results in a product with 6 ppm HCN residue, whereas pulp dried at 40°C and 22°C produces gari containing 13 ppm and 143 ppm HCN residue, respectively (Figure 1.10).

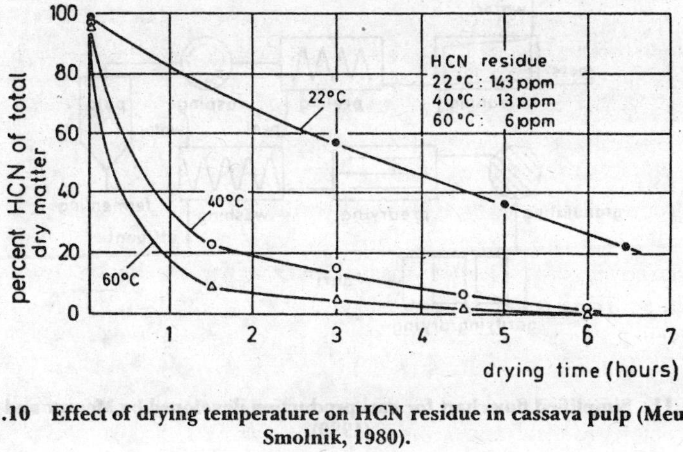


**Figure 1.10 Effect of drying temperature on HCN residue in cassava pulp (Meuser and Smolnik, 1980).**

**Table 1.2 Effects of different drying techniques on the hydrocyanic acid content of cassava**

| Drying technique | | HCN (ppm dry matter) |
|---|---|---|
| Freeze drying | (pulp) | 439 |
| Flash drying | (slices) | 432 |
| Air drying at 40°C | (chips, pulp) | 13 |
| Heated air drying | (chips) | 14 |
| at 180°C | (fermented pulp) | 77 |
| Drum drying | (pulp) | 8 |
| | (fermented pulp) | 121 |
| HCN of pulp | (free and bound) | 900 |

Source: Meuser and Smolnik, 1980.

## Mechanization of gari production

In more recent times, most of the unit operations in gari manufacture have been mechanized, and a couple of mechanized gari plants have been established (Purcel and Williams, 1973; Wadwa, 1973; Cock, 1985). The mechanized process is similar to the traditional method. The fresh cassava is weighed, washed, peeled, grated and pressed. After pressing, the pulp is sieved, roasted and dried. The final product, gari, may be separated by sieving into coarse, medium and fine fractions.

Based on laboratory and industrial pilot plant experiments, Meuser and Smolnik (1980) developed a mechanized process for gari manufacture. A flow-chart for the plant using this process is shown in Figure 1.11. The cassava roots are cleaned and peeled mechanically. After peeling they are rasped to a pulp, which is then fermented anaerobically at 37°C, with the sugars being decomposed to lactic acid, acetic acid and a small amount of ethanol. After

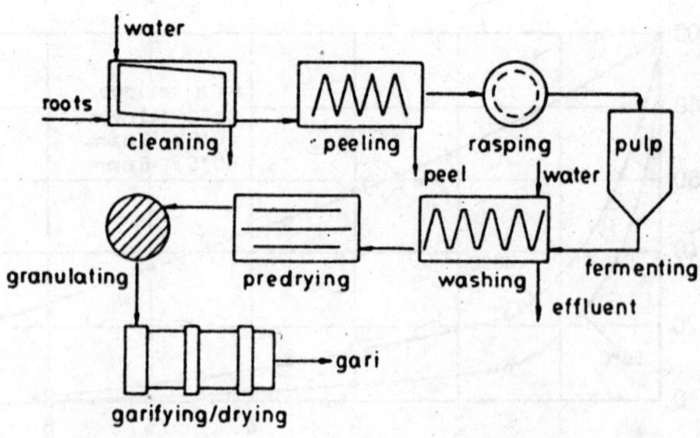


**Figure 1.11 Simplified flowchart for gari production developed by Meuser and Smolnik (1980).**

fermentation, which usually takes 3–5 days, the pulp is washed and drained mechanically to remove the cyanide and other soluble substances. The drained pulp is then predried and granulated. The product is dried again and may be separated into grits of any granular size. From this manufacturing process, gari with a moisture content of less than 12 per cent and a residual hydrocyanic acid content of 10 ppm, is reported to have been obtained. The product is storable for more than a year without any change in taste and aroma. The washing of the pulp causes a loss of about 50 per cent of the nitrogen containing substances and 40 to 70 per cent of the minerals.

### *Farinha de mandioca and cassava bread*

Farinha de mandioca is the most common form in which cassava is consumed in Brazil and many other South American countries. The preparation of farinha de mandioca consists of peeling, grating, pressing, mixing, heating and roasting. The cassava roots are usually peeled with a knife and then grated. The grater used by South American Indians consists of a board upon which a number of flints are embedded in a thin layer of wax. Graters consisting of a wooden board with wooden teeth are also in use. Hand-driven rotating rasps, similar to the raspers in rural mills, are used to do the rasping in relatively more developed regions. After rasping or grating, the cassava pulp is packed in leaves and pressed under a heavy stone, sometimes with the aid of pole leverage, or in a wooden screw press. The traditional equipment for this purpose in Latin America is known as a tipiti. A tipiti is a basket cylinder which has been specially woven so that it can take on both a long and thin as well as a short and bulky form. In the latter form the basket is packed with fresh cassava pulp and hung from the bough of a tree. This 'sock' is pulled until it is long and thin; during this operation the pulp is subjected to considerable pressure, which drains off a substantial proportion of the cyanide-containing juice.

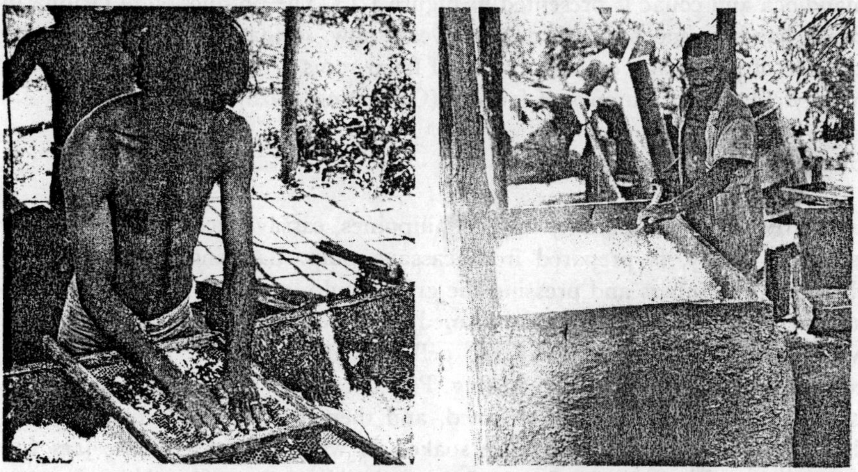

**Figure 1.12 Preparation of farinha de mandioca in Brazil, (a) sieving and (b) baking.**

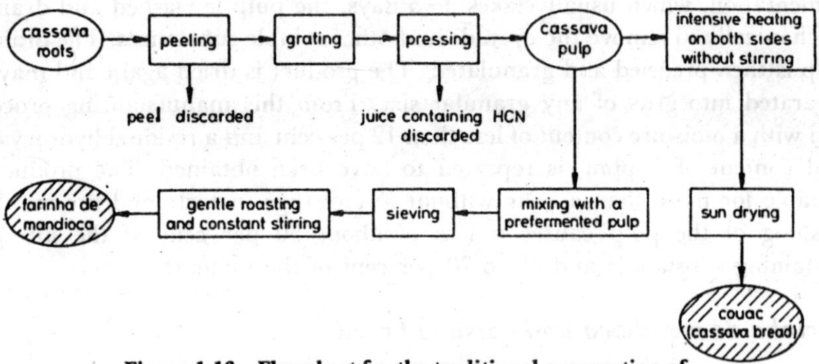


**Figure 1.13 Flowchart for the traditional preparation of farinha de mandioca and couac.**

The drained pulp from the tipiti may be used either to prepare farinha or couac (bread). For farinha the pulp is mixed with 3 day-old prefermented pulp, pounded together and rubbed through a sieve (Figure 1.12a). The result of this operation is a slightly damp meal which is then heated in an open pan. The pulp is turned continuously with a wooden rake during the 3 to 4 hours of baking, which yields a granular, only slightly roasted product (Figure 1.12b). Farinha de mandioca is thus very much like gari in its mode of preparation. The only major difference is that the pressing and fermentation times for farinha are considerably shorter. In Latin America, farinha has the same status as an excellent cereal and is usually eaten with meat and gravy.

Couac (bread) is obtained by heating the pulp more intensively, without stirring, until the mass is slightly brown on one side. After treating the other side, the product is then dried in the sun, and in this state it can be stored for a long time. This cassava bread, or couac, is very hard, and is usually eaten after being dipped in gravy. A flowchart for the preparation of farinha de mandioca and couac is presented in Figure 1.13. The production of farinha in Brazil has been mechanized for some years now, so that a processing plant is capable of processing over 30 tonnes of fresh roots a day. One tonne of fresh roots yields 0.3–0.4 tonnes of farinha (Onwueme, 1978).

## *Cassava rice – landang*

In Southeast Asia, especially in the Philippines, cassava rice, or landang as it is called locally, is prepared from cassava pulp. Landang is prepared by shredding the tubers and pressing the grated pulp in a cloth until most of the juice is squeezed out. Pellets are formed after whirling the drained mass in a winnowing basket. The size of the pellets is determined by their moisture content and the speed of the whirling. Pellets of more or less uniform size are isolated by sifting. They are steamed, and then dried in the sun for several days. Alternatively the tubers are soaked in water in eathernware jars (no contact with metal) until, after 5–7 days, they begin to soften. Next they are macerated, the fibre is removed by hand and the mass is air-dried before being formed into pellets. The end product is a substitute for rice or maize and can be kept for up to six months before being damaged by moulds (Grace, 1977).

### *Cassava chips*

Cassava chips are pieces of dried, sliced or chipped roots not exceeding 6 cm in length, with a starch content of 70 per cent or more. The method of processing chips consists essentially of preparing the roots, slicing, and then drying the slices until they have a storable moisture content of about 12 per cent.

Preparing the roots, that is trimming, peeling and washing, may be done either manually or mechanically. By mechanical means a washing-peeling machine is used. A machine of this kind consists essentially of a rotating slotted drum and water jet. Water enters the drum through its axis and washes the tubers, while the friction and rubbing action of the roots accomplish the peeling. The working principle of such a washer-peeler is shown in Figure 1.14.

Villa *et al.* (1979) report that a washer-peeler of this design, with a drum

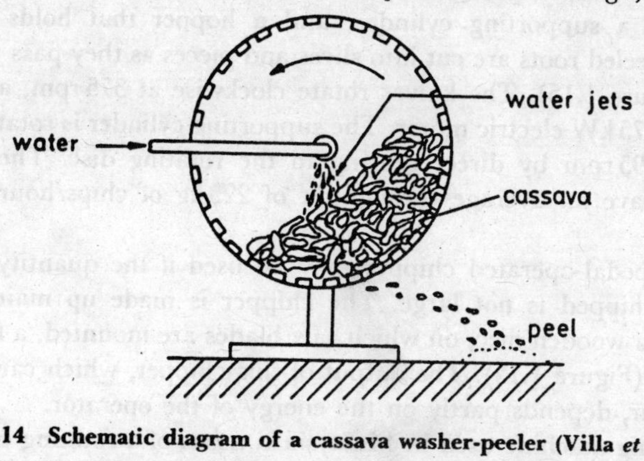


**Figure 1.14 Schematic diagram of a cassava washer-peeler (Villa *et al.* 1979).**

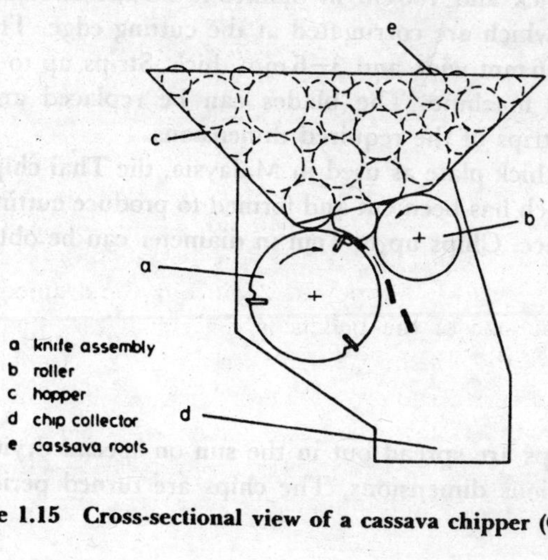


**Figure 1.15 Cross-sectional view of a cassava chipper (Odigboh, 1976).**

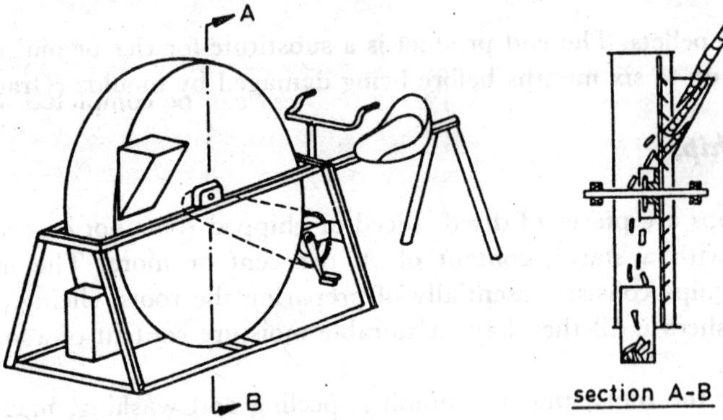


**Figure 1.16 Pedal-operated cassava chipper.**

diameter of 1 m rotating at 40 rpm, and a residence time of 60–90 seconds, peels over 90 per cent of the root surface without producing excessive losses. From the washer-peeler the tubers may be taken to an engine-driven chipping machine which consists essentially of an assembly of knives mounted on a rotating disc, a supporting cylinder, and a hopper that holds the peeled tubers. The peeled roots are cut into slices and pieces as they pass through the machine (Figure 1.15). The knives rotate clockwise at 375 rpm, and are belt driven by a 0.75 kW electric motor. The supporting cylinder is rotated counter-clockwise at 95 rpm by direct gearing to the rotating disc. The chipper is reported to have an average throughput of 225 kg of chips/hour (Odigboh, 1976b).

A simple, pedal-operated chipper may be used if the quantity of cassava roots to be chipped is not large. The chipper is made up mainly of three components; a wooden disc, on which saw blades are mounted, a frame and a bicycle drive (Figure 1.15). The output of this chipper, which can be as high as 250 kg/hour, depends partly on the energy of the operator.

The chipping machine used in Malaysia consists of a rotating circular steel plate about 12 mm thick and 100 cm in diameter. To the circular plate are fixed six steel blades which are corrugated at the cutting edge. The machine produces strips about 6 mm wide and 3–6 mm thick. Strips up to 25 cm long can be made by the machine. The blades can be replaced and may be adjusted to produce strips of the required dimensions.

Instead of a heavy thick plate as used in Malaysia, the Thai chipper uses a thin circular plate which has been cut and formed to produce cutting elements across the whole surface. Chips up to 3 cm in diameter can be obtained from the Thai chipper.

## Drying cassava chips

### *Traditional sun drying*

Traditionally, the chips are spread out in the sun on special drying areas or concrete floors of various dimensions. The chips are turned periodically for even drying.

In ideal climatic conditions, that is on sunny days with low relative humidity, drying to a moisture content of 12–13 per cent can be completed within 3–5 days. Long drying periods usually result in attacks of bacteria and mould.

Experiments in Madagascar suggest that the concentration of chips during sun drying should not exceed 10–15 kg/m$^2$, the required drying area being about 150 m$^2$ for each tonne of dried roots produced per day (Grace, 1977).

More recent research at the Asian Institute of Technology, the Tropical Products Institute and CIAT has shown that drying rates can be considerably improved by placing the cassava chips on inclined or suspended trays. This arrangement allows more air to flow through the chips. The need to turn chips periodically is eliminated. However, the capital cost of the trays is reported to be relatively high per unit of dried cassava since each tray can normally only dry one batch of cassava every 2 days (Cock, 1985).

Villa *et al.* (1979) report on drying cassava chips with a naturally ventillated vertical drier, consisting mainly of two 1.5 × 2.0 m vertical and parallel frames covered with galvanized wire screens (8 mm spacing between the 1 mm wires). The vertical drier allows drying of at least 20 kg/m$^2$ without deterioration and can be fitted with a roof to protect the product from rain.

*Solar drying*

Villa *et al.* (1979) report on drying cassava chips, using a solar drier (baracaca drier, Figure 1.17).

The drying unit consists essentially of a solar collector, a fan and a perforated bin to accommodate the chips. The solar collector (Figure 1.18) is made up of

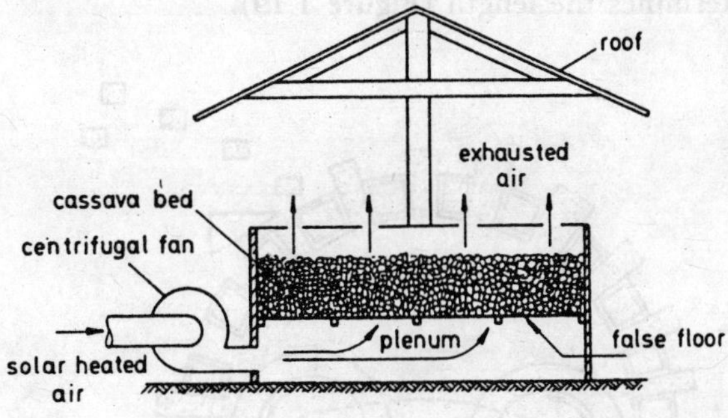


**Figure 1.17 Schematic diagram of a solar drier (baracaca drier) (Villa *et al.*, 1979).**

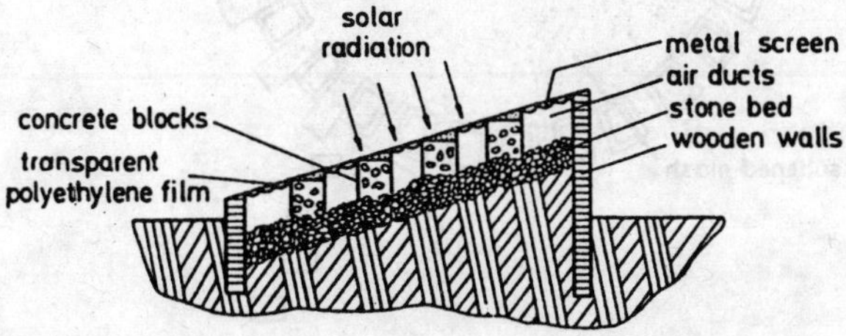


**Figure 1.18 Solar collector (Villa *et al.*, 1979).**

a black painted stone bed and bricks that collect and store the solar energy, wooden lateral walls and a white painted metal screen which supports a transparent polyethylene film located at the top of the collector.

Using the solar drier shown in Figure 1.15, 3 tonnes of cassava roots, with a moisture content of about 62 per cent when fresh, were washed and cut into chips and then dried to a moisture content of 13 per cent. The drying operation took 30 hours. The quality of the product was reported to have been considered excellent by a feed manufacturer.

A centrifugal fan was used to move 60 $m^3$/min of air through the perforated floor. The air was preheated in a 60 $m^2$ (3 × 20) plane solar collector. The average temperature and the relative humidity of the drying air were 35.2°C and 42.7 per cent, respectively. On average the increase in temperature and the decrease in relative humidity due to the solar collector were 6.6°C and 18.1 per cent, respectively.

## *Cassava pellets*

Pellets are produced by grinding the dried tubers and compressing the powder into cylindrical shapes. The cylinders are between 2 and 3 cm long and are 0.5–1 cm in diameter. A pelleting plant consists of one or a linked series of small pelleting units. The ground cassava is carried to a hopper that feeds a pelleting press via a feed worm with facilities for the input of steam. In the press, the ground material is pushed through the perforations of a horizontal die by rollers, and the material is cut into pellets as it emerges from the other side. The perforation size determines the diameter of the pellets, and the depth of the die determines the length (Figure 1.19).

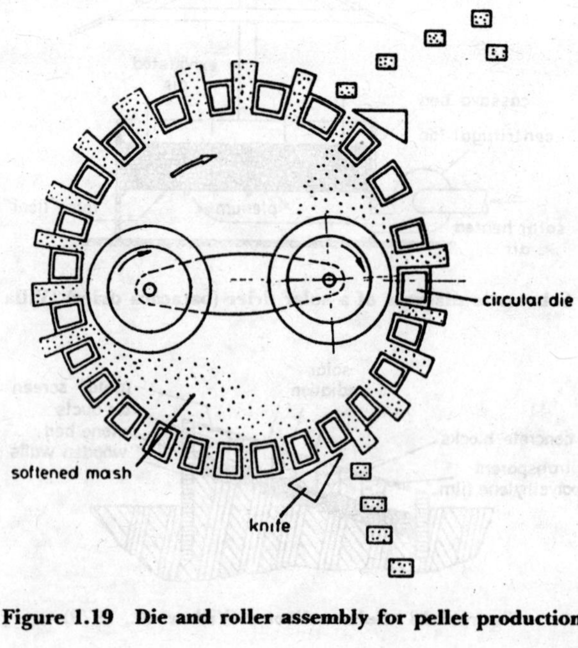


**Figure 1.19 Die and roller assembly for pellet production.**

## Baked cassava products

Tapioca is a commercial term applied to baked cassava products. In Malaysia and some other regions in Southeast Asia these products are commonly known as sago. Tapioca, in the form of flakes (irregular lumps), pearls and seeds (small round beads), is made from partly gelatinized cassava starch by heat treatment of the moist starch in shallow pans. When heated, the wet granules gelatinize, burst, and stick together. Stirring is carried out for even roasting and to prevent scorching.

In the preparation of tapioca flakes, the moist starch, with a water content of about 45 per cent, is reduced in size using a mill. The lumps produced are then rubbed through a sieve of about 20 mesh (units per square inch) to produce a coarse-grained moist flour, ready for gelatinization.

Raw starch is made up of microscopic granules held together in a network of molecules by hydrogen bonds so strong that these granules are insoluble at room temperature. At higher temperatures the insoluble starch product becomes an amorphous substance, miscible with water, yielding solutions with considerable viscosity, which after cooling set to a semi-solid elastic mass – a jelly, or gel. This process (gelatinization) may be accomplished by the action of chemicals or by heating in an aqueous medium. It is accompanied by a change in molecular arrangement and a certain amount of swelling.

With cassava starch, gelatinization sets in at about 60°C, and the process is completed at about 80°C. In the manufacture of baked products, the treatment is carried out at moderate temperatures so as to cause gelatinization only in the surface layer of the lumps of moist starch. The product therefore consists of a raw starch core enclosed by a thin, tough, gelatinized layer.

For hand-worked flakes, gelatinization is carried out in shallow pans about 60–90 cm in diameter and 20–25 cm deep, which have previously been smeared with oil and which are then placed in holes in a brick oven and heated over a moderate fire for about 3–5 minutes. The mass is continuously stirred to prevent burning and to ensure uniform gelatinization. The flakes are then dried in enclosed driers at about 50°C until they have a storable moisture content of about 12 per cent.

For the preparation of pearls and seeds, the sieved damp starch is put into a long cylindrical bag of twill cloth which is held at each end by one man. The two men throw the mass of starch lumps from one end of the bag to the other, subjecting the mass to rhythmic jerking movements. After some minutes of this treatment, the irregular lumps change into seeds or pearls of various sizes and cohesive strength. This operation is known by its Indonesian name as the gangsor method. After gangsoring, the granules are graded according to size by screening them between plates with circular holes. The final product is obtained by roasting the granules for about 15 minutes in shallow pans which have been smeared with coconut or groundnut oil, followed by drying.

In Malaysia the moist starch is fed into open, cylindrical, rotating pans 0.9 m in diameter and 1.2 m deep. The rotation forces the starch grains to stick together in the form of small particles or beads. As is the case for flakes, gelatinization and subsequent drying complete the process.

## Cassava starch

The traditional method of processing cassava starch consists essentially of peeling, washing, grating, filtering and drying.

Freshly harvested tubers are hand peeled, washed and then grated into a pulp. The pulp is put into a basket over a pan or bucket covered with a piece of clean cloth. Water is poured over the basket, the starch is washed out and is filtered through the cloth into a pan. It is left overnight and the liquid is poured off in the morning. This is repeated until all the starch has been removed by rinsing, after which it is sun-dried.

Another traditional method is to put the grated tubers in bags and to pour enough water over them to soak the contents. The bags are then squeezed and a white liquid is expressed, which is poured into buckets. More water is added to the grated cassava, and the process is repeated until the liquid is no longer white. The liquid is left for several hours so that the starch can settle and the supernatant can be poured off. The remaining starch is washed 3–4 times until the liquid is clear. It is finally poured off and the starch is sun-dried.

The methods described here for cassava starch production are simple, labour intensive and cheap in terms of equipment. Yet productivity is low, and a further problem not to be overlooked is the time required to complete settling. Up to 24 hours may be required, allowing undesirable contamination and fermentation through contact between the starch and the fruit water. If economically feasible, technically more efficient processing methods have certain advantages.

For small to medium-scale production, processing of starch from cassava tubers may be carried out as shown in Figure 1.20.

Peeling and washing are accomplished in equipment similar to that used in the production of gari and chips. The peel has to be removed in order to facilitate the rasping or grinding of the tubers. In large factories only the outer skin, and not the cortex, is removed.

After washing and peeling, the tubers are conveyed to the rasping or disintegrating section to produce pulp or slurry. The purpose of rasping is to rupture the cell walls in order to release the starch granules. This is an important

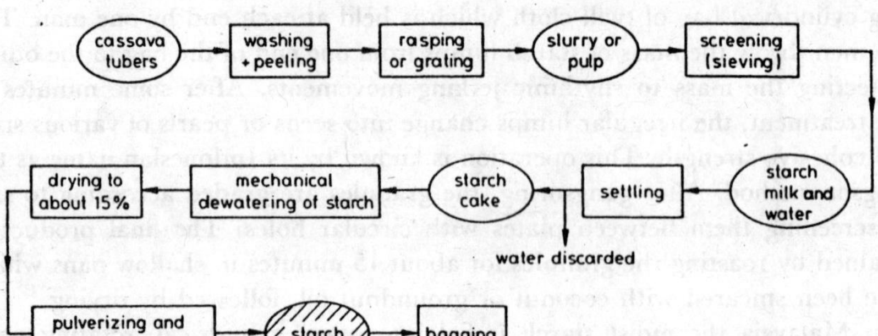


**Figure 1.20 Simplified flowchart for cassava starch processing.**

stage in processing, since the starch yield is dependent on efficient rasping.

From the crusher, disintegrator or rasper, the pulp is pumped to the extraction stage where the starch milk is separated from the fibrous material. This process is generally known as screening. A wide variety of designs for screens exist, differing in the complexity of their construction and their efficiency of operation.

Basically, rotating, vibrating screens and pusher-type centrifuges are used. A simple form of rotating screen consists of a conical hardwood frame, fixed on a hollow, horizontal axis, at least 3 m long, covered with cloth. The pulp is fed into the cone at the narrow end. By the rotation of the screen, approximately 50 rpm, the pulp slowly moves to the other end. At the same time, water is sprayed on the pulp from a number of openings in the hollow shaft, against the flow along the screen and under pressure. As a result, by the time the pulp reaches the lower end of the cone, it has been more or less completely washed.

The principle of a vibrating screen lies in the application of a vibrating movement to a flat sieve surface. The sieves are mounted one above the other, resting in a sieve box frame set at an inclined angle. The fresh pulp, previously mixed with water, is brought to the higher end of the screen. The pulp on the top of the screen is slowly pushed downwards by the vibrating motion during screening.

A pusher-type centrifuge consists essentially of an inlet funnel, a pusher disc and a basket. The starch-pulp slurry is introduced through a conical funnel and the cake is formed between the flange and the end of the basket. By means of a reciprocating pusher, the pulp is moved along the surface of the basket, and the liquid passes through the holes. Figure 1.21 shows schematically the parts of a pusher-type centrifuge.

On leaving the extractor the starch milk is stored in settling tanks where the starch sinks to the bottom. This stage of the process must be completed in as short a time as possible because of the chemical changes in the solution –

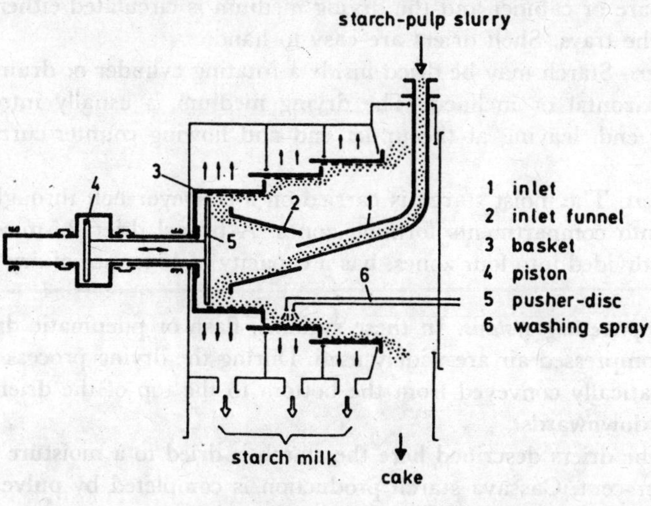


**Figure 1.21 Schematic diagram of a pusher-type centrifuge.**

the formation of complexes between starch and proteins and fatty substances – and the increase in micro-organisms, which eventually leads to fermentation, whereby alcohols and organic acids are produced, particularly butyric acid. These biochemical changes exert negative effects on the quality of the end product.

To reduce the settling time, the traditional settling method (rectangular and round tanks or basins) is replaced partly by the so-called settling table. The settling table is a flat wooden trough about 50 m long, which may be horizontal or gently sloping from the end where the starch liquid enters. The liquid flows in at a controlled rate, and initially the larger particles of starch separate, so that the starch can be graded along the length of the table. As the starch settles it builds up and increases the slope of the incline. The build-up of starch is then periodically removed.

In some cases centrifugal separators, which perform the separation process faster than the settling tables, are used.

After separation and refining, the starch milk is ready for drying. A preliminary mechanical dewatering is generally undertaken to reduce the moisture content to about 40 per cent. Centrifugal driers or vacuum filters are used for this purpose.

Drying is carried out by various methods. Starch can be sun-dried on trays. The trays may be placed directly on the ground or may be laid on racks above the ground. One advantage of sun drying is the bleaching effect of ultra-violet rays. At the same time, however, sun drying has certain disadvantages, such as contamination of the starch with dust particles and a certain degree of chemical degradation.

Ovens with wood-fired furnaces or drying floors, under which a wood fire provides the heat for drying, are widely used in Malaysia and other parts of the Far East. In larger factories, oil-fired or engine powered driers are used to reduce the drying time. Some of these include:

*Tray or shelf driers*. The material to be dried is placed on trays or shelves within an enclosure or cabinet and the drying medium is circulated either across or through the trays. Shelf driers are easy to handle.

*Rotary driers*. Starch may be dried inside a rotating cylinder or drum, which is either horizontal or inclined. The drying medium is usually introduced at the lower end, leaving at the upper end and flowing counter-current to the material.

*Tunnel driers*. The moist starch is carried on a conveyor belt through a tunnel divided into compartments forming zones. A tunnel drier 2.5 m wide, 10 m long and divided into four zones, has a capacity of 15 tonnes of dry starch per 24 hours.

*Continuous processing systems*. In these systems, flash or pneumatic driers using heated, compressed air are widely used. During the drying process the starch is pneumatically conveyed from the bottom to the top of the drier and then deflected downwards.

In all the driers described here the starch is dried to a moisture content of 10–15 per cent. Cassava starch production is completed by pulverizing and dry-screening the dried starch before it is bagged.

## Cassava flour

Until recently cassava flour only played a minor role in the diet of people in cassava-producing countries, compared with other processed cassava products such as gari and farinha de mandioca.

The flour from cassava has received, and is still receiving, considerable attention, as bread consumption and consumption of confectionery products increase in the developing countries where, for economic reasons, wheat flour must be wholly or at least partly substituted in bread making by locally available flour.

Research work on composite flours in bread making has shown that cassava flour can be substituted for up to 20 per cent of the wheat or cereal flours used in baking without significantly changing the final product or the processing methods (Edwards, 1974).

Traditionally, cassava flour is made from peeled, washed, sliced and sun-dried cassava tubers. The sun-dried chips are ground with a mortar and pestle into flour. In most cases flour prepared by this method is not of good quality due to considerable contamination with dirt. Also a certain degree of fermentation occurs during the relatively long drying period. It is therefore important to avoid long drying times, since fermentation imparts a dark colour and an acid taste to the product. Cassava flour of acceptable quality is obtained through the accelerated drying of the partially dehydrated pulp. In this process, manually or mechanically peeled cassava tubers are washed, chipped

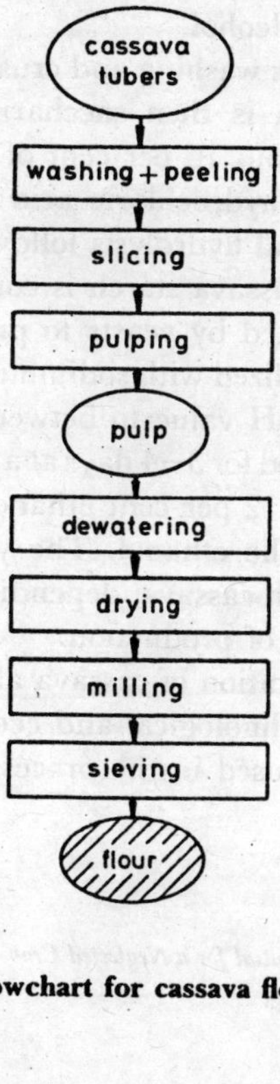


**Figure 1.22 Flowchart for cassava flour processing.**

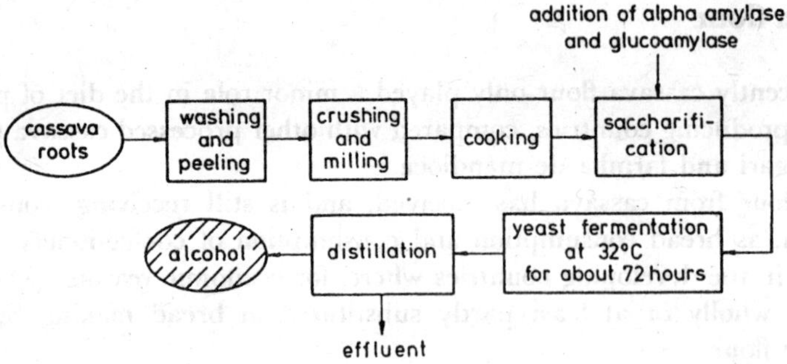


**Figure 1.23 Simplified flowchart for processing alcohol from cassava.**

and crushed or grated. Some water is removed in a centrifuge to give a moist pulp, which is then dried with hot air driers. The dried product is milled in a hammer or roller mill and then sieved to give cassava flour. The sequence of processing for cassava flour production is illustrated in Figure 1.22.

## Cassava alcohol

Cassava in its fresh form contains as much as 30 per cent carbohydrates. With this property it presents one of the richest fermentable raw materials for the production of alcohol.

Ethyl alcohol (ethanol) production from cassava is a sequence of chemical processes designed to saccharify the cassava starch, ferment the resulting solution and then distil the alcohol.

The overall process includes washing and crushing the tubers, and screening and cooking the pulp which is then saccharified. Saccharification, that is raising the total sugars to about 16 per cent of the content, is carried out by either mixing the pulp with hydrochloric acid or sulphuric acid in pressure cookers, or by means of partial hydrolysis followed by enzyme conversion. As a result of saccharification, cassava starch is converted to fermentable sugars, which can be readily fermented by yeasts to produce alcohol (Figure 1.23).

The mixture may be neutralized with sodium carbonate ($Na_2CO_3$) to remove the free acid and bring the pH value to between 5.0 and 7.0.

Yeast fermentation is allowed for 3–4 days at a suitable temperature, resulting in a solution containing 6 to 12 per cent ethanol. This solution is distilled by heat treatment to draw off the ethanol. The yield is about 70–110 litres of pure alcohol per tonne of fresh cassava, depending mainly on the variety used, its maturity and the method of production.

The widespread implementation of cassava alcohol production is associated with a number of serious technological and economic problems, one of them being the amount of energy used in the processing.

### *References*

**Cock, J.H.** (1985). *Cassava. New Potential for a Neglected Crop.* Westview Press/Boulder and London. 26–31.

**Collard, P.** and **Levi, S.** (1959). A two-stage fermentation of cassava. *Nature,* **183**: 620.

**Conn, E.E.** (1980). Cassava Toxicity. In (Bourdoux, P., Mafuta, M., Hanson, A. and Ermans A.E., eds), *The role of linamarin in the etiology of endemic goitre and cretinism,* Ottawa, Canada, IDRC – 136e.

**Coursey, D.G.** (1973). Cassava as food: Toxicity and Technology. In (Nestel, B. and MacIntyre, R., eds), *Chronic cassava toxicity.* Proceedings of an International Workshop, London, England, 29–30 January, 1973. IDRC, Ottawa, Canada, IDRC–010e, 27–36.

**Edwards, D.** (1974). *The Industrial Manufacture of Cassava Products. An economic study.* Tropical Products Institute, London.

**Ejiofor, M.A.N.** and **Okafor, N.** (1981). *Tropical Root Crops: Research Strategies for the 1980s.* Proceedings of the first triennial root crops symposium of the International Society for Tropical Root Crops – Africa Branch, 8–12 September, 1980, Ibadan, Nigeria.

**Erman, A.M., Mbulamoko, N.M., Delange, F.** and **Ahluwalia, R.** (1980). *Role of cassava in the etiology of endemic goitre and cretinism.* International Development Research Centre, Ottawa, Canada, IDRC – 136e, 1–182.

**FAO Production Yearbook**, Rome, 1980.

**Grace, M.R.** (1977). *Cassava Processing,* FAO, Rome.

**Meuser, F.** and **Smolnik, H.D.** (1980). Processing of cassava to gari and other foodstuffs. *Starch/staerke,* 32, **4**: 116–122.

**Ngaba, P.R.** and **Lee, J.S.** (1979). Research Note: Fermentation of cassava. *Food Science,* **44**: 157.

**Odigboh, E.U.** (1976a). Mechanization of Nigerian cassava production and processing, research needs and interests. *Journal of Agric. Engng.*, **31**: 20.

**Odigboh, E.U.** (1976b). A cassava peeling machine; development, design and construction. *Journal of Agric. Engineering Research*, **3**: 361.

**Oke, O.L.** (1968). Cassava as food in Nigeria. *World Rev. Nutr. Diet.* **11**: 170.

**Oke, O.L.** (1979). Some aspects of the role of cynogenic glycosides in Nutrition. *World Rev. Diet.* **33**: 70–103.

**Oke, O.L.** (1982). *Processing and Detoxification of Cassava*: Proceedings of a workshop held in Ottawa, Canada, 31 May–2 June, 1982.

**Onwueme, I.C.** (1978). *The Tropical Tuber Crops. Yams, Cassava, Sweet Potatoes, Cocoyams.* John Wiley and Sons Ltd, 145 pp.

**Osuntokun, B.O., Monekosso, G.L.** and **Wilson, J.** (1969). Relationship of a degenerative tropical neuropathy to diet. Report of a field survey. *British Medical Journal.* 1: 547–550.

**Phillips, T.P.** (1982). *An Overview of Cassava Consumption and Production*: Proceedings of a workshop held in Ottawa, Canada, 31 May–2 June, 1982.

**Purcel, J.G.** and **Williams, D.H.** (1973). *History of the development of the first mechanized continuous gari manufacturing plant,* 3rd Int. Symp. Trop. Root Crops, Ibadan, Nigeria.

**Villa, L.G., Roa, G.** and **Braunbeck, O.** (1979). *Natural and convected air drying of cassava chips*: Paper presented at the 1979 Summer Meeting of ASAE and CSAE, Paper No. 79–4037.

**Wadhwa, N.D.** (1973). *Cassava gari plant,* 3rd Int. Symp. Trop. Root Crops, Ibadan, Nigeria.

**Williams, C.E.** (1979). *The Role of Women in Cassava Processing in Nigeria*: In (Plucknett, D.L., ed.), *Small-scale Processing and Storage of Tropical Root Crops.* Westview Press, Colorado.

# 2

# Cocoa

## Classification and production

Cocoa, *Theobroma cacao* L., is one of the 22 species that constitute the genus *Theobroma*, a member of the family Sterculiaceae. Their centre of origin is considered to be the lower eastern equatorial slopes of the Andes where the greatest range of variation in natural populations exists.

In practical plant production, three varieties of cocoa are distinguished:
*Criollos* from Central America to Columbia.
*Forasteros* from the upper Amazon, with the group amelonados originating from Guyana and the lower Amazon.
*Trinitarios*. These hybrid populations result from natural crosses between criollo and forastero types. Trinitario was the term applied originally to the Trinidad population, however the name would now cover some of the cocoas in such countries as Sri Lanka, Indonesia and Papua New Guinea, where criollos have crossed with forasteros.

For many years cocoa production was confined to the western hemisphere. It was introduced into Ghana in 1879 from Fernando Po, giving rise to the large-scale West African production. Today the main cocoa producing countries are Ghana, Nigeria, the Ivory Coast and Brazil. By far the greatest volume of production, the so-called bulk cocoa of commerce, is produced from the amelonado forasteros of West Africa and Brazil. Amelonado cocoa provides about 80 per cent of the world's supplies. Production of criollo cocoa is now so insignificant that it is of little importance in world trade.

Yields vary between 1 and 1.5 tonnes of dry beans/hectare/year. Twelve pods produce about 450 g of dry cocoa and 45 kg of wet beans give 18 kg of dry beans (Irvine, 1974). World production of cocoa is over 2 million tonnes, Africa producing over 50 per cent of the world's raw cocoa (FAO, 1986). Cocoa beans of commerce may be classified into:

1. Fine cocoa, such as criollo cocoa, of which negligible quantities are now produced.
2. Flavoured cocoas, as are produced in Central America and the West Indies. The flavour of Trinidad cocoa is probably due to the mixture of a large number of genotypes in the trinitario variety, and is somewhat astringent.
3. Ordinary forastero or amelonado cocoa as is produced in West Africa and Brazil, which usually has a bland flavour.

### *Uses*

Cocoa is of ancient cultivation in Central America. The Indians believed it to be of divine origin, hence the generic name *Theobroma* given by Linnaeus, meaning the 'food of the gods'. Indians made a thick beverage by pounding roasted cocoa beans with maize and capsicum. Cocoa is produced by comparatively few countries, all of them tropical, but the product is processed and consumed mainly in temperate countries. Cocoa is used to manufacture chocolate bars by grinding the cocoa mass with sugar and extra cocoa fat. Lecithin may also be added to increase fluidity. In making milk chocolate, dried or powdered milk is incorporated.

Cocoa butter is used in cosmetics and pharmaceutical preparations. In Brazil, cocoa jelly is made on a small scale from sweatings, and in Mexico sweatings are sometimes allowed to ferment to produce cocoa wine. The cocoa shell (testa) is used in stock feeds or as manure; it is a source of theobromine, shell fat and vitamin D. The pod is rich in potash and is used in soap making.

### *Cocoa butter*

Cocoa butter is the fat of beans from *Theobroma cacao*. Fermented and roasted beans are shelled, milled and then pressed to recover the fat. Cocoa butter has an ivory colour, a hard brittle consistency and a pleasant aroma. It dissolves in ether, chloroform and boiling 90 per cent alcohol.

The total amount of triglycerides in cocoa butter (palmitin 24 per cent and stearin 35 per cent) outweighs that of the fatty acids (oleic acid 38 per cent and linoleic acid 2 per cent). The triglyceride composition gives cocoa butter its characteristic melting properties; the short melting range of 32°–34°C is the most desirable property (Ullmans Enzyklopaedia, 1976).

## Physical characteristics of cocoa

The cocoa pod is usually considered to be a drupe. It is variable in size, 10–32 cm long, nearly spherical to cylindrical, pointed or blunt, smooth or warty, with or without furrows. The pericarp (husk) is fleshy and thick with the mesocarp varying in its degree of lignification. Pods attain full size 4–5

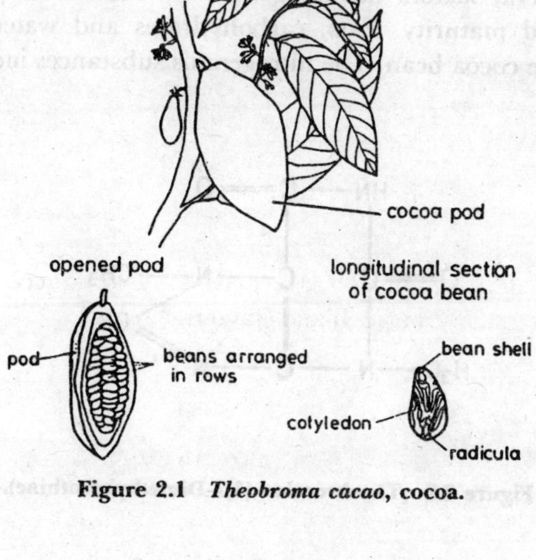


**Figure 2.1** ***Theobroma cacao*, cocoa.**

months after fertilization and require a further month for ripening. The cocoa pod is composed of about 42 per cent beans, 2 per cent sweatings (mucilage) and 56 per cent husk.

A pod contains 20–60 beans. Forastero cultivars have more beans than criollo cultivars, usually arranged in 5 rows; they are variable in size (2–4 × 1.2–2 cm) and shape (ovoid or elliptic), and have varying shades of purple cotyledons. The cotyledons of criollo varieties are normally white while those of criollo × forastero crosses have white to deep purple cotyledons. The cocoa bean consists essentially of a shell (testa), which represents 10–14 per cent of the dry weight, and the kernel (cotyledon) which represents the remaining 86 to 90 per cent (Figure 2.1).

Fresh beans are surrounded by a mucilaginous, whitish, sugary, acid pulp which develops from the outer integument of the ovule. The pulp is removed during fermentation and drying. The most important differences between typical criollo and forastero cocoas are shown in Table 2.1.

**Table 2.1 The most important differences between criollo and forastero cocoas**

| Characteristic | Criollo | Forastero |
|---|---|---|
| Colour of pod | Yellow-red, freckled | Green-yellow |
| Form of pod | Oblong, longitudinal, pointed | Oval |
| Surface of pod | Warty, deeply furrowed | Smooth, slightly furrowed |
| Pod shell | Thin and soft | Firm and tough |
| Size of bean | Large, round | Small, flat |
| Number of beans in pod | 20–40 | 30–60 |
| Colour of cotyledon | White-cream | Purple |
| Aroma | Strong | Weak |
| Yield | Low | High |

## Composition of cocoa beans and pods

The chemical composition of the cocoa bean and the pod husk will vary according to several factors including variety, cultivation practices, region of cultivation and maturity. Fats, carbohydrates and water are the major components of the cocoa bean. The nitrogenous substances include many non-

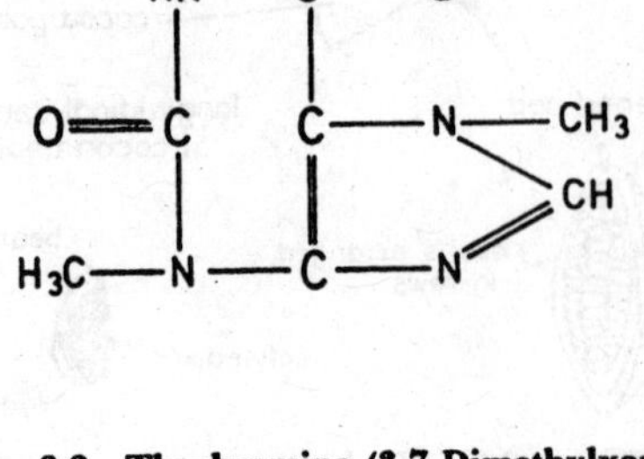


**Figure 2.2 Theobromine (3,7-Dimethylxanthine).**

protein compounds such as theobromine (Figure 2.2) and caffeine (both alkaloids) as well as traces of ammonia and asparagine. Theobromine is a stimulant, increasing muscular activity. Symptoms such as excessive stimulation of the kidneys, the heart and smooth muscles have been associated with a high intake of theobromine (Heimann, 1976). The carbohydrates in cocoa consist mainly of starch, pectins and fibre. Table 2.2 gives a more detailed description of the composition of fermented and unfermented cocoa bean cotyledons after drying.

**Table 2.2 Chemical composition of the cotyledons of fermented and unfermented cocoa beans after drying**

| Constituent | Unfermented beans (%) | Fermented beans (%) |
|---|---|---|
| Water | 3.65 | 2.13 |
| Fat | 53.05 | 54.68 |
| Ash | 2.63 | 2.74 |
| Total nitrogen | 2.28 | 2.16 |
| Protein nitrogen | 1.50 | 1.34 |
| Ammonia nitrogen | 0.028 | 0.042 |
| Amide nitrogen | 0.188 | 0.336 |
| Theobromine | 1.71 | 1.42 |
| Caffeine | 0.085 | 0.066 |
| Glucose | 0.30 | 0.10 |
| Sucrose | Nil | Nil |
| Starch | 6.10 | 6.14 |
| Pectins | 2.25 | 4.11 |
| Fibre | 2.09 | 2.13 |
| Cellulose | 1.92 | 1.96 |
| Pentosans | 1.27 | 1.21 |
| Mucilage and gums | 0.38 | 1.84 |
| Acids | 0.3 | 0.43 |

Source: Opeke, 1982.

**Table 2.3 Approximate chemical composition of cocoa pod husks**

| Constituent | Cocoa pod husk (%) |
|---|---|
| Water | 57.75 |
| Total dry matter | 42.25 |
| Crude protein | 9.69 |
| Fatty substances | 0.15 |
| Ash ($SiO_2$ free) | 10.80* |
| Crude fibre | 33.90 |
| Nitrogen-free extracts | 42.21 |
| Glucose | 1.16 |
| Sucrose | 0.18 |
| Pectin | 5.30 |
| Theobromine | 0.20 |

*Contains the silica component.
Source: Opeke, 1982.

The use of cocoa pod husks as a feed is of special interest in West Africa, in view of livestock development (Owusu-Domfeh, 1972). Table 2.3 gives the composition of cocoa pod husks. The cocoa pod husk has a rather low alkaloid content, while the tannin content is virtually nonexistent. The crude fibre content is low, and since it is unlignified it seems that pod husks should be digestible by all classes of livestock.

## Harvesting and pod breaking

Harvest time varies from one region to another. For countries north of the equator, the main cocoa harvesting time is usually from September to December, while the lean season is from April to June.

Harvesting involves plucking or detaching ripe pods from the trees and

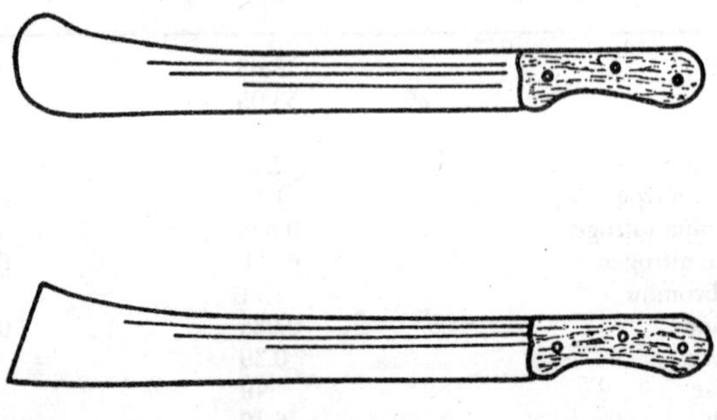

**Figure 2.3 Cutlasses for various operations in cocoa production.**

**Figure 2.4 Opening cocoa pods with a knife to remove the beans.**

opening them to extract the wet beans. On ripening, the pods change colour, green pods becoming orange-yellow and red pods turning orange. Pods are plucked from the trees using various forms of knife. A short-handled cutlass (Figure 2.3) or machete is often used for pods within reach and special hooked knives on long poles for pods out of reach.

After plucking, the pods are opened by cutting them diagonally with a knife which is then twisted, thereby breaking off a portion of the pod (Figure 2.4). The beans are removed from the pod with the fingers. This method of pod opening using a knife usually results in some damage to the beans, sometimes as many as 5 per cent of the beans receiving a cut testa. To circumvent this drawback, pods can be opened by cracking them on a stone or with a wooden mallet or on a triangular strip of wood fixed to a table. Over 3000 pods a day may be broken by an experienced worker.

## Fermentation of cocoa beans

Fermentation is probably the most important operation in the processing of cocoa beans. It involves keeping a mass of cocoa beans well insulated so that heat is retained, while at the same time air is allowed to pass through the mass. Fermentation normally involves the conversion of sugar to alcohol and alcohol to acetic acid. However, in the case of cocoa, as well as these fundamental changes, other changes also take place inside the bean.

The process of fermentation, which lasts up to 7 days, is dependent, among other factors, on the activities of a succession of micro-organisms. These may be described as follows (Powell 1959, Irvine 1974, Wood, 1975). When the beans are removed from the pod they are covered with mucilage or pulp, forming about 15–20 per cent of the weight of the wet beans. The pulp consists of 80 per cent water, 10–15 per cent glucose and fructose, up to 0.5 per cent non-volatile acids – largely citric, and small amounts of sucrose,

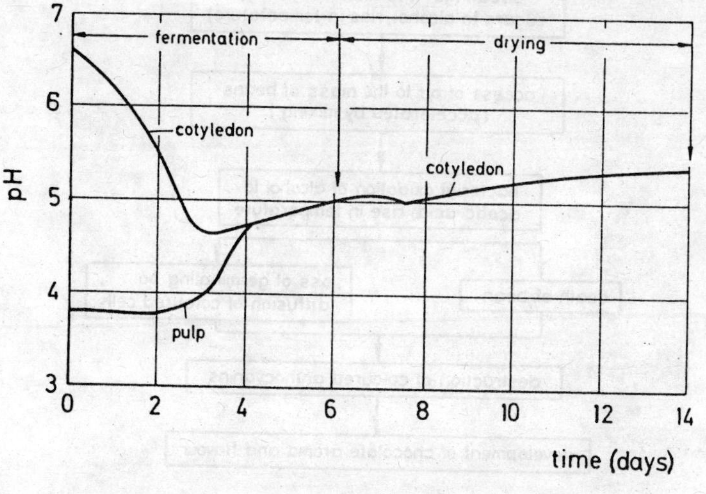


**Figure 2.5** **Changes in pH during box fermentation of cocoa beans (Wood, 1975).**

starch and salts. The presence of sugar and the acidity (pH 3.5) provide the conditions necessary for the growth of micro-organisms.

During the first stage of fermentation, yeasts proliferate and convert the pulp sugars into alcohol with the evolution of carbon dioxide. Chemically this may be expressed as:

$$\underset{\text{sugar}}{C_6H_{12}O_6} \rightarrow \underset{\text{ethyl alcohol}}{2C_2H_5OH} + \underset{\text{carbon dioxide}}{2CO_2}$$

The conversion of sugars into alcohol with the evolution of $CO_2$ occurs under relatively anaerobic conditions. These conditions allow the development of lactic acid bacteria, which contribute in the breakdown of sugars. When the sweatings are drained, the conditions become more aerobic. The acidity is also reduced by the removal of citric acid. As the pulp cells collapse, air enters the mass and there is oxidation of the alcohol to acetic acid by *Acetobacter* spp., which become active as soon as the ethyl alcohol content reaches a sufficient level. Heat is produced by the oxidation.

According to Kenten and Powell (1960) the reactions in which the micro-organisms are involved produce much heat, raising the temperature of the mass of beans. The pH of the fresh pulp is 3.5; at the same stage the pH of the cotyledon is 6.6. As the testa is permeable to acetic acid this penetrates the cotyledon, kills the bean and lowers the pH to 4.8 by the third day of

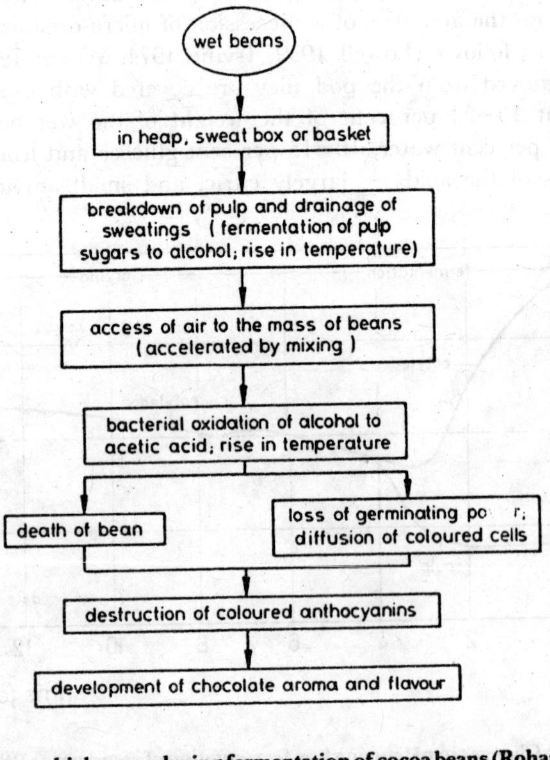


**Figure 2.6 Changes which occur during fermentation of cocoa beans (Rohan 1958a; 1963).**

fermentation. Thereafter, the pH rises gradually during the remaining period of fermentation and drying, and is usually 5.5 in the dried bean (Figure 2.5).

The death of the bean, mainly caused by the penetration of alcohol and acetic acid into the cotyledon tissues, results in the breakdown of the internal cell structure, allowing various enzyme reactions to occur involving the conversion of some of the polyphenolic compounds. The anthocyanins are hydrolysed to give cyanidin and two reducing sugars; other polyphenols undergo chemical changes or diffuse through the testa, both processes contributing to the reduction in bitterness and astringency.

The proteins in the cotyledon are hydrolysed to amino-acids and converted to insoluble forms by reactions with the polyphenols. These changes are reflected in the flavour, aroma and colour of the beans. The changes which take place, both outside and inside the cocoa bean during fermentation, are

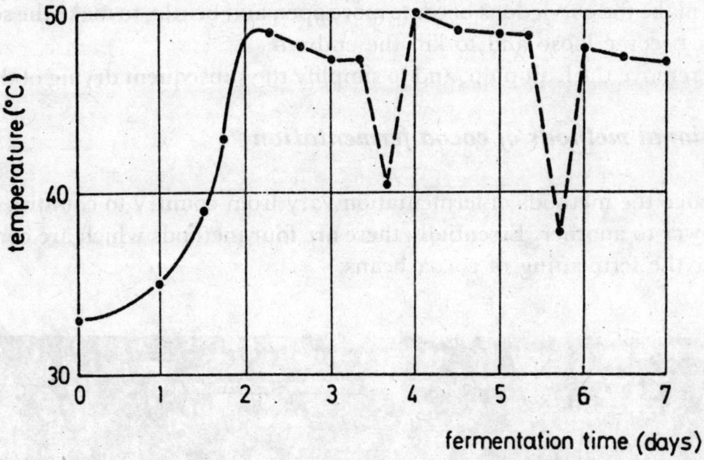


**Figure 2.7 Temperature curve of a box fermentation of cocoa beans (Wood, 1975).**

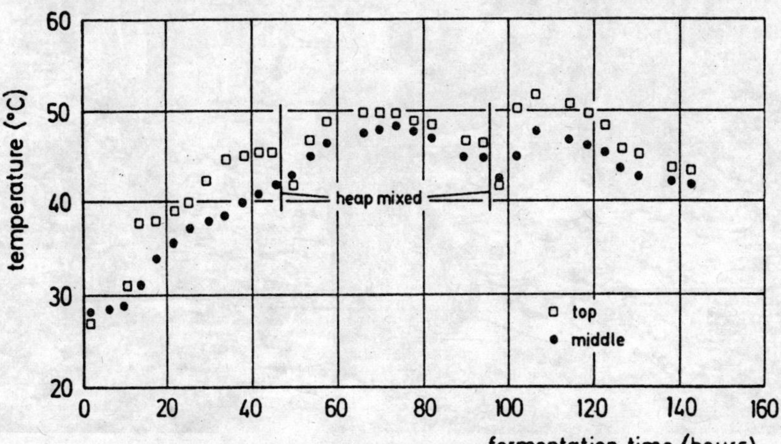


**Figure 2.8 Temperature curve of a heap fermentation of cocoa beans (Wood, 1975).**

demonstrated in a simplified form in Figure 2.6.

Temperature curves of box and heap fermentations are shown in Figures 2.7 and 2.8. The temperature rises steadily during the first 48 hours of fermentation, reaching 42–48°C by the end of that time. After the first mixing, the temperature rises to 50°C. Thereafter it falls slowly, but will rise again after the beans are turned. After 6 days the temperature is about 48°C.

The changes accompanying over-fermentation can increase the temperature to 50°–52°C. When beans are over-fermented, unpleasant, fishy, rancid smells develop in the fermenting mass. These are the result of the breakdown of proteins and the production of ammonia. On the other hand, if the temperature does not rise above 45°C the beans will be underfermented.

The objectives of fermentation are:

1. To reduce the bitterness of the unfermented beans and to obtain cocoa beans which on roasting will produce chocolate with the desired aroma and flavour.
2. To make the cotyledons become more open and brittle, to make the seed coat or testa become loose and to kill the embryo.
3. To remove the fruit pulp, and to simplify the subsequent drying of the bean.

## *Traditional methods of cocoa fermentation*

In practice the methods of fermentation vary from country to country, or from one grower to another. Essentially there are four methods which are commonly used for the fermenting of cocoa beans:

**Figure 2.9 Fermenting cocoa beans in heaps on plantain leaves.**

1. The heap method.
2. The box method.
3. Fermenting in baskets.
4. Curing on a drying platform.

## Heap fermentation

This is the most popular method of fermenting cocoa in West Africa. The wet beans are heaped on a layer of banana or plantain leaves, and covered with the same material (Figure 2.9). If a layer of wooden sticks, arranged like the spokes of a wheel, is first and put underneath the heap, the sweatings to drain away more readily. Sometimes a trench is dug around the heap to carry away the by-products of fermentation. The leaves are turned up at the end so as to cover the heap completely and are kept in position by pieces of wood or stones (Figure 2.10).

The beans are turned after 48 hours and once more before the end of the 6 day fermentation period, so that the outside beans are turned into the centre of the heap, and the whole heap is more or less evenly fermented.

## Box fermentation

Fermentation in boxes involves the use of strong wooden boxes measuring 1.2 × 1.2 × 0.9 m, holding up to 1.5 tonnes of wet beans. The boxes may be arranged side by side or in tiers. The tiers arrangement facilitates mixing, which can be effected by simply transferring the beans from one box to the next one in line below. Drainage holes 15 mm in diameter are provided in the bottom of each box to facilitate the removal of sweatings and to permit aeration. In carrying out fermentation, the wet beans are put in the boxes to a depth no greater than 0.75 m.

The side by side arrangement consists essentially of two boxes, raised from the ground, with sliding tops, open fronts, and holes drilled in the bottom.

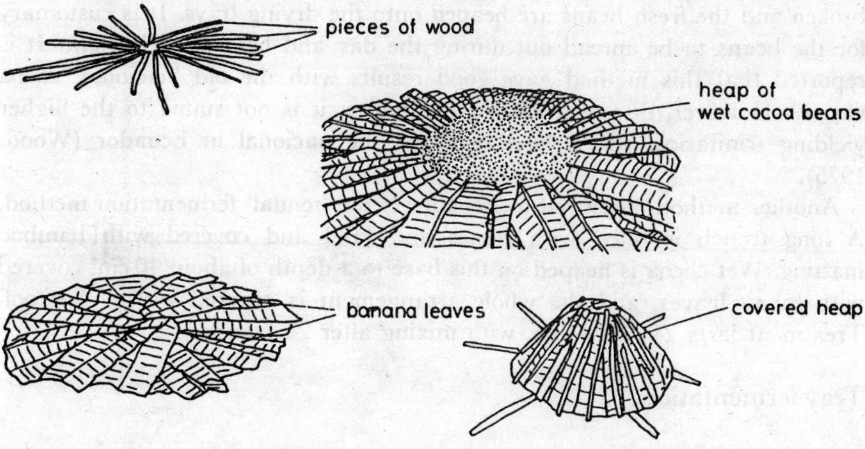


**Figure 2.10 Heap fermentation.**

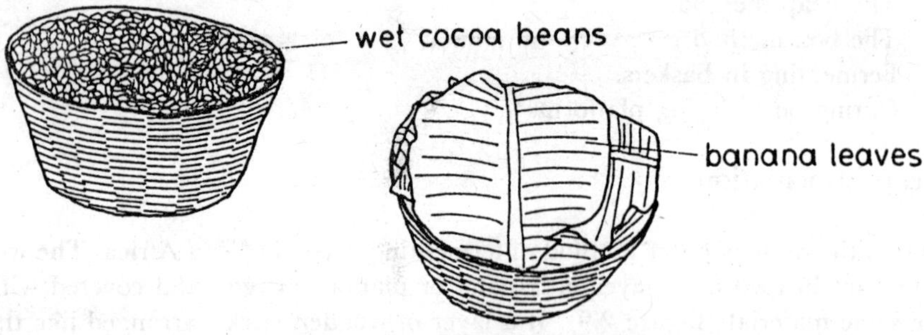


**Figure 2.11 Basket fermentation.**

One box is filled with beans, and covered with plantain leaves or sacks and a board weighted with stones. When it is time to turn or mix the beans they are shovelled into the second box and later transferred back into the first box.

The box fermentation method is used mainly in the West Indies and many countries in Latin America, although it is not unknown in West Africa. It is more applicable on large farms or plantations where relatively large quantities of cocoa are harvested. A typical box fermentation lasts for six days, and during this time the beans are mixed twice after 2–4 days.

### Basket fermentation

Woven baskets are used in this method. They are filled with the wet beans, then covered with leaves (Figure 2.11). The sweatings drain through the sides and bottoms of the baskets and mixing is effected by transferring the beans from one basket to another. The size of the baskets varies; in Ghana and Nigeria the range is from about 10–150 kg wet beans capacity.

### Curing on drying platforms

This method is confined almost entirely to Ecuador. The cocoa pods are broken and the fresh beans are heaped onto the drying trays. It is customary for the beans to be spread out during the day and heaped each night. It is reported that this method gave good results with the old 'nacional' cocoa variety. However, due to a decline in quality, it is not suited to the higher yielding trinitarios, which have replaced the nacional in Ecuador (Wood, 1975).

Another method employed in Ecuador is the 'tendal' fermentation method. A long trench is filled with stones and sand and covered with bamboo matting. Wet cocoa is heaped on this base to a depth of about 40 cm, covered with bihau leaves, and the whole arrangement is covered with a tin roof. Treatment lasts 24–36 hours, with mixing after 24 hours.

### Tray fermentation

Tray fermentation has evolved from Rohan's studies of heap fermentation

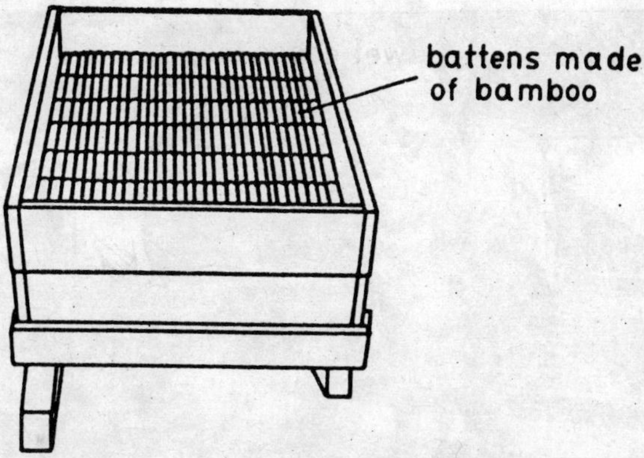


**Figure 2.12 Tray (box) for fermenting cocoa beans.**

(Rohan, 1958a). About 10 cm layers of beans are fermented in trays stacked one on top of the other. The trays are usually 0.9 × 0.6 m × 13 cm deep. Battens are fixed across the bottom of the trays and matting rests on the battens to hold the beans in place. Allowing for the battens and matting, the effective depth of the tray is 10 cm. Such a tray will hold 45 kg of wet beans (Figure 2.12).

After opening the pods, the beans are put into trays which are stacked up to 12–14 trays high. The bottom tray is often left empty to allow easy drainage of sweatings and to improve aeration. After 24 hours, the stack of trays is covered with sacks to retain heat. Fermentation in trays has been reported to be quicker than fermentation in heaps or boxes; a four-day tray fermentation of amelonado or Amazon cocoa in Ghana resulted in a product superior to that produced by normal heap fermentation (Rohan, 1958a; Allison and Rohan, 1958).

Since the fermentation period is shorter, the output from the tray method is higher. The saving in labour is in the order of 20 per cent because mixing during tray fermentation is eliminated. However, the appearance of the beans after tray fermentation is less uniform than after box fermentation.

## Drying of cocoa beans

At the end of fermentation the moisture content of cocoa beans is about 55 per cent and this must be reduced to 6–7 per cent for safe storage. The purpose of drying, apart from reducing the moisture content, is to prevent the growth of moulds and to complete the chemical changes taking place within the bean.

Cocoa beans may be dried in the sun (sun drying) or by means of mechanical driers (artificial drying). Artificially dried beans will normally be more acid than beans dried in the sun but artificial drying over a period of 24–72 hours is unlikely to result in beans with unacceptable acidity (Wood, 1975). Well-dried, fermented beans are partly pale purple and partly chocolate coloured. Dry beans give a crackling sound when rubbed in the hand and powder easily.

The beans may be polished after drying; this is done mechanically or by

**Figure 2.13 Polishing cocoa beans by trampling with bare feet.**

foot. Polishing by foot is known as 'dancing the cocoa' in Trinidad. The beans are wetted and trampled with bare feet, care being taken not to crush the beans but rather to rub one against the other (Figure 2.13).

## *Sun drying of cocoa beans*

In West Africa and in the West Indies, where the main harvest coincides with the dry season, cocoa beans are usually dried in the sun. In the West Indies sun drying takes place on wooden drying floors with movable roofs or alternatively on wooden trays on rails which can be pushed under a fixed roof. In West Africa beans are dried on mats raised off the ground (Figure 2.14); the

**Figure 2.14 Cocoa beans drying on raised meats in the sun in Cameroon.**

**Figure 2.15 A wooden drying tray on rails, with a fixed roof.**

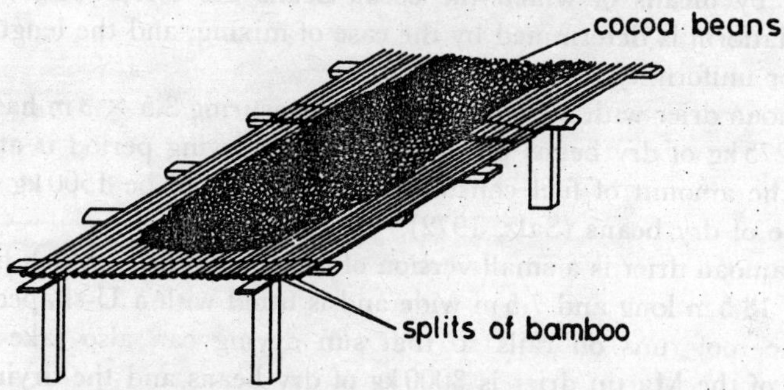


**Figure 2.16 Sun drying of cocoa beans on a mat.**

mats are made of strips of bamboo, or split rachides of oil or raphia palms, with spaces between the slats so that air can circulate from below (Figures 2.15 and 2.16).

The area required for sun drying is about 3 $m^2$/50 kg of dry beans, the normal drying (floor) area being 18 × 6 m for a load of 2250 kg dry beans. The beans are spread out on such floors to a depth of 5 cm. Under favourable conditions drying is completed in 8 days, whilst under less favourable conditions, drying may take as long as 14 days (Wood, 1975).

## *Mechanical drying*

In areas with little sunshine and high atmospheric (air) humidity, cocoa beans are dried by mechanical means. In this process, drying should also be carried out slowly and at moderate air temperatures (60–70°C), in order to complete the enzymic oxidation. Some of the factors involved in determining the most suitable conditions for drying include the depth of beans, the rate of air flow, and the temperature of the drying air. The effect of these factors was investigated by Shelton (1967). He showed that a deep layer of beans (25 cm), an air flow of 3 $m^3$/min and air temperatures of 60–65°C gave the most economical conditions.

In Papua New Guinea a method known as interrupted drying is practised by some growers. Cocoa beans are dried for 12 hours, rested for 12 hours and then given a final drying. The purpose of the rest period is to allow the moisture within the beans to migrate to the surface. A reduction in acidity and the prevention of case-hardening are reported to be two of the advantages of this method. A further purpose of the rest periods between the periods of drying is economy of fuel use.

## Samoan drier

The Samoan drier consists of a flue in a plenum chamber with a permeable drying platform above. Air inlets are provided to allow easy flow of convection currents, by means of which the cocoa beans are dried. The width of the drying platform is determined by the ease of mixing, and the length is limited to 6 m for uniformity of drying.

A Samoan drier with a drying platform measuring 3.5 × 3 m has a capacity of 225–275 kg of dry beans per batch and the drying period is about 36–48 hours. The amount of fuel consumed is estimated to be 1500 kg of firewood per tonne of dry beans (Salz, 1972).

The Samoan drier is a small version of the Martin drier which has a drying platform 13.5 m long and 7.5 m wide and is fitted with a U-shaped flue and a roof. The roof runs on rails so that sun drying can also take place. The capacity of the Martin drier is 3000 kg of dry beans and the drying period is 20 hours.

These simple driers are suitable for small-scale cocoa growers and can be cheaply run in areas where wood fuel is readily available. To avoid smoke escaping and contaminating the beans, the flues, which corrode fairly quickly, must be attended to quite often. On large estates and plantations, drying in the sun or with the use of simple driers may have to be replaced in the future by more powerful driers.

## Platform driers

The platform drier consists of a drying platform, a fan heater unit and a duct. The drying platform is long and narrow to ensure uniform distribution of the hot air and ease of mixing the beans, a typical installation measuring 9 × 2.5 m (Wood, 1961). The platform is made of nylon-covered expanded metal or perforated aluminium sheets, having holes not exceeding 10 mm in diameter. The plenum chamber beneath the platform is about 1 m deep and is connected to the fan-heater unit by the duct. A platform drier (6 × 2.7 m), coupled to a fan-heater unit providing 70 m$^3$/min of air at a temperature of 50°C above ambient, has a capacity of 1 tonne of dry beans and a drying period of 32–34 hours. The fuel consumption is about 180–230 litres of fuel (diesel oil) per tonne of dry beans. The fan heater unit is driven by an internal combustion engine using diesel fuel. Installing the drier in an enclosed space and recirculating the air leads to the production of 1 tonne of dry beans within 24 hours at a fuel consumption of 155 litres/tonne (Allison and Kenten, 1964).

In some areas the drying platform has been replaced by a drying bin of similar dimensions, in which case the depth of beans is about 45 cm.

## Other driers for cocoa

Rotary driers may be used for the later stages of drying cocoa, when the beans are externally dry. However, the drum should revolve slowly at about 0.25 to 0.5 rpm, to prevent breakage of shells and beans. The perforations in the metal cover of the drum should be 8–10 mm in diameter.

The Buttner drier is a continuous drier which can be used for cocoa beans. It consists of an endless belt of small trays, each holding about 40 kg of wet beans. The trays pass down a drying tower through which hot air is blown. The drier can handle 9 tonnes of wet beans when fully charged.

## Cocoa powder and chocolate production

Dried cocoa beans serve as the raw material for cocoa powder and cocoa butter is the raw material for chocolate.

First the beans are cleaned to remove any foreign matter and to separate small and broken beans. Cleaning and separating is followed by roasting. The roasting process consists of heating the beans to 100–140°C for 45–90 minutes.

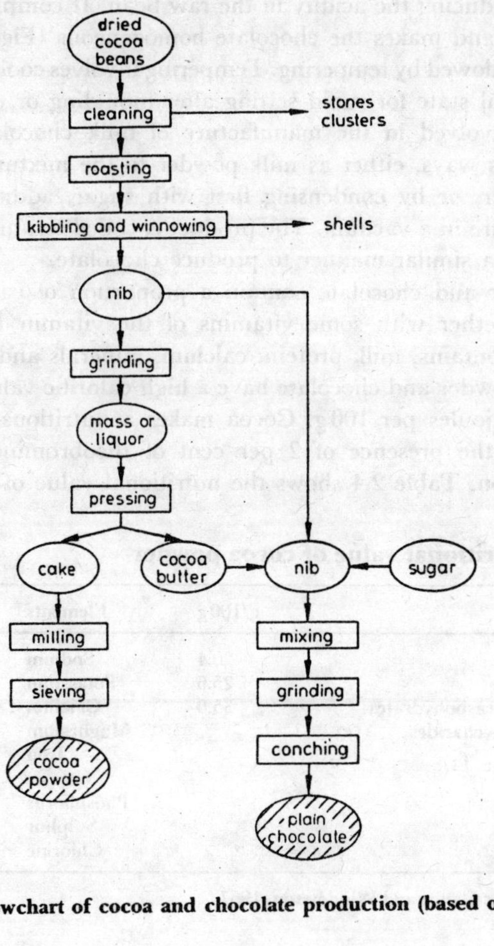


**Figure 2.17 Flowchart of cocoa and chocolate production (based on Wood, 1975).**

Roasting develops flavour and colour, and facilitates easy removal of shells.

In the next step the roasted beans are cracked or kibbled in a mill, after which the shell and nib (the roasted cotyledon) are separated by winnowing. The separated nib is ground to a liquor or mass which has a fat content of 55–58 per cent. For the production of cocoa powder, a substantial proportion of the fat is removed in hydraulic presses at pressures ranging from 300–400 bars. The cocoa powder comes from the press as a cake, with a fat content of 20–23 per cent. The cake is then crushed in a mill. The resulting powder is sieved through a fine silk, nylon or wire mesh. In some cases, cocoa powder is made from mass which has been treated with an alkali, with the purpose of controlling its dispersability.

If chocolate production is the objective then the cocoa butter, nib and sugar are mixed to enable the chocolate to be easily moulded. The mixture is then ground to such a consistency that the chocolate is smooth to the palate. In the past, the grinding process was carried out in melangeurs; these are heavy granite rollers on a revolving granite bed. Grinding nowadays is in a series of rollers. After grinding the mass is conched. During the conching process, which may last for several hours, the chocolate is heated to drive off volatile acids, thereby reducing the acidity in the raw bean. It completes the development of flavour and makes the chocolate homogeneous (Figure 2.17).

Conching is followed by tempering. Tempering involves cooling and obtaining the right physical state for rapid setting after moulding or enrobing. Similar processes are involved in the manufacture of milk chocolate. The milk is added in various ways, either as milk powder to the mixture of mass, sugar and cocoa butter, or by condensing first with sugar, adding the mass and drying the mixture in a vacuum. The product is called 'crumb', and is ground and conched in a similar manner to produce chocolate.

Cocoa powder and chocolate contain a proportion of carbohydrates, fats and protein together with some vitamins of the vitamin B complex. Milk chocolate also contains, milk protein, calcium, minerals and vitamins.

Both cocoa powder and chocolate have a high calorific value, cocoa powder producing 1898 joules per 100 g. Cocoa makes a nutritious drink with milk and sugar and the presence of 2 per cent of theobromine gives a mildly stimulating action. Table 2.4 shows the nutritional value of cocoa powder.

**Table 2.4 Nutritional value of cocoa powder**

| Nutrients | g/100 g | Elements | mg/100 g |
|---|---|---|---|
| Protein | 20.4 | Sodium | 650 |
| Fat | 25.6 | Potassium | 534 |
| Available carbohydrates | 35.0 | Calcium | 51.2 |
| as monosaccharides | | Magnesium | 192 |
| | | Iron | 14.3 |
| | | Copper | 3.40 |
| | | Phosphorus | 685 |
| | | Sulphur | 160 |
| | | Chlorine | 199 |

Source: Mc Cance and Widdowson, 1967.

## Grading

Cocoa may be graded by taking at random representative samples from the bulk. A final selection of 300 beans is made from the harvest and weighed. The 300 beans taken from the main crop should weigh more than 300 g. The weight as stated is affected mainly by the variety of cocoa. There are two grades of cocoa based on the following criteria (Opeke, 1982):

*Grade I cocoa*: Less than 3 per cent slaty beans
Less than 3 per cent mouldy beans
Less than 3 per cent other defectives

*Grade II cocoa*: Less than 5 per cent slaty beans
Less than 4 per cent mouldy beans
Less than 5 per cent other defectives

Slaty beans are defined as having a slaty colour on half or more of the surface exposed by a test cut made lengthwise through the centre. A slaty bean is usually the result of drying a bean before any of the initial changes associated with fermentation have taken place. Such a bean has therefore been killed by drying instead of by the effects of acid and heat.

Mouldy beans are defined as beans in which mould is visible to the naked eye on the internal parts. This is the most serious defect of cocoa beans and should be avoided since the presence of mouldy beans affects the flavour of the finished cocoa product.

Defective beans are infested beans, germinated beans and smoky beans. Infested beans result from attack by several stored product pests, the major one being the tropical warehouse moth, *Cadra cantella*. Other pests include *Araecerus fasciculatus* (the coffee bean weevil) and *Tribolium casteneum* (the rust red flour beetle). Germinated beans are considered to be defective because the hole in the testa left by the emerging radicle offers an opening for moths and insects to invade the beans; smoky beans have a smoky smell or taste due to contamination with smoke during drying (wood-fired drying) or storage.

## *Storage*

Cocoa beans can be stored relatively well in temperate countries but in the tropics, where high temperatures and high relative humidities usually prevail, the rapid development of moulds can easily take place. Cocoa beans and cocoa powder are hygroscopic and will absorb moisture under humid conditions.

For safe storage, the moisture content of cocoa beans should be between 6 and 7 per cent; above 8 per cent there is danger of moulds developing within the beans; below 5 per cent the beans are very brittle and can result in bean breakages.

Cocoa stores should be built with particular attention being paid to temperature, humidity and cleanliness. The relative humidity in cocoa stores should not exceed 80 per cent for any length of time. Also of importance is air circulation in cocoa stores; any impedance to free circulation by covering the stack with a moisture-proof material, could lead to damp conditions and mould development at the top of the stack. Where humid conditions prevail or where the interval between bagging and shipment is long, properly dried beans (6 to 7 per cent moisture content) can be protected against the absorption of moisture and consequent enhancement of mould development by the use of sealed polythene linings inside jute bags. Polythene is impervious to moisture and will prevent the absorption of moisture by properly dried beans.

## *References*


**Allison, H.W.S.** and **Kenten R.H.** (1964). Mechanical drying of cocoa. *Tropical Agriculture*, **41**: 115–120.

**Allison, H.W.S.** and **Rohan T.A.** (1958). A new approach to the fermentation of West African Amelonado cocoa. *Tropical Agriculture*, **35**: 279–88.

**FAO** (1987). *FAO Mon. Bull. Statist.* June 1987.

**Heimann, W.** (1976). *Grundzüge der Lebensmittelchemie*, Steinkopff Darmstadt.

**Irvine, F.R.** (1974). *West African Crops*, University Press, Oxford, London.

**Kenten, R.H.** and **Powell B.D.** (1960). Production of heat during fermentation of cocoa beans. *Journal Science Food Agric.*, **11** (7): 396–400.

**Mc Cance, R.A.** and **Widdowson, E.M.** (1967). The composition of foods. *Medical Research Council, Spec. Rep. Series*, No. 297.

**Opeke, L.K.** (1982). *Tropical Tree Crops*, John Wiley and Sons Ltd., 108–119.

**Owusu-Domfeh, K.** (1972). The future of cocoa and its by-products in the feeding of livestock. *Ghana J. Agric. Sci.*, **5**: 57–64.

**Powell, B.D.** (1959). Cocoa fermentation: A note on the anaerobic and aerobic phases. *Chem. and Ind.* 991–2.

**Rohan, T.A.** (1958 a). *Observations on the fermentation of West African Amelonado cocoa.* Proc. Cocoa Conf., London, 1957, 203–7.

**Rohan, T.A.** (1963). *Processing of raw cocoa for the market*, FAO Rome.

**Salz, A.G.** (1972). *Cocoa processing — A practical approach to fermentation and drying cocoa and coconuts in Malaysia*, Proc. Conf. Incorp. Soc. Planters. Kuala Lumpur. 181–218.

**Shelton, B.** (1967). Artificial drying of cocoa beans. *Tropical Agriculture*, 44, 125–32.

**Ullmanns Enzyklopädie der Technischen Chemie**. (1976). 4. *Auflage, Bd.* 11.

**Wood, G.A.R.** (1961). Experiments on cocoa drying in the Cameroons, *Tropical Agriculture* **38**: 1–11.

**Wood, G.A.R.** (1975). *Cocoa*. 3rd Edition; Chapter 12, Tropical Agricultural Series, Longman.

# 3

# Coconut

## Production and distribution

The coconut palm, *Cocos nucifera* L., is a monocotyledon (Figure 3.1). It has neither a tap root nor a root-stock with divided branches; instead, many thousands of long roots uniform in thickness, emerge from the base of the swollen bole or stem. Coconuts are by far the most important nut in the world (Popenoe, 1969). They are grown for different purposes in various parts of the world. Due to their wide distribution, the original habitat of the coconut palm has, for a long time, been a matter of considerable controversy. One school of thought expounds a New World origin with subsequent dispersal to Polynesia and Asia. Others suggest an Indo-Pacific origin. The distribution of the coconut palm extends over the islands and coasts of most of the tropics and even in some places beyond the tropics. Today the main coconut growing areas of the world lie geographically within about 22°N and S of the equator. Outside these regions the palm generally tends to grow more slowly, is mainly vegetative, and has little or no commercial value.

The annual coconut crop of the world, estimated at approximately 17 million nuts, comes from an area of 3.5 million hectares distributed mainly over the tropical coasts of India, Indonesia, the Philippines, Sri Lanka, Malaysia, Central America and Africa. Over 80 different varieties have been described in the literature (FAO, 1970); many of these are reported to be simply 'ecological

**Figure 3.1 A fruiting coconut palm.**

types' which have evolved as a result of adaptation to certain environments. Recent classifications distinguish between definite genetic characteristics, in particular the mode of pollination, with further breakdown into types within these varieties. Accordingly, there are tall and dwarf varieties. The tall varieties are allogamous, or cross-pollinating, and the dwarf varieties are autogamous, or self-pollinating. The latter varieties are referred to as dwarf because they generally have a vegetative structure of short or reduced size compared with the former. Named tall and dwarf varieties of the coconut palm based on fruit and nut characteristics have been reported in the literature (FAO, 1970).

For most producing countries, statistics regarding production are available for copra entering world trade, with no allowance for local consumption of fresh coconuts, processed coconuts and non-food products. In many cases only export figures are available. A convenient source of estimates of world coconut and copra output is FAO. World coconut and copra production in 1984 was 31 million tonnes and 3.7 million tonnes, respectively. The largest producer of coconut is Indonesia accounting for nearly 31 per cent, followed by the Philippines producing 21 per cent of world output. The Philippines are the leading copra producer, accounting for nearly 38 per cent of world copra production (FAO Production Yearbook, Vol. 35, 1984). The local processing of coconut oil, the chief commercial product from the coconut palm, in coconut producing countries is expanding. The Philippines (53 per cent) and Indonesia (16.4 per cent) exported over 69 per cent of the coconut oil exported in 1985 (FAO, Mon. Bull. Statist., 1986). Total world export of coconut oil was 1.2 million tonnes in 1985.

## *Uses of the coconut palm*

Burkill (1935) describes the coconut palm as 'one of nature's greatest gifts to man' because of its numerous uses. Of the more than 100 products made directly or indirectly from coconuts, seven are important in world trade. These are:

1. Whole coconuts (tender and green as well as dry and mature).
2. Copra – dried kernel of the nuts used for shredding and pressing oil.
3. Coconut oil – extracted from copra or directly from the meat.
4. Coconut oil cake – the residue after extraction of oil.
5. Desiccated shredded coconut.
6. Coconut skim milk and coconut protein.
7. Coir – fibre from the husks.

Coconuts provide food, oil, medicine, fibre, timber, thatch, mats, fuel and domestic utensils. The oil is used for cooking, anointing the body, illumination, lubrication and making soap. It plays a prominent role in the customs and folklore of many people. Coconut oil also finds a number of uses in industry, including the manufacture of detergents and resins. Because it has a high melting point, coconut stearine is a valuable confectionery fat; it is also used as a substitute for cocoa butter and in the manufacture of candles. Coconut cake is widely used as food for cattle and poultry. Desiccated coconut, prepared by shredding and drying the fresh endosperm, is widely used in confectionery and bakery products. Coconut milk, produced by squeezing the freshly grated endosperm through a sieve, is widely used in curries, sweets and other cooking.

The water in the cavity of the fresh green unripe fruit is commonly drunk as a beverage throughout the tropics. It is cool, slightly aromatic and somewhat aerated as it is saturated with carbon dioxide derived from the respiration of the internal tissues.

The shell – the hard stony endocarp – accounts for about 25 per cent of the weight of the husked nut. Shells are used as fuel, particularly in copra kilns. Half shells are also used for bowls, cups, scoops and ladles. The shells take a high polish and can be carved and decorated. They are also used for making buttons, combs, bangles and musical instruments. Shells are used in the preparation of charcoal, and on distillation they yield wood tar and furfural.

Coir is obtained from the husks of the mesocarp which is removed in the preparation of copra. In southern India much of the coir is made into yarn; superior quality yarn is manufactured into mats, matting, rugs and carpets, while inferior grades are used for ropes and twine. Coir is used for filtering olive oil. Coir fibre has natural resilience and durability and is water resistant. It is useful in the making of cables.

The leaves of the coconut palm are plaited and used for thatching, mats, screens, hats, baskets and other articles. The midribs are made into brooms and baskets. The trunks are used for building.

## Structure of the coconut fruit

The fruit of the coconut palm is more or less ovoid, its exact shape depending upon variety, which also determines its size. Beneath the smooth outer skin, or exocarp, is a fibrous covering, or mesocarp. Together these form the husk. Beneath this, covered by a very hard shell, or endocarp, is the nut itself (Figure 3.2). The coconut consists of four parts; about 35 per cent husk, 12 per cent shell, 28 per cent meat and 25 per cent water (Aten *et al.* 1953). When unripe, the nut is almost completely filled with a watery liquid, called coconut

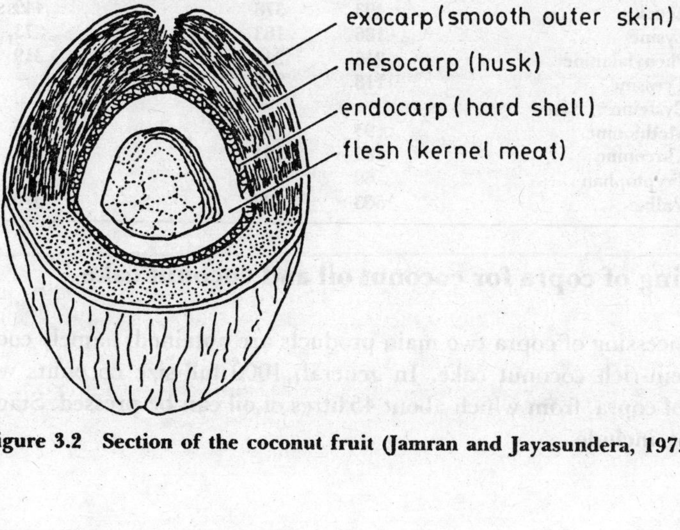


**Figure 3.2 Section of the coconut fruit (Jamran and Jayasundera, 1975).**

water, the amount and composition of which changes as development and ripening occur. When the nut is completely ripe, this water has almost disappeared, having been absorbed into the white endosperm. A small amount of water is essential if the nut is to germinate; a dry coconut will not germinate.

The 'meat', or endosperm, of the coconut is thin and jellylike when young, but a thickness of about 1 cm or more develops as the nut ripens. During development, the fruit attains its full size 6 months after fertilization. The shell begins to harden after about 7 months. After 10 months the endosperm is fully formed and hardened. The fruit is ripe and the endocarp fully hardened at 12 months, the average weight being between 2 and 3 kg.

Coconut is rich in phosphorus and calcium. Arginine, leucine, valine and phenylalanine are the main essential amino-acids found in coconut meat (Tables 3.1 and 3.2).

**Table 3.1 Mineral and vitamin content of coconut**

| Minerals and vitamins | Content (mg/100 g of coconut concentrate) |
|---|---|
| Calcium | 25 |
| Phosphorus | 75 |
| Iron | – |
| Carotene (vitamin A) | – |
| Ascorbic acid (vitamin C) | 2.0 |
| Thiamine | 0.15 |
| Riboflavin | 0.1 |
| Niacin | 0.1 |

**Table 3.2 Essential amino-acid content of coconut**

| Amino-acid | Content (mg/g) | | |
|---|---|---|---|
| | Coconut seed meat | Copra | Coconut seed protein (globulin) |
| Arginine | 673 | 688 | 1046 |
| Histidine | 98 | 97 | 151 |
| Isoleucine | 295 | 331 | 277 |
| Leucine | 493 | 378 | 449 |
| Lysine | 186 | 163 | 273 |
| Phenylalanine | 246 | 244 | 319 |
| Tyrosine | 118 | 250 | – |
| Cysteine | – | – | – |
| Methionine | 94 | 78 | 126 |
| Threonine | 183 | 194 | 254 |
| Tryptophan | 60 | 47 | 41 |
| Valine | 303 | 322 | 370 |

## Processing of copra for coconut oil and coconut cake

In the processing of copra two main products are obtained, namely coconut oil and protein-rich coconut cake. In general, 1000 full-size coconuts will yield 226.5 kg of copra, from which about 45 litres of oil can be pressed. Stages of oil processing include:

1. Pretreatment, including all treatments of the copra prior to extraction, such as drying, cleaning, crushing or size reduction and cooking.
2. Extraction – the separation of the raw material into oil and residue.
3. Postextraction treatment, consisting of the purification of the oil and the treatment of the cake.

## *Copra drying*

Copra is the name for dried coconut meat or kernel. It contains about 65 per cent oil which is extracted by crushing and pressing. The largest part of the world's coconut yield is dried to produce copra which is processed into coconut oil and coconut cake. On average, about 6000 nuts are required for one tonne of copra. Fresh coconut meat contains between 50 and 55 per cent water, most of which is removed during drying to yield a product containing less than 7 per cent moisture. Drying must be carried out as quickly as possible since coconut meat deteriorates very rapidly, being an excellent substrate for the growth of moulds and bacteria.

Three drying methods are in general use:
1. Sun drying.
2. Smoke curing or drying over an open fire in a direct drier or kiln, where the combustion gases produced come into contact with the drying coconut meat.
3. Indirect hot-air drying, where the combustion gases do not come into direct contact with the drying coconut meat.

In addition to these three methods, a combination of preliminary sun drying followed by kiln drying is also frequently used. The drying period should be as short as possible to prevent the coconut meat from deteriorating. One of the factors that can considerably shorten the drying period is high temperature. On the other hand, if the drying is done at too high a temperature the texture of the meat is altered and browning may occur, resulting in discoloured oil. Moreover, the copra cake derived from it may be so badly burned as to be unsuitable even for cattle feed. Another disadvantage of drying at too high a temperature is that the copra becomes 'case-hardened' and does not dry thoroughly. As a result a hard dry 'shell' encloses poorly dried copra. Copra dried in this way will have a rubbery consistency, which may cause difficulties in milling, and the copra cake will contain a high residual oil content.

Grimwood (1975) summarizes recommended drying practices as follows:
1. The moisture content must be reduced from 50–55 per cent to 35 per cent, preferrably within 24 hours.
2. During the subsequent 24 hours the moisture content should be reduced to about 20 per cent.
3. In the next 24 hours the moisture content should be reduced to about 5–6 per cent.

Figure 3.3 shows the moisture reduction rates for various drying methods (Kiln drying, sun drying, and forced hot air drying) with different drying cycles. Three types of drying cycle are shown in Figure 3.4. Curve K shows the rate of drying in a conventional direct-fired kiln, in which the moisture is

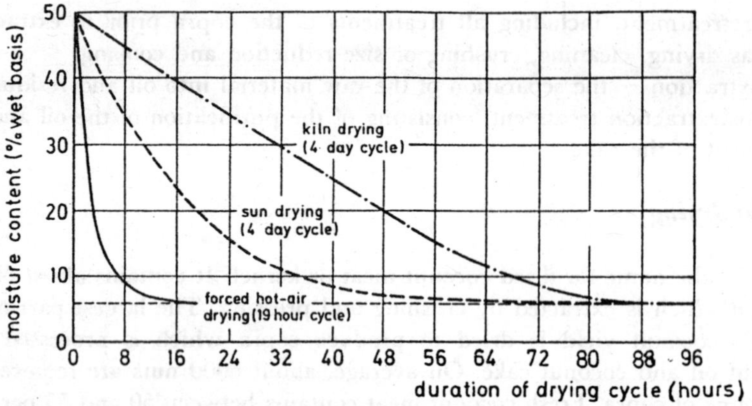


**Figure 3.3 Moisture reduction rates from various drying cycles (Salz, 1971).**

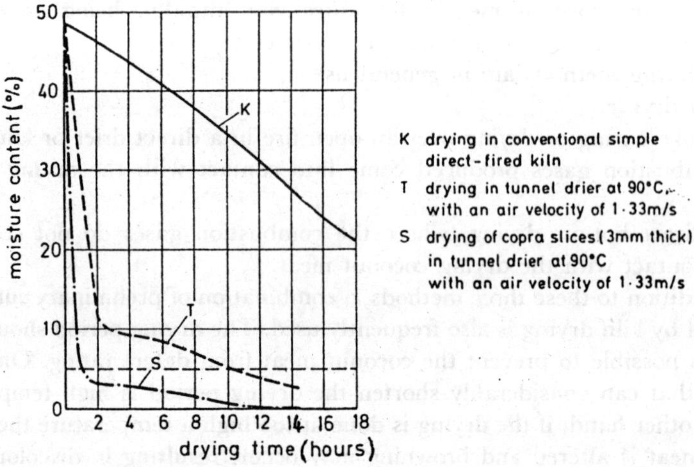


**Figure 3.4 Drying curves for copra (Des Vos, 1956).**

reduced from about 50 per cent to 5 per cent in 48 hours. Curve T shows the results obtained from a test with a tunnel drier, where the air temperature is kept at 90°C with an air velocity of 1.33 m/s. The drying period required to reduce the moisture to 5 per cent is 11 hours. Curve S shows a total drying period of only one hour when the fresh coconut meat is cut into slices approximately 3 mm thick, with drying carried out under the same conditions as curve T.

It is reported that the oil pressed from copra dried in this manner has good flavour and does not become rancid after eight months of storage.

The effect of the size of copra pieces (Figure 3.5) and temperature on drying rate were investigated by Palmer (1968). Palmer also suggests that an initial high temperature of 90°C can be used if the temperature is lowered to 70°C after two hours to prevent case-hardening.

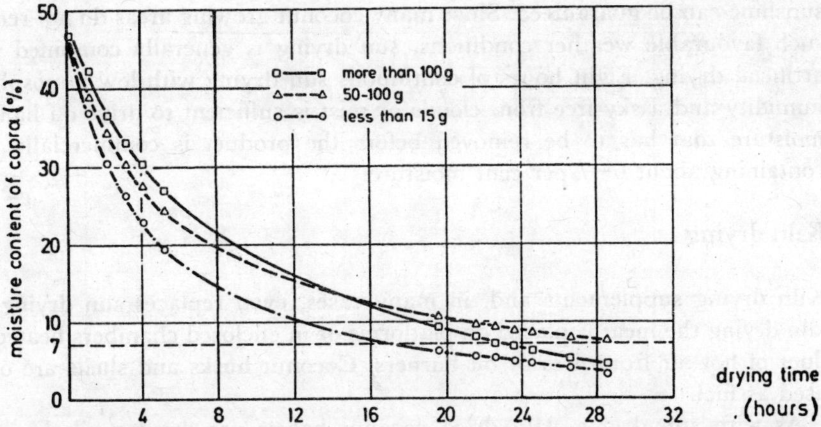


**Figure 3.5 Comparative drying rates for copra pieces of different sizes (Palmer, 1968).**

## Sun drying of copra

The first step in sun drying copra is to remove the husk from the mature nut. The husked nut is then opened by striking it at the 'equator' with a heavy knife; this operation splits the nut in two halves (Figures 3.6 and 3.7). The halves, after being left to drain, are laid on racks or concrete floors with the opened side turned to the sun. The average dimensions of such racks are approximately 1.2 × 2 m, so that they can easily be carried by two men.

After about 2–3 days drying, the kernels (meat) are detached from the shell, if necessary by means of a thin wooden lever. The kernel of copra is again dried for another 4–5 days. During the process of drying, the copra yard is covered with large screens or tarpaulins at night or during rainfall to protect the copra from damp. To obtain good quality cup copra it is necessary to dry the kernels in the sun continuously for 6–8 days.

This method of drying is, of course, only successful when long periods of

**Figure 3.6 Men preparing copra from coconuts in Papua New Guinea.**

**Figure 3.7 Coconuts split in half, ready for drying.**

sunshine can be guaranteed. Since many coconut growing areas do not receive such favourable weather conditions, sun drying is generally combined with artificial drying. Eight hours of continuous sun drying with low atmospheric humidity and a sky free from clouds or mist is sufficient to drive off half the moisture that has to be removed before the product is commercially dry, containing about 6–7 per cent moisture.

## Kiln drying

Kiln drying supplements and, in many cases, even replaces sun drying. In kiln drying the meat is placed on platforms or in enclosed chambers heated by flues of hot air from fires or oil burners. Coconut husks and shells are often used as fuel.

As with sun drying, kiln dried coconut halves are dried until the meat loosens, then the shells are discarded and drying is continued until a moisture content of 5–7 per cent has been reached. There are many different designs of copra kilns. The efficiency of the drying, in whatever type of kiln, depends upon the temperature, humidity and movement of the air. The optimum economic load for farm drying kilns is 200 coconuts, or 400 halves, arranged on four grills and heated to 50–59°C (Grimwood, 1975).

*Direct kiln driers*

In direct kiln driers, the coconut meat is heated directly by the smoke and hot combustion gases of the fuel, usually resulting in a heavily smoked product. In its simplest form the kiln consists merely of an open grill made from wood or bamboo slats to form the drying platform, 2–3 m wide, raised on posts above ground level (Figure 3.8).

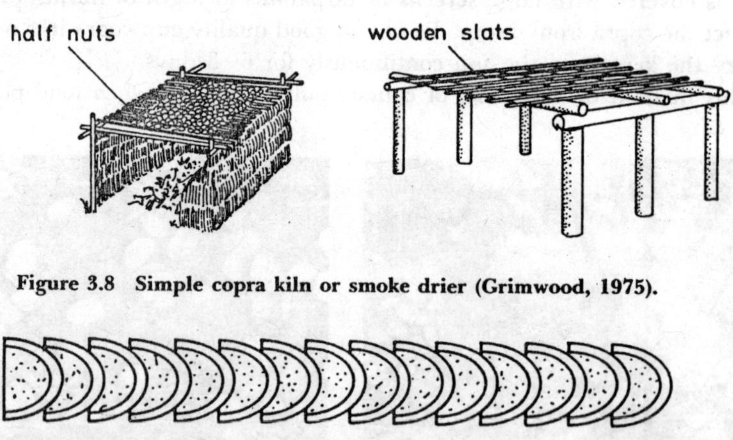


**Figure 3.8 Simple copra kiln or smoke drier (Grimwood, 1975).**

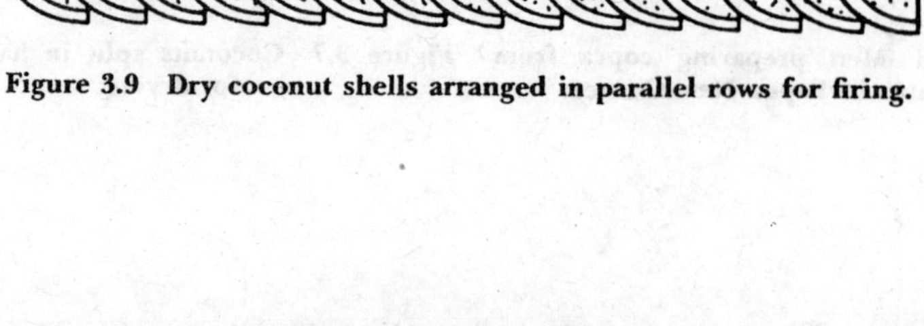

**Figure 3.9 Dry coconut shells arranged in parallel rows for firing.**

The split coconuts are placed on the grill and a mixture of combustible rubbish is burned underneath, either continuously or intermittently. Kilns of this type are often called *smoke kilns*. The most common drawbacks of such kilns are considerable heat loss, serious risk of fire, inadequate control of temperature and smoke, and a poorly dried end product. To protect the copra and the fire against eventual rain and wind, the smoke kiln is usually covered with a roof.

The drying process is more manageable when only dry coconut shells, fitted into each other in parallel rows as shown in Figure 3.9, are fired. The proper placement of the half nuts to be dried on a grate is illustrated in Figure 3.10. By suitable design, and the use of dry coconut shells as fuel, it is possible to produce satisfactory copra in kilns using direct heat. Of the numerous existing types of direct driers, the standard Sri Lanka kiln is reported to be one of the

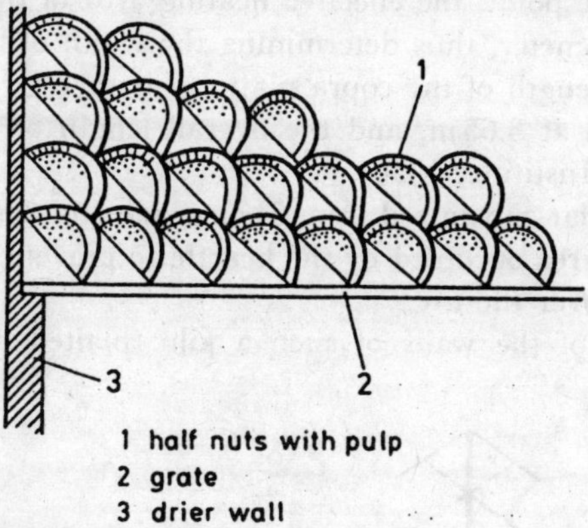


**Figure 3.10 Proper arrangement of half nuts on a grate (Tergast, 1937).**

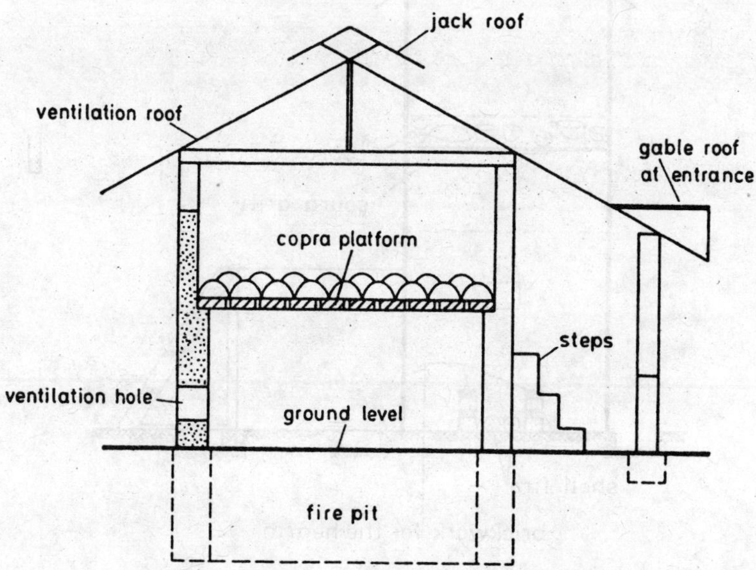


**Figure 3.11 Improved Sri Lanka copra kiln (Coconut Research Institute, 1956).**

best for general use. The copra obtained is of good quality and compares favourably with that obtained by indirect drying.

The standard kiln consists essentially of a fire pit about 0.6 m deep, a drying grill or platform 1.2 m above ground level and a corrugated iron roof fitted with a jack roof for good ventilation. The kiln has a total capacity of 10000 nuts, allowing a daily input of 2500 nuts (Figure 3.11).

*The smallholder's kiln (Cooke type)*

This is a modified version of the Sri Lanka kiln (Figure 3.11), designed for use on smallholdings of eight hectares or less (Figure 3.12). The design of the kiln is based on the travelling cone of heat produced at the end of a double chain of coconut shells 3.65 m long, lit at one end and burning at the rate of about 60 cm an hour, depending on the size of the shells. When the copra grill is 2.13 m above the fire point, the effective heating area of the cone is approximately 1.52 m in diameter, thus determining the width of the copra grill and the kiln itself. The length of the copra platform is therefore fixed at 3 m, the length of the hearth at 3.65 m, and the overall length of the kiln at 4.88 m (Coconut Research Institute, 1956).

A heat-spreader, flat-galvanized iron, 51 cm wide and 3 m long, is nailed to the rafters over the area occupied by the hearth, to prevent overheating of the copra immediately over the fire.

For construction of the walls of such a kiln, plaited coconut leaves are

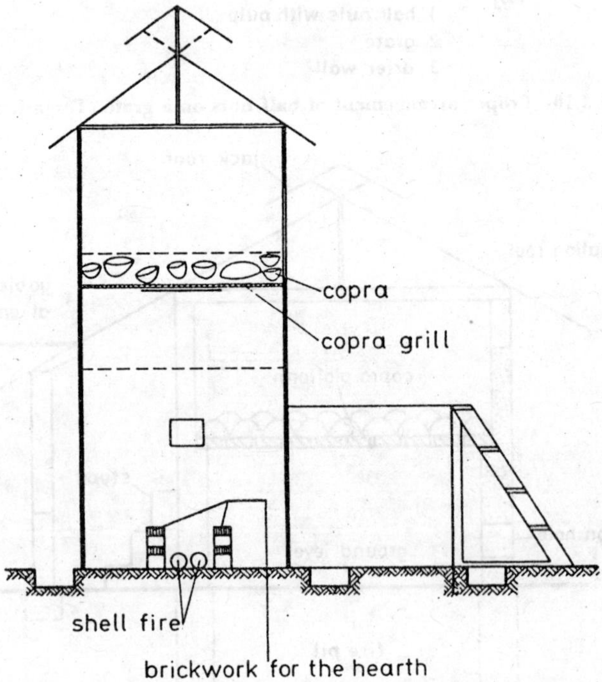


**Figure 3.12 Copra kiln (Cooke type) (Coconut Research Institute, 1956).**

reported to be preferable to sheet iron or brickwork, because plaited coconut leaves allow air to filter evenly into the fire without any of the wind disturbance that is normally associated with ventilation holes. It is also reported that the material keeps the air inside the kiln warm, so that no moisture condensation and chilling occur on cold nights as is the case with galvanized iron.

Superior in operation, the Cooke type kiln is reported to produce better quality copra in half the normal time, as little as two and a half days, without forcing the drying process in any way.

Other direct driers include the charcoal-fired oven, the Malaysian '30-acre' kiln and the IST-Nacoco drier. Detailed accounts, designs and modes of operation for these driers are described by Grimwood (1975).

## Indirect driers

In indirect driers, the coconut meat to be dried does not come into contact with the hot combustion gases and smoke from the fuel. For this reason the copra obtained from the indirect driers is usually of better quality. The simplest types of such driers consist in principle of an iron plate under which a fire pit is arranged.

*The Samoa drier*

The Samoa drier dries the copra by means of a convection current. The drier consists of a flue in a plenum chamber with a perforated drying platform above (Figure 3.13). Air inlets are provided to allow the convection current to flow. The coconut meat to be dried may be placed on iron mesh or coarse coir matting laid over a grating. The Samoa drier is suitable for small farms or small co-operative societies.

The disadvantage of the drier is that the flues, which are made from sheet iron and are approximately 30 cm in diameter, must be carefully maintained; this simple flue will corrode fairly quickly, which can lead to smoke escaping and contaminating the copra.

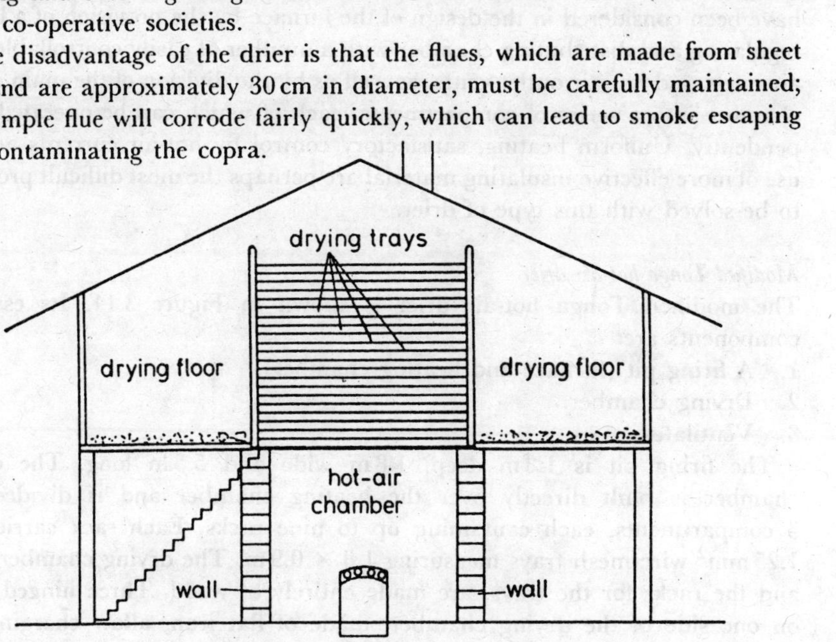


**Figure 3.13 Samoa drier (Grimwood, 1975).**

*The Comoro drier*

The Comoro drier consists essentially of a furnace, drying plates and a chimney. The furnace produces hot air which circulates inside it and at the bottom of the kiln, either through a sheet-iron pipe or under sheet-iron plates supported by brickwork baffles. Above, a number of plates are placed on slides about 18 cm apart. Each plate has a drying space of 1.3 × 0.4 m on which coconut meat is laid. It is reported that in this kiln 500 kg of meat can be dried in not more than 48 hours. Wood is the customary fuel. The kiln should not be heated too quickly, as oil may fall on the sheet-iron plates, causing an outbreak of fire; in any case, it is safer to put 3–4 cm of sand on the bottom plates to absorb any dripping oil.

*Fiji drier*

The Fiji drier consists of a furnace and an air heating system, a hot-air mixing and regulating chamber and a copra drying chamber, which is inclined and built to one side of the hot-air regulating chamber. It is divided into six separate compartments, each fed by hot air from one of the controlled apertures. Each compartment is fitted with a door, hinged at the top, with adjustable vents, and has six angle iron runners, spaced at 15 cm intervals, for carrying trays of coconut meat to be dried.

The advantages of the inclined chamber drier include a low risk of fire to the drier, since very little timber is used in its construction. Even if by accident any part of the hot-air regulating chamber or any copra drying compartment catches fire, that part can be isolated immediately by closing the hot-air dampers, and the fire can then be extinguished by completely closing all apertures. The questions of economic fuel consumption and heat conservation have been considered in the design of the furnace by the provision of a hot-air regulating and distributing chamber with a number of easily controllable cold-air inlets and hot-air outlet vents, as well as by the division of the main drying chamber into a series of compartments, each of which can be operated independently. Uniform heating, satisfactory control by hot-air currents and the use of more effective insulating material are perhaps the most difficult problems to be solved with this type of drier.

*Modified Tonga hot-air drier*

The modified Tonga hot-air drier is shown in Figure 3.14. Its essential components are:

1. A firing pit (furnace and heating chamber).
2. Drying chamber.
3. Ventilator.

The firing pit is 1.8 m deep, 1.8 m wide and 5.5 m long. The drying chamber is built directly over the heating chamber and is divided into 3 compartments, each containing up to nine racks. Each rack carries two 1.25 $mm^2$ wire-mesh trays measuring 1.8 × 0.9 m. The drying chamber itself, and the racks for the trays, are made entirely of wood. Three hinged doors on one side of the drying chamber, made of flat iron, allow charging and discharging of trays, and on the same side an open lean-to shelter provides

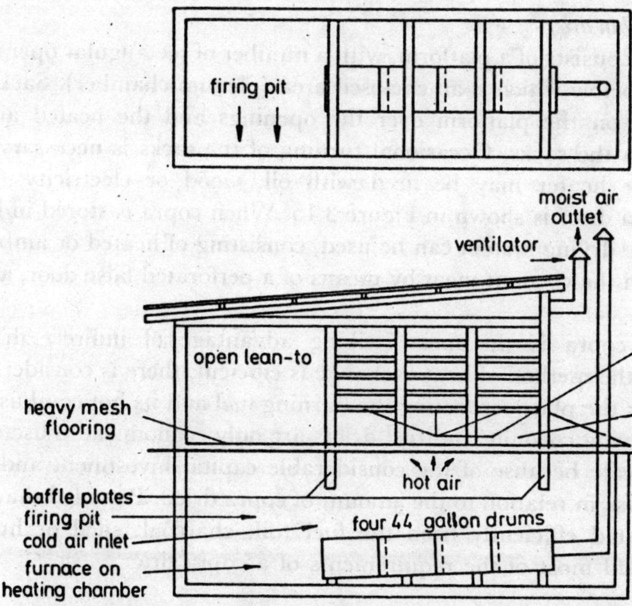


**Figure 3.14 Modified Tonga hot-air drier (Coconut Bulletin, 1964).**

protection during rainfall. A ventilator is placed high up in the chamber to discharge all the moist air.

Two baffle plates at the ends of the drying chamber project from ground level to about 0.6 m into the pit. These are used to direct the hot air up into the drying chamber.

The furnace and heating chamber consist of four 44 gallon drums, the ends of which are inserted one into another for about 15–23 cm, to make a single chamber. The fire, fuelled by coconut husks and shells, is set in the end portion of the furnace. The flooring of this chamber is made of heavy 5 cm wire mesh.

After firing, and when the drier is sufficiently hot, the whole nuts are split in half and stacked in the chamber, the first layer faces up and the remainder face down.

Predrying is carried out for 12 hours. Next morning the half nuts are removed, and the meat is cut into two or three pieces and put on the trays above for further drying. After this stage of drying, usually for two days, the copra is taken out of the drier and spread out in the sun for another two or three days.

The chamber has a capacity of 800 nuts per drying session, that is, 150–250 kg of dry copra. The amount of time required for drying is 12 hours for predrying, 24 hours for drying after removing the copra from the husks, and finally two days of sun drying or, alternatively, 36 hours of kiln drying.

Condensation will not occur if the furnace has been fired and the heating chamber is sufficiently heated for about an hour prior to stacking the half nuts in the drier.

*In-sack drier and in-bin drier*

The in-sack drier consists of a platform, with a number of rectangular openings of the same size, below which is an enclosed area (plenum chamber). Sacks of copra are placed on the platform over the openings and the heated air is forced up through the sacks. Occasional turning of the sacks is necessary for even drying. The heater may be fired with oil, wood or electricity. The principle of such a drier is shown in Figure 3.15. When copra is stored in bins or sheds, a simple drying system can be used, consisting of heated or ambient air blown through the coconut meat by means of a perforated false door, a fan and a heater.

Good quality copra is the most striking advantage of indirect driers. However, unless the method of heat exchange is efficient, there is considerable loss of heat across the plate separating the burning fuel and its hot combustion gases from the drying coconut. Indirect driers are only economical to use on a relatively large scale because of the considerable capital investment and the high rate of fuel use in relation to the amount of copra dried. Provided that the hot air is generated efficiently from the fuel (oil, charcoal, shell or husk), indirect driers fulfil most of the requirements of a copra drier.

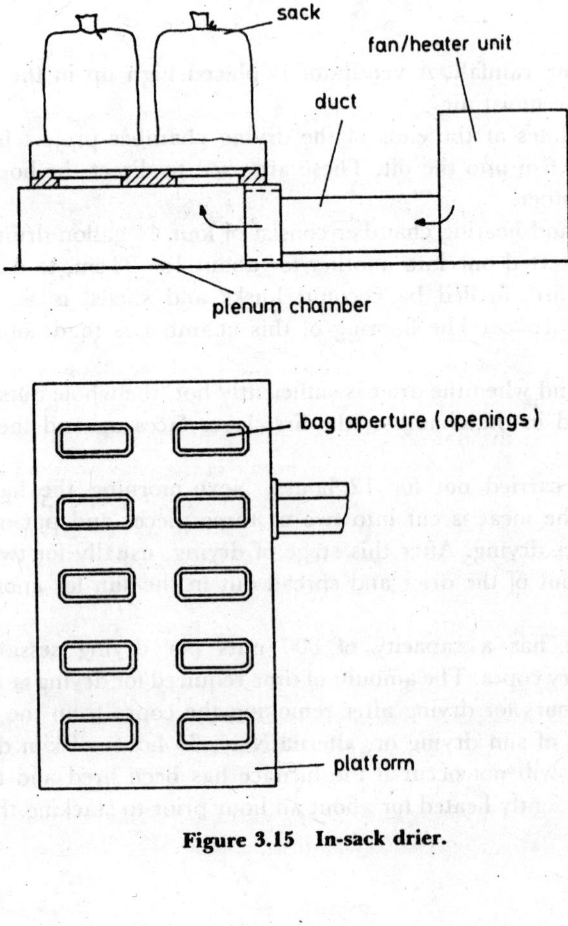


**Figure 3.15 In-sack drier.**

*Selected continuous indirect convection driers*

In this part of the chapter, selected continuous indirect driers still used in commercial practice are described. These are:

1. The Lister reversible air-flow system. This drier essentially consists of a heavy-duty air-cooled diesel engine driving a high-output axial fan, two perforated drying bins and an oil-fired heater to provide additional heat if necessary. In operation, heated air is forced up through one bin and down through the other. After the content of the first bin has been dried, it is removed and replaced by a fresh wet batch. Then the air is reversed to go up through the second bin and down through the freshly loaded bin. After the content of the second bin has been dried, it is emptied and reloaded, and the process is repeated. Copra can thus be dried to a 5 per cent moisture content. The reversible bin system is used with a drying unit and no additional heat for quantities not exceeding two tonnes dry copra/day. For larger outputs (3–7 tonnes) drying units with additional heater units are used. The major advantages claimed by the manufacturer are:

(a) The copra produced is of high quality compared with sun or kiln dried copra.

(b) Bins, which may be constructed using locally available materials, are relatively cheap.

(c) Low running costs.

(d) Low labour requirements.

The major disadvantage is the high capital cost of the machinery.

2. The commercial double-chamber drier. This type of drier has two chambers, each with a drying floor of 15 perforated plates and a capacity of about 25 tonnes. The coconut meat is placed in these chambers in bags. The drying process normally takes 40 hours, and loading and unloading can be effected in four hours. At the beginning of the drying operation the first chamber is charged on the first day and the second chamber on the second day. During the drying process, hot air is sucked through the channels of the steam heater into the first chamber and, later on, into the second chamber. After two days the drying process in the first chamber is completed, so the copra can be removed. By means of a valve system, hot air is then circulated in the opposite direction, so that the hottest air passes first through the second chamber and then through the first. Using this system one of the chambers can be discharged and loaded each day, regularly changing the direction of the hot air.

3. The tropical drying system (TDS). This system consists of a divided bin, a furnace and a fan. Air is drawn across the furnace, where it is heated, and passes up through the coconut meat in the first (left-hand) bin. It then flows over and down through the coconut meat in the second (right-hand) bin and out through the fan. The air returns across the furnace for recirculation or is expelled via the exhaust duct. By altering the position of the louvres, the air direction is reversed so that air can flow from right to left. The tropical drying system has the advantage of eliminating cold spots and giving even drying over the entire copra bed.

4. Chula driers. These solid-fuel driers are available in three basic types with different capacities (800 kg copra in 24 hours type, 1275–2050 kg copra in

24 hours and 680–6140 kg copra in 24 hours). The Chula drier consists essentially of two parts: the drying chamber and the air heater. The drying chamber rests at one end on the air heater and at the other end on steel supports. The air heater consists of a furnace and flues. The drying air enters around the sides of the furnace and is heated by passing through heating chambers. It is then carried through the flues along the bottom of the drying chamber and rises through the drying trays. When a tray of dried copra is taken out, it is replaced by a tray of wet coconut at the cool end, and the intervening trays are pushed along the steel rails provided; thus the trays gradually progress towards the hot zone of the drier, as charging and emptying are continuous. Approximately six trays of copra can be unloaded every four hours.

The 800 kg capacity drier has a natural-draught furnace while the other two have in addition a power-driven fan for assisting air circulation. Chula driers provided with power-driven fans and two-way valves ensure more uniform drying of the copra, especially during the drying of surface moisture because the air flow can be reversed.

## *Cleaning*

The next pretreatment after drying is cleaning the copra to remove stones, dirt and other foreign matter. In small processing plants cleaning is done by hand; foreign matter is picked out or the material is sieved through bamboo screens. In bigger plants, mechanical devices such as revolving or shaking screens, magnets or aspirators are common. Reels, rotary screens and revolving screens are designed to remove impurities such as dust, sand, pieces of shell, twigs and stalks from the copra. They have the shape of round drums or 6–8 sided cages. In the case of round drums, the screens are usually made from curved plates of perforated metal while the polygonal cages consist of frames with flat screens. The drum or cage is inclined to the horizontal to ensure continuous movement of the material. They are suitable for small-scale operations.

In recent years preference has been given to flat screens as cleaning equipment due to their simple construction, they are also easily accessible and controllable. Flat screens are fixed on a horizontal or slightly inclined frame which is either suspended or supported on springs and is subjected to a shaking motion. Shaking screens work at about 200–400 strokes/minute, whilst vibrating screens or vibrators work at about 3000 vibrations/minute.

Vibrating screens or vibrators are suspended freely on steel springs, enabling them to vibrate or oscillate freely in the vertical plane, or to perform circular or elliptical vibrations. With up to about 3000 oscillations/minute – more than the resonance of the system – they have a very short, fast, vigorous stroke. Vibrators are more popular than shakers because of their efficiency and their smooth action.

For the removal of foreign matter from the copra, a current of air may be used. This is based on the principle of winnowing, as used for the separation of chaff from grain. The device for cleaning copra in this way is called an aspirator or an air lift. An aspirator consists essentially of a fan, an adjustable valve and a feed roll. Copra is fed into the aspirator through an opening at the top. The air current passes vertically upwards to the descending product, carrying away the impurities. The cleaned copra leaves the aspirator at the bottom, against the current of the incoming air. Aspiration is usually combined with screening, in shakers or vibrators. In the simplest form of such a combination, a suction fan is mounted on the top of a shaking screen, and the fine impurities are sucked off the end of the screens by means of broad nozzles.

In the process of cleaning, iron particles can only be removed by magnetic separators. Simple permanent magnets have the advantages of simplicity and small size, and can therefore be easily attached to a cleaner; on the other hand their strength is limited. Electromagnets can be made in almost any size and strength, but they require a direct current and their power consumption is rather high.

## *Crushing*

Crushing of copra into small particles is done to facilitate easy extraction of the oil and consequently to increase the yield. Crushing is carried out in two stages. In the first stage a preliminary reduction occurs in special copra breakers, hammer mills or in equipment of similar construction. The final reduction prior to extraction is usually accomplished in roller mills, disk mills or disk beater mills.

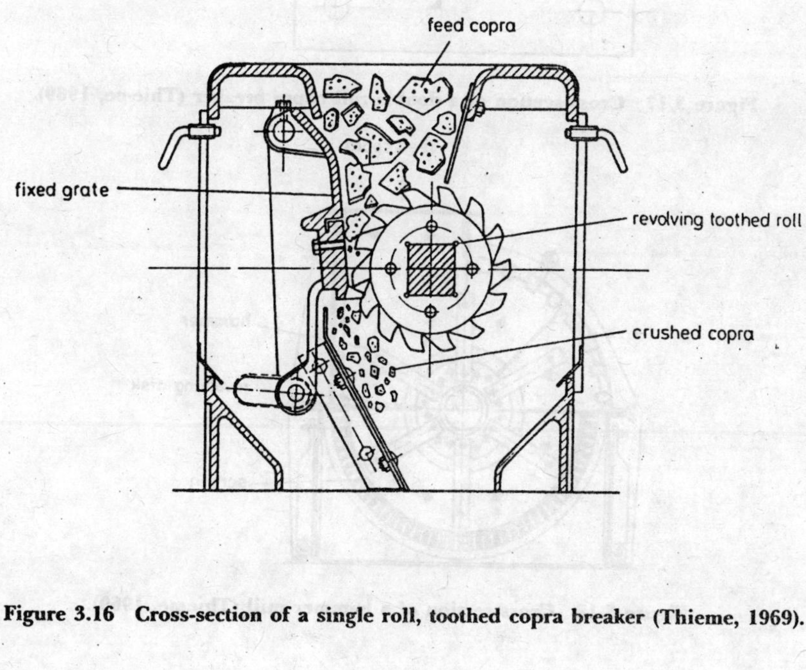


**Figure 3.16 Cross-section of a single roll, toothed copra breaker (Thieme, 1969).**

## Copra breakers

Copra breakers consist essentially of a feed hopper and either horizontal shafts on which blunt knives, teeth or beaters are fastened, or fluted or toothed rollers. Single roll or double roll breakers are the two main types of breakers (Figure 3.16 and 3.17). The former has only one toothed roll that works against a fixed grate. The latter, on the other hand, has two rolls with meshing teeth to crush the copra.

Hammer mills are also used for copra crushing. The beaters fixed on a shaft work at 2000–4000 rpm. Crushing takes place mainly due to the impact of the hammers, combined with some cutting and rubbing action. Below the hammers, a screen in the form of a trough is mounted, through which the crushed copra passes. A hammer mill, unlike the copra breaker, can be adjusted to grind to a fine powder. A cross-section of a typical hammer mill is shown in (Figure 3.18).

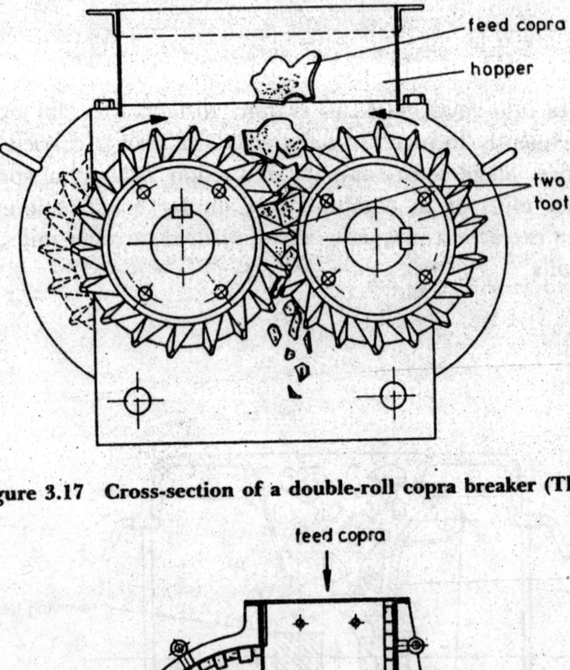


**Figure 3.17 Cross-section of a double-roll copra breaker (Thieme, 1969).**

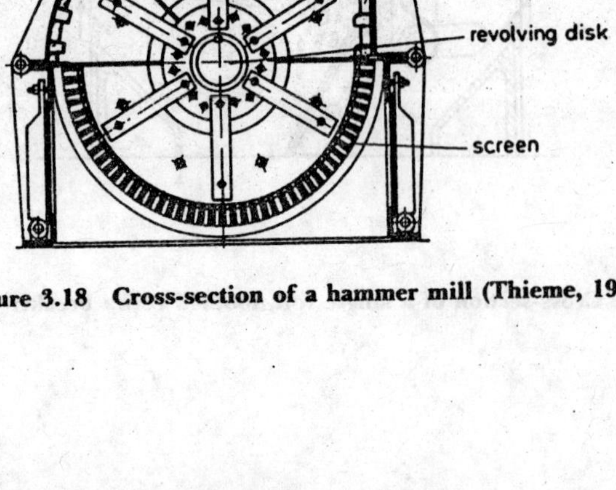


**Figure 3.18 Cross-section of a hammer mill (Thieme, 1969).**

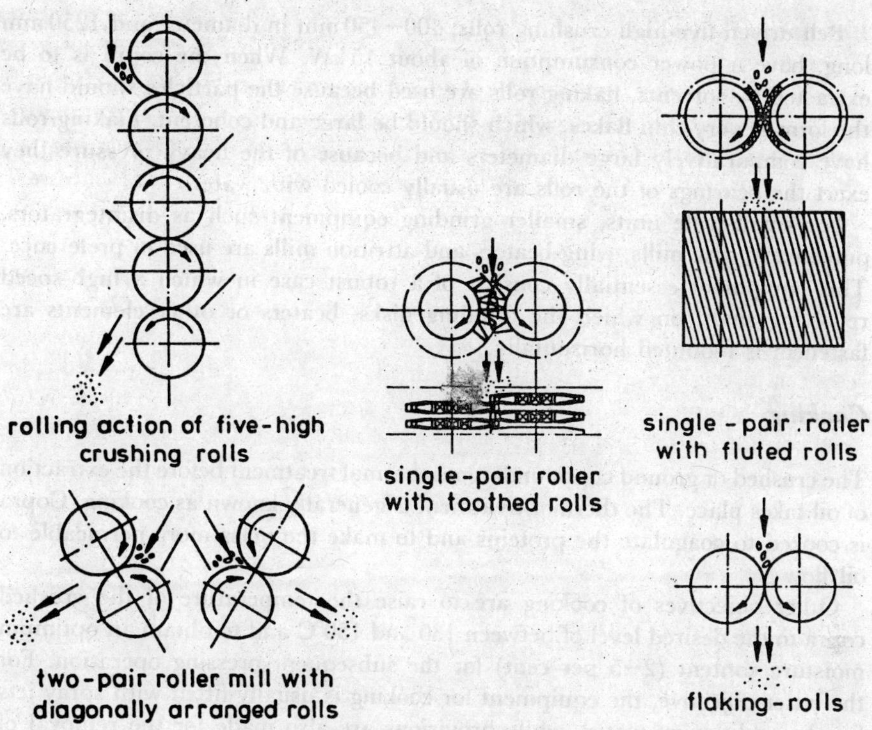


**Figure 3.19 Various types of roller mills.**

After preliminary crushing of the copra in a breaker or a hammer mill, the material is first of all ground to a fine powder between rolls or so-called seed-crushing rolls. Two types of mills are distinguished, depending on the arrangement of the rolls, the roller mills and the Anglo-American rolls. In both types, crushing takes place by pressure and friction between the surfaces of the rolls, which can either be smooth or fluted; the pressure is produced either by the weight of the rolls themselves, by means of spring-loaded bearings or hydraulically. In the roller mills, the rollers are used in pairs, either single pairs or 2–4 pairs; each pair having its shaft either in a horizontal plane or slightly inclined (Figure 3.19).

The so-called Anglo-American rolls have their shafts arranged in a vertical plane and consist therefore of a series of single rolls placed one above the other. Pressure is produced by the weight of the rolls. A common version of the Anglo-American rolls is the 'five-high crushing-rolls', consisting essentially of five rolls and a stripper plate.

For medium-sized and small mills, the diameter of the rolls may vary from 25–80 cm and the length from about 30–100 cm. Usually, alternate rolls have different diameters as this is considered to prolong the life of the rolls; also a certain degree of 'slip' occurring between rolls – resulting from differences in speed – is believed to provide the desired grinding action. The mills make between 120 and 200 rpm, depending on the diameter of the rolls, and have a capacity of 150–4000 kg/hour.

Belt driven five-high crushing rolls, 400–450 mm in diameter and 1250 mm long, have a power consumption of about 15 kW. When the copra is to be extracted by solvents, flaking rolls are used because the particles should have the form of very thin flakes, which should be large and coherent. Flaking rolls have comparatively large diameters and because of the heavy pressure they exert the bearings of the rolls are usually cooled with water.

For small-scale units, smaller grinding equipment such as disintegrators, pulverizers, disc mills, wing-beaters and attrition mills are used in preference. This equipment essentially consists of a robust case in which a high-speed revolving shaft, on which the grinding disks, beaters or other elements are fastened, is mounted horizontally.

## *Cooking*

The crushed or ground copra undergoes thermal treatment before the extraction of oil takes place. The thermal treatment is generally known as cooking. Copra is cooked to coagulate the proteins and to make the mass more permeable to oil flow.

Other objectives of cooking are to raise the temperature of the crushed copra to the desired level of between 130 and 150°C and to obtain an optimum moisture content (2–5 per cent) for the subsequent pressing operation. For the latter objective, the equipment for cooking is usually fitted with spray jets for the addition of water, while provisions are also made for the removal of excess moisture by means of an exhaust pipe and fan.

The equipment for cooking is shaped like a pan or kettle. The machines are variously called *cookers*, *scorchers*, *heaters* or *mill kettles* and consist basically of flat pans with a stirrer (Figure 3.20). Heating may be done by means of an open fire or steam jackets at the sides or the base. There are two types of cooker, namely the open-type and the closed-type. The simple, open-type cooker is suitable for rural areas, where steam boilers are not available. The cooker has a depth of 300 mm and a diameter of 700–1200 mm, and a capacity of 75–125 kg/hour of crushed copra. Based on 2–3 charges/hour, the power requirement is more than 0.75 kW (Thieme, 1969). For a higher capacity of

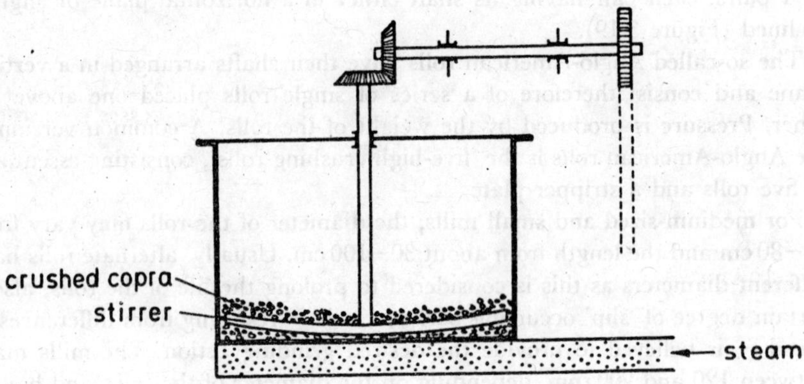


**Figure 3.20 Simple copra cooker.**

100 tonnes or more per 24 hours, the so-called three-high open-type kettle may be used. It has steam-jacketed bases, sides and stirrers with a central shaft passing through all the pans.

### Closed-type cookers

Closed-type cookers are also often called *stack cookers*. They consist of 3–6 closed round pans arranged one on top of the other, independently jacketed for steam heating. Each pan is equipped with rotating arms driven by a central shaft. Crushed copra is fed from the top of the cooker into the highest pan, dropped from there through an automatically operated gate at the bottom of the pan into the pan below, and so on until it eventually leaves the cooker at the bottom.

## *Oil extraction from copra*

Oil may be removed from the heated, ground copra by mechanical pressing, centrifuges or solvent extraction. Mechanical pressing is the simplest method, in which oil is extracted by mechanical devices using levers, wedges, screws or hydraulic appliances. Centrifuges and solvent extraction plants are only used in special cases because of their high investment costs. Mechanical pressing is universally used and may be continuous or batchwise.

In the batch method, the copra is spread on press cloths and metal racks. These are stacked, with space between them, and pressed, with about 96 000 kPa applied by hydraulic lift, causing the oil to flow to the edge of the tier and down the sides into the collecting trough. The pressure which can be applied in open batch presses is limited. For the same material under the same conditions of temperature and draining time, the decisive factor is the applied pressure; the higher yields achieved with modern equipment are largely attributed to the higher pressures applied. The equipment for oil extraction may be divided into three groups: batch presses, expellers, and solvent extraction equipment.

### Batch presses

Stump and screw presses belong to the oldest type of batch presses used in the production of coconut oil. Screw presses have screws to produce pressure. The copra to be pressed is placed in an enclosed cage and by turning the screw, pressure is exerted. The oil runs through the perforations of the cage into a collecting pan. A diagram of a screw press is included in the section on palm oil extraction.

*Chekkus*

Chekkus are widely used in India and Sri Lanka for the pressing of copra to yield oil. They consist essentially of a stone or wooden mortar 75–100 cm in height. Inside the mortar, which has an outlet for the oil, is a pestle made from hardwood approximately 150 cm long and 60–80 cm in circumference.

It is held obliquely in position by a cone-like construction; this enables the pestle to revolve, while at the same time it is forced inside the mortar by means of a lever about 5 m long which at its extremity is loaded with stones. The pestle is slowly turned round by means of a pair of oxen fastened to the free end of the lever. Two men are required to operate it – one attending the oxen and one the mortar. The capacity is about 100 kg of copra per day in six charges.

Power driven chekkus are called *ghanis*. The mortar itself revolves against the pestle; it is driven by a gear from a shafting, so that a battery of mortars can be run from one engine. The amount of oil expressed per kg of copra is higher than in a chekku press. However, power consumption is high in relation to output, and wear and tear on the mortars is considerable.

## Continuous working screw presses – expellers

An expeller consists essentially of a main shaft carrying the screw assembly which pushes the copra to be pressed continuously forward in a horizontal cage or barrel, while simultaneously exerting sufficient pressure to force the oil through the openings of the cage. At the discharge end, a movable cone or choke controls the operating pressure by changing the width of the annular space through which the press cake must pass.

The simplest expellers resemble household machines and are mostly hand-driven; their capacity is up to 20 kg per hour. Other single engine-driven

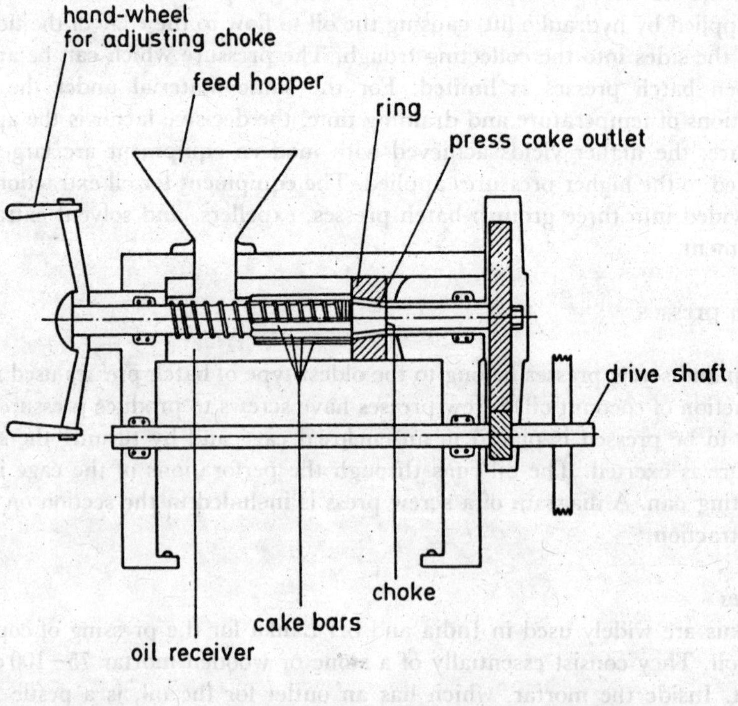


**Figure 3.21 Continuous working screw press – expeller.**

expellers have capacities between 40 and 200 kg copra per hour and require 2.5–7.5 kW (Figure 3.21). Using single expellers, copra has in some cases to be pressed twice – prepressing and finish pressing – for satisfactory extraction.

## Oil extraction by solvent

Oil extraction by means of expellers is naturally associated with losses which can be as high as 10 per cent or more; a certain percentage of oil always remains in the cake, even with the most efficient expeller. With solvent extraction it is possible to reduce the residual oil in cake or meal to less than 1 per cent. For raw materials which originally contain only a low percentage of oil, such as rice bran, cotton seed and soya beans, solvent extraction is therefore the ideal extraction method. The use of solvent extraction is less appropriate for raw materials with a high percentage of oil, such as copra.

Most solvent extraction plants have relatively high capacities, the minimum capacity being approximately 25 tonnes of raw material per day. Compared with expellers and presses, the solvent plant is a complicated system that requires skilled technicians to maintain it and keep it running economically. Solvent extraction plants may be divided into two main groups: batch extraction plants and continuous extraction plants.

### *Batch extraction*

A solvent extraction plant working batchwise consists basically of the following parts:

1. Extractor.
2. Feed hopper.
3. Calandria heater (evaporator).
4. Solvent tank.
5. Miscella tank.
6. Condenser.
7. Water separator.

The raw material is fed from the hopper into the extractor. From their respective tanks, solvent and weak miscella are fed into the extractor, where they dissolve the oil. They pass as enriched miscella through a filter within the extractor to the chamber which is connected with the evaporator. Circulating rapidly, the solvent immediately begins to evaporate here. The solvent vapours are condensed in the condensers, from where they are fed into the extractor. The concentrated miscella is then passed back from the calandria heater to a chamber. The oil is collected in the lower part of the calandria heater (evaporator), where the last traces of solvent are also driven out by steam. The process is repeated automatically until the material no longer contains extractable oil. The working principle is based on the so-called *Merz system*.

### *Continuous extraction*

In all continuous extractors, the raw material moves through the vessel continuously, against the flow of the extracting medium – pure solvent and weak miscella. The various types of continuous extractors include:

*Basket-type extractors* are also known as Bollmann extractors. The material rests in the bucket or basket and is moved vertically in a *bucket elevator*, horizontally in a *basket band* or in a rotary motion. The solvent/miscella mixture is sprayed on the buckets. The advantages of this system are good enrichment of the miscella, the use of only a minimum number of pumps and an almost negligible power requirement.
*Semibasket-type extractors* are similar in construction to the basket-type, except that the walls and perforated bottoms of the baskets are independent of each other and can be moved separately. This system, which can only be used horizontally or in a rotary motion, offers the advantage of pouring and mixing the material as it passes from the higher stage to the lower one.
*Extractors with horizontal conveyors* contain solvent which moves horizontally through the bath against the flow of the material; the screw, which pushes the raw material forwards, causes a certain amount of disintegration in the material.
*Extractors with vertical conveyors* consist essentially of two vertical screws interconnected at the bottom by a horizontal screw. They move the material first downwards, then across and finally upwards through the solvent bath.

## *Postextraction operations*

The postextraction treatment includes cleaning the oil and powdering or pressing the residual copra cake. Coconut oil extracted from copra is cleaned or refined to remove free fatty acids, odours, flavours and other impurities such as traces of water, undesirable colours and gum.

Refining consists of adding a calculated amount of alkali to the oil to remove free fatty acids, subjecting it to steam under a vacuum to remove odours and flavours, and filtering it with carbon to decolourize it. In the process, a precipitate is formed which is removed as *foots*. Coarse impurities are removed by screening, by filteration or by centrifugal separation; water may be removed by evaporation; for degumming settling tanks are used.

### Simple oil cleaning equipment

Most expellers are built with a screen (strainer) under the barrel to prevent coarse particles from entering the main collecting vessel. For higher capacities of oil, a separate shaking or vibrating screen with up to 3000 oscillations per minute may be installed on the top of the receiving tank for the extracted oil. Perforated plates with up to 324 holes per 6.45 $cm^2$, or woven wire cloth with up to about 100 x 100 mesh can be used.

A piece of equipment which should also be present in the postextraction operation is the filter press. It is used for filtering the oil through filter cloth. Filter presses consist of filter elements arranged in parallel, which are pressed together hydraulically or by means of a screw (Figure 3.22).

The two main types of filter presses are the *chamber presses* and the *plate* and *frame presses*. The plates of the chamber press have a raised rim on both sides, by means of which chambers are formed between the plates. The plates have

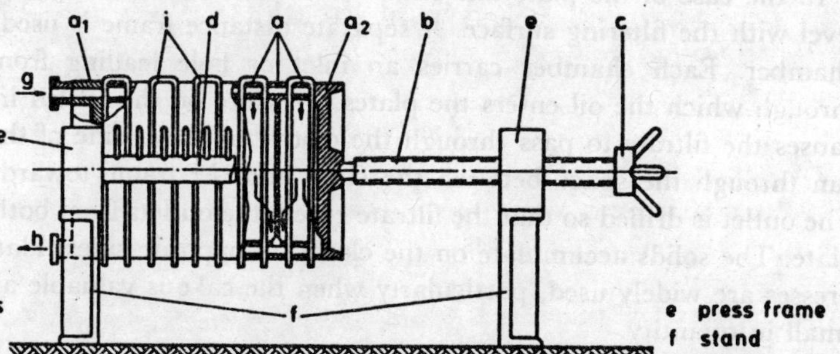


**Figure 3.22 Filter press (Ullrich, 1967).**

feed holes in the centre with corresponding holes in the filter cloths. The oil to be filtered is put into the press, filling the chambers between the plates and eventually filtering through the pores of the cloth. The filtered oil can drain into passages at the base of the plate and from there it can be collected (Figure 3.23).

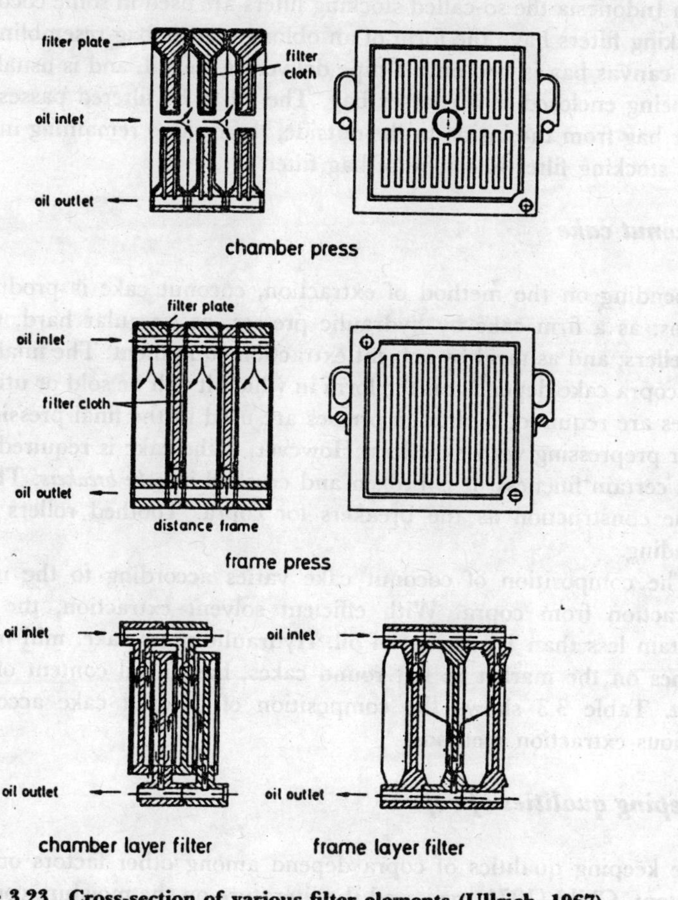


**Figure 3.23 Cross-section of various filter elements (Ullrich, 1967).**

In the case of the plate and frame press the rims of the plates are almost level with the filtering surface. A separate distance frame is used to form the chamber. Each chamber carries an inlet or hole leading from a conduit through which the oil enters the plates. Pressure on the oil fed into the press causes the filtrate to pass through the cloths on either side of the plates and run through the space between the cloth and the frame towards the outlet. The outlet is drilled so that the filtrate enters the outlets from both sides of the plate. The solids accumulate on the cloths on opposite sides. Plate and frame presses are widely used, particularly when the cake is valuable and relatively small in quantity.

Filter cloths are made from natural as well as from artificial fibres. A combination used frequently consists of a rather thick porous cloth as an underlayer and a thin, densely woven twill cloth on top of it. The filtering takes place through the thin cloth to avoid any mechanical damage and to facilitate the flow of oil after it has passed through it.

Other types of filter used are bag filters, leaf filters and rotating filters, which work without interruption, the solid material being removed automatically and continuously.

In Indonesia the so-called stocking filters are used in some coconut oil mills. Stocking filters have the form of an oblong canvas bag resembling a stocking. The canvas bag is tied onto a pipe delivering the oil, and is usually supported by being enclosed in a coarser bag. The oil to be filtered passes through the filter bag from the inside to the outside, the residue remaining inside the bag. The stocking filter is a type of bag filter.

### *Coconut cake*

Depending on the method of extraction, coconut cake is produced in three forms: as a firm cake by hydraulic presses; as irregular hard, thin slabs by expellers; and as meal by solvent extraction equipment. The final treatment of the copra cake depends on the form in which it will be sold or utilized. If hard cakes are required, hydraulic presses are used in the final pressing operation, after prepressing with expellers. However, if the cake is required as a powder of a certain fineness, it is broken and crushed in *cake breakers*. These have the same construction as the breakers for copra. Toothed rollers are used for grinding.

The composition of coconut cake varies according to the method of oil extraction from copra. With efficient solvent extraction, the cake should contain less than 1 per cent of oil. Hydraulic press cake, mill poonac, which comes on the market as flat round cakes, has an oil content of about 6 per cent. Table 3.3 shows the composition of coconut cake according to the various extraction methods.

### *Keeping qualities of copra*

The keeping qualities of copra depend among other factors on its moisture content. Child (1974) reviewed the literature on the moisture content of copra

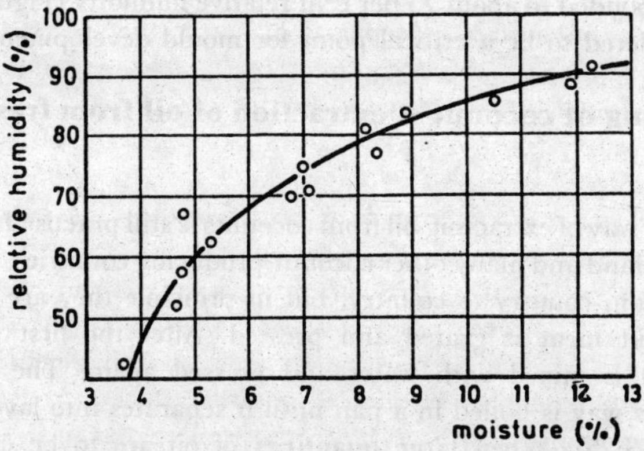


**Figure 3.24 Correlation between humidity and moisture content of copra (Jamaican Research Department, 1961–1962).**

and its keeping qualities, and concluded that copra, dried to 6–7 per cent moisture and stored in a well-ventilated dry place, does not seriously deteriorate.

The Jamaican Research Department made studies on the equilibrium moisture content of copra at various relative humidities of ambient air. Good quality copra was stored in jars in which the relative humidity was controlled. After five weeks the samples were removed and their moisture content was determined. A further five weeks storage caused insignificant changes, indicating that the samples had reached equilibrium. A moisture content of 7

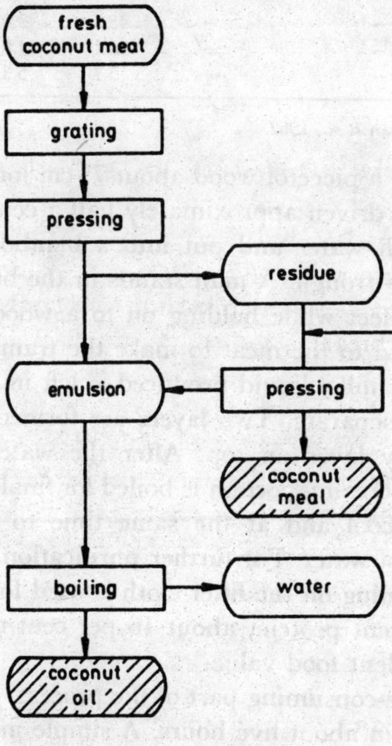


**Figure 3.25 Flowchart for traditional wet processing of coconut oil.**

per cent corresponded to about 75 per cent relative humidity (Figure 3.24), and this was considered to be a critical point for mould development.

## Wet processing of coconut – extraction of oil from fresh coconuts

The traditional way of extracting oil from coconuts is still practised in Indonesia, Malaysia, Thailand and many other coconut-producing countries. The methods vary slightly from country to country, but in principle they are the same.

Fresh coconut meat is grated and pressed. After the first pressing, the grated coconut is mixed with water and pressed again. The milky liquid obtained in this way is boiled in a pan until it separates into layers of oil and water (Figure 3.25). When large quantities of oil are to be extracted, the pieces of fresh coconut meat are soaked overnight and grated the following morning with a grater.

**Table 3.3 Composition of coconut cake**

| Constituent | Solvent extracted meal (%) | Blend of solvent extracted with hydraulic press meal (%) | Hydraulic press (%) | Expeller (%) | Ghani (%) | Chekku (inefficiently pressed) (%) |
|---|---|---|---|---|---|---|
| Moisture | 10.95 | 10.82 | 11.0 | 10.0 | 9.4 | 13.3 |
| Fat | 0.53 | 3.26 | 6.0 | 10.0 | 12.9 | 26.7 |
| Protein (6.25 × N) | 20.38 | 19.38 | 19.8 | 19.1 | 20.5 | 14.3 |
| Carbohydrate | – | – | 45.3 | 43.8 | 40.1 | 32.8 |
| Fibre | – | – | 12.2 | 11.8 | 11.9 | 8.9 |
| Ash | – | – | 5.7 | 5.3 | 5.2 | 4.0 |

Source: Nathanael, 1960; Rajaschharan *et al.*, 1962.

The traditional grater is a piece of wood about 75 cm long and 10 cm wide, into which pins have been driven approximately half a centimetre apart. The grated meat is mixed with water and put into a bamboo basket, which is placed in a square wooden trough. A man stands in the basket and tramples the grated meat with his feet while holding on to a wooden support. From time to time water is added to the meat to make the trampling easier and to increase the oil yield. The milky liquid produced is left in a vessel for several hours so that the oil can separate. Two layers are formed – a watery layer underneath and a creamy layer on top. After the watery layer has been drained off, the remaining creamy portion is boiled for final separation (Figure 3.26). The oil is left to cool and at the same time to 'dry', through the evaporation of the traces of water. For further purification it may be filtered.

The slimy matter remaining on the filter cloth is used for human consumption. With about 50 per cent protein, about 16 per cent oil and 12 per cent carbohydrate, it has excellent food value.

Grating is the most time-consuming part of the process; two skilled workers can grate about 500 nuts in about five hours. A simple motorized grater can grate up to 200 coconuts in ten minutes. Approximately 150 nuts yield 17 kg

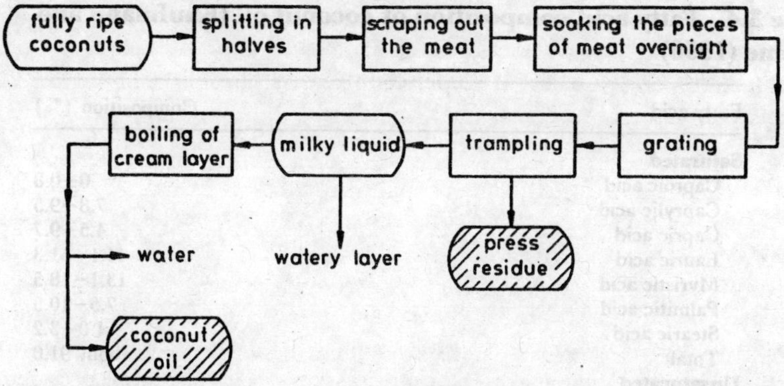


**Figure 3.26 Traditional method of extracting oil from coconuts.**

of oil. Pressing the meat from 500 coconuts by trampling takes about five hours and the subsequent separation of the oil by boiling takes 3–4 hours (Thieme, 1969).

The traditional method for the processing of coconut oil is characterized by the following features:

1. Oil is extracted from the fresh undried meat (endosperm) of coconuts.
2. Very simple equipment is used in the operation.
3. Oil losses in the residue are high.
4. The operation is labour intensive.
5. The oil is of very good quality but it still has too high a moisture content, which prevents it from being stored for a long time.

## Chemical and physical properties of coconut oil

Coconut oil is classified as an edible industrial oil. Unrefined, in the country of origin, it is a colourless to brownish yellow oil, depending on the quality of the raw material and the method of processing. In temperate climates it appears as a greasy, somewhat crystalline, white or yellowish solid fat, which has a melting point of between 20° and 26°C. The flavour and taste of coconut oil varies from pleasant, like that of fresh coconuts, to rancid and disagreeable if improperly prepared and stored. When properly refined, it becomes odourless, tasteless and water-white. Unrefined coconut oil, like other vegetable oils, will keep much longer than refined coconut oil. The reason for this is that unrefined oils contain tocopherol, which acts as a natural antioxidant, protecting the oil against atmospheric oxidation and rancidity.

Generally, fats and oils are classified according to their characteristics and consequently their utilization. The main fatty acids in coconut oil are lauric, myristic, palmitic, oleic, caprylic, capric and stearic. Table 3.4 gives the fatty acid composition of coconut oil. The oil is normally solid below 20°C and more than 90 per cent of it is glycerol esters of saturated fatty acids. Lauric acid is the main constituent with smaller amounts of myristic and palmitic acids. The unsaturated oleic and linoleic acids make up no more than 9 per cent of the oil.

**Table 3.4 Fatty acid composition of coconut oil (Kaufmann and Thieme (1956)**

| Fatty acid | Composition (%) |
|---|---|
| **Saturated** | |
| Caproic acid | 0–0.8 |
| Caprylic acid | 7.8–9.5 |
| Capric acid | 4.5–9.7 |
| Lauric acid | 44.1–51.3 |
| Myristic acid | 13.1–18.5 |
| Palmitic acid | 7.5–10.5 |
| Stearic acid | 1.0–3.2 |
| Total | about 91.0 |
| **Unsaturated** | |
| Oleic acid | 5.0–8.2 |
| Linoleic acid | 1.0–2.6 |
| Total | about 9.0 |

Due to its high content of lauric and myristic acids, the oil has a very high saponification value and the soap lathers freely in hard and salty water. The low content of unsaturated acids makes coconut oil resistant to oxidative rancidity, thus adding to the keeping quality of baked food and fillings made with this oil.

Coconut oil represents a very complex mixture of glycerides. The oil contains 84 per cent trisaturated glycerides. Of the other glycerides, 12 per cent contain two saturated and one unsaturated fatty acid, and 4 per cent contain only one saturated and two unsaturated fatty acids. The glycerides contain 50–60 per cent caprylo-lauro-myristin and up to 20 per cent myristo-dilaurin, as well as smaller amounts of lauro-dimyristin.

Besides glycerides and free fatty acids – usually expressed as the percentage of lauric acid – coconut oil contains about 0.2–0.6 per cent unsaponifiable matter, that is the material that does not react with caustic soda to form soap. The unsaponifiable matter contains phosphatides (lecithin and cephalin), gums, sterols, tocopherols and lipochromes. About one third of the unsaponifiable matter in coconut oil consists of sterols (Bezard *et al.*, 1971).

In Table 3.5 the most important chemical and physical characteristics of coconut oil are presented.

**Table 3.5 Some chemical and physical properties of coconut oil**

| Property | Usual range |
|---|---|
| Acid value | 1.0–10.0 |
| Saponification value | 251.0–264.0 |
| Iodine value | 7.0–10.0 |
| Unsaponifiable matter (%) | 0.15–0.6 |
| Refractive index at 40°C | 1.448–1.450 |
| Specific gravity at 40°C | 0.907–0.913 |
| Melting point | 20–28°C |
| Solidifying point | 22–23°C |

Source: Kaufmann and Thieme, 1956.

## Desiccated coconut

Desiccated coconut is the dried, disintegrated white kernel of the coconut, processed under strict hygienic conditions for human consumption. It retains, for the most part, the original oil and protein of the unprocessed fresh nuts. The manufacture of desiccated coconut is a process involving husking, shelling, removing the testa, washing, sterilizing, disintegrating, drying, sieving and packing (Figure 3.27).

Whole coconuts are shelled by hand with the aid of a small hatchet or knife. After shelling they are pared; paring is the removal of the testa from the kernel with a special knife, rather like a potato peeler. The kernels are cut open to release the coconut water and then thoroughly washed in fresh water to remove invert sugars from the inner surfaces. To ensure elimination of *Salmonella* and drastically reduce all other microbiological counts, the washed coconut is sterilized.

In Sri Lanka a sterilizing bath is incorporated into the process and the coconut pieces are immersed for about one and a half minutes in boiling water. In the Philippines the disintegrated nut is subjected to live steam treatment in stainless steel blanchers at a temperature of about 88°C for five minutes.

Sterilization is followed by granulating or shredding, depending on the thickness desired, in a disintegrating machine. This is essentially comprised of a main rotor and a ring-adjustment mechanism. A typical disintegrator for this purpose has an output of 250–1000 kg per hour. After disintegrating, the wet pieces are desiccated or dried to a final moisture content of 2–2.5 per cent.

In Sri Lanka the wet meat is spread out to a depth of about 4 cm on trays, approximately 1.5 $m^2$ and made of a fine metal mesh supported by a wooden frame. Each tray has a capacity of 12 kg of wet meat. This meat is dried at 88°C for 45 minutes in a hot air drier. During the process the trays are moved twice and the content is raked over to ensure uniform drying and to break up any lumps that might have formed. In the Philippines, continuous conveyor driers are exclusively used in the manufacture of desiccated coconut. In this

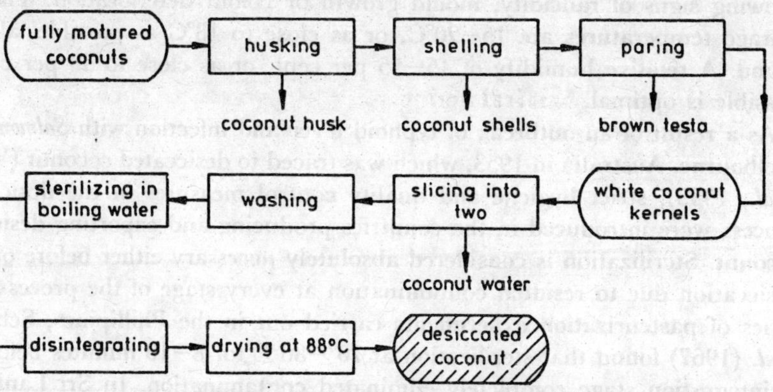


**Figure 3.27 Flowchart for processing of desiccated coconut.**

type of drier a continuous fine mesh conveyor belt some 3 m across, carrying the wet disintegrated meat at a depth of 7–8 cm, passes through a tunnel at a constant rate. Hot air of a uniform velocity, produced by blowing air over steam heated coils, is circulated through the bed of the coconut. About halfway through the tunnel, the coconut is usually transferred by a tumbling action onto a second conveyor belt placed beneath the first, effectively turning the coconut over in the process. In this type of drier the inlet temperature of the first stage of drying is about 115°C, while at the second stage it is reduced to 105°C. A tunnel of 40 m has an output of about 1350 kg/hour.

For the manufacture of desiccated coconut, it is necessary to process nuts on the basis of yield and quality. Yield will vary from variety to variety. A variety with a thick layer of endosperm is essential for economic yields.

Rajasekharan *et al.* (1962) showed the advantages of seasoning (maturing) in desiccated coconut production by demonstrating over a period of five months – from the ninth to the fourteenth month (from unripe to fully ripe nuts) – the changes in water and oil content; the oil content increased from 22.2 to 39.1 per cent, while the coconut water decreased from 57.0 to 42.9 per cent. The yield of desiccated coconut increased from 87 to 139 kg during the same period. Good desiccated coconut should be crisp, snow white in colour, with a sweet, pleasant and fresh taste. The oil content ranges from 68–72 per cent, but the oil should not contain more than 0.1 per cent of free fatty acids.

## Storage of desiccated coconut

The keeping qualities of desiccated coconut, like most processed, fat-rich foods, depend to a large extent on the moisture content. Desiccated coconut, with a moisture content of 6.73 per cent and kept in sealed containers, shows considerable growth of green, pink and brown moulds, and also has a strongly rancid odour after two months of storage. On the other hand, under similar conditions, desiccated coconut with 1.8 per cent moisture shows no trace of mould and does not have a bad odour after a two month storage period (Child, 1940). If it is kept under favourable conditions (temperature, relative humidity, ventilation), desiccated coconut can be stored for months without showing signs of rancidity, mould growth or colour deterioration. The best storage temperatures are 15–20°C, or as close to 18°C as possible, all year round. A relative humidity of 45–55 per cent, or as close to 50 per cent as possible is optimal.

As a result of an outbreak of typhoid fever and infection with *Salmonella* in Melbourne, Australia in 1953, which was traced to desiccated coconut (Wilson *et al.*, 1955), strict hygiene and quality control measures in the desiccating process were introduced in the countries producing and exporting desiccated coconut. Sterilization is considered absolutely necessary either before or after desiccation due to residual contamination at every stage of the process. In a series of pasteurization experiments carried out in the Phillipines, Schaffner *et al.* (1967) found that sterilization at 70°–80°C for 8–10 minutes before the disintegration stage completely eliminated contamination. In Sri Lanka the sterilization process is carried out at 100°C for one and a half minutes.

A draft of code of hygienic practices for the manufacture of desiccated coconut has been published by the Joint FAO/WHO Food Standards Programme, Codex Alimentarius Commission (FAO/WHO, 1970). Importing countries of desiccated coconut have their own set of standard specifications. These generally refer to the moisture, fat and free fatty acid contents, the absence of bacteriological contamination, and the amount of extraneous material present. For example the standard specifications drafted by the UK Cocoa, Chocolate and Confectionery Alliance, among others, refer to freshness, moisture content (not exceeding 3 per cent), fat content (should not be less than 65 per cent of the total weight), acidity (not exceeding 0.3 per cent of the free fatty acids calculated as lauric acid) and flavour. The specifications issued by Franklin Baker, USA, for desiccated coconut refer to similar physical and chemical qualities (Franklin Baker, 1971).

## Coconut milk extraction

Coconut milk is obtained by comminuting or grating the coconut endosperm into pulp, with or without the addition of water, and pressing or dewatering the pulp. The pressed or extracted milk is first of all centrifuged to separate the cream from the watery and solid portion. The cream is then mixed with 1.5–2 times its weight of water and pasteurized for about 15–30 minutes. The pasteurized cream is mixed thoroughly with a dairy stabilizer, homogenized and then heated almost to boiling point. The cream obtained in this way is poured into cans or bottles while still hot.

The most commonly used equipment for reducing the pulp into a suitable mash for milk extraction is the hammer mill. A hammer mill equipped with a

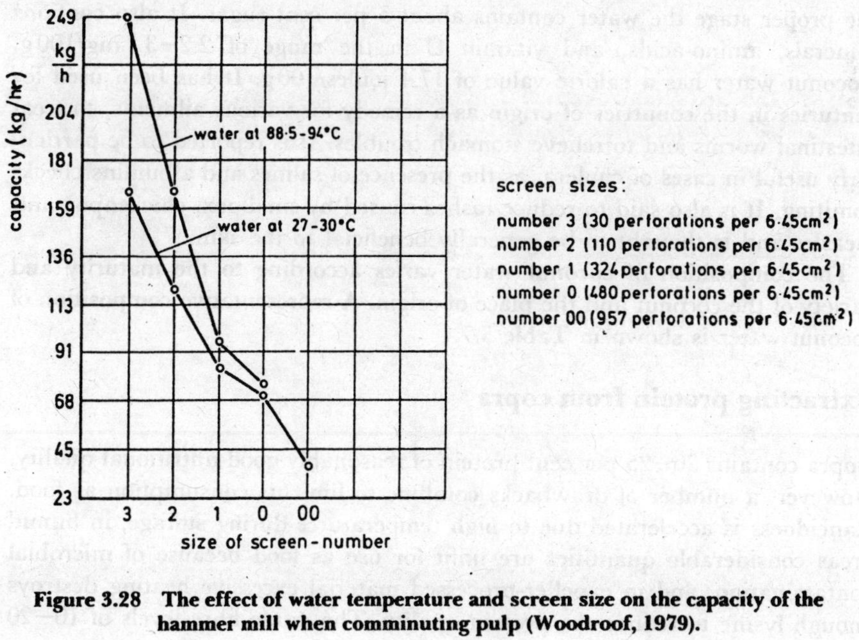


**Figure 3.28 The effect of water temperature and screen size on the capacity of the hammer mill when comminuting pulp (Woodroof, 1979).**

3.7 kW motor, a 25.4 cm diameter and 13.9 cm wide rotor, and a speed of 4600 rpm, may grind 170 kg of coconut pulp/hour using a number three screen (30 perforations/6.45 $cm^2$) (Woodroof, 1979). The performance of the hammer mill in comminuting coconut pulp is shown in Figure 3.28. Also shown is the effect of water temperature on the capacity of the mill.

The composition of coconut milk varies according to the variety of coconut used, its maturity and the environment in which it has grown. Processing conditions also influence the composition and quality of the milk (Nathanael, 1964). The chemical composition of coconut milk is shown in Table 3.6.

**Table 3.6 Composition of coconut milk**

| Constituent | Composition (%) | | | |
|---|---|---|---|---|
| Moisture | 54.1 | 50.0 | 52.0 | 47.0–53.0 |
| Fat | 32.2 | 39.8 | 27.0 | 39.8 |
| Protein | 4.4 | 2.8 | 4.0 | 2.6–2.9 |
| Sugars | – | 3.0 | – | 2.8–3.2 |
| Total solids | – | 10.4 | – | 10.3–10.5 |
| Ash | 1.0 | 1.2 | 1.0 | 1.1–1.3 |
| Carbohydrates | 8.3 | – | – | – |
| Starch | – | 0.09 | – | 0.08–1.0 |

Sources: Nathaniel, 1964; Popper *et al.*, 1966; Clemente and Villacorte, 1933.

## Coconut water

Coconut water is an important beverage in the coconut producing areas. About 1000 coconuts contain on average about 140 litres of coconut water. At the proper stage the water contains about 5 per cent sugar. It also contains minerals, amino-acids, and vitamin C in the range of 2.2–3.7 mg/100 g. Coconut water has a caloric value of 17.4 joules/100 g. It has been used for centuries in the countries of origin as a remedy for various ailments, to treat intestinal worms and to relieve stomach troubles. It is reported to be particularly useful in cases of cholera, as the presence of salines and albumins checks vomiting. It is also said to reduce rashes caused by smallpox, chickenpox and measles and is thought to be generally beneficial to the skin.

The composition of coconut water varies according to the maturity and variety of the coconut and the place of origin. A representative composition of coconut water is shown in Table 3.7.

## Extracting protein from copra

Copra contains 20–25 per cent protein of reasonably good nutritional quality. However, a number of drawbacks combine to limit its consumption as food. Rancidness is accelerated due to high temperatures during storage; in humid areas considerable quantities are unfit for use as food because of microbial contamination, and in expeller-processed material excessive heating destroys enough lysine to reduce the protein quality. The crude fibre levels of 10–20

**Table 3.7 The composition of coconut water**

| Constituent | Composition (%) |
|---|---|
| Water | 95.5 |
| Nitrogen | 0.05 |
| Phosphoric acid | 0.56 |
| Potassium | 2.61 |
| Calcium oxide | 0.69 |
| Magnesium oxide | 0.59 |
| Iron | 0.5 |
| Total solids | 4.71 |
| Reducing sugars | 0.80 |
| Total sugars | 2.08 |
| Ash | 0.62 |

Source: Pandalai, 1958.

per cent reduce its usefulness in infant feeding. The combination of all these drawbacks results in the wastage of much of the supply of protein from coconut copra.

The amount of available protein for food purposes may be increased by the use of enzymes. Enzymes capable of improving the yield of protein from coconut flour from 49 to 85 per cent have been derived from *Pestalotiopsis westerdijikii*. During enzyme treatment about half of the crude fibre is destroyed. Optimal conditions for extracting protein from coconut flour are reported to be 40°C for 3 hours at pH 5.6–6.0, using approximately 50 mg of enzyme/g of coconut residue (Woodroof, 1979).

## Tapping the coconut palm

Tapping is the term used to describe the method of obtaining sap from the tender unopened inflorescence of the palm. The art of tapping is very old, having been passed down for generations. Child (1974) comprehensively describes the method of tapping. The unopened flower spathe is prepared by bruising it slightly with gentle taps using a small mallet. After three weeks, when the spathe is nearly ready to produce sap, a strip of about 7 cm is cut from the end to prevent opening. During the preparation the spathe is gradually bent over, so that by the time the sap flows, a receptacle can be placed over the end for its collection. The flow of juice gradually increases; when it is in full flow the tapper usually changes the pot twice daily, at the same time he shaves a thin slice from the end of the spathe, tapping slightly with the mallet, and smearing on a mixture containing bruised leaves, which according to Child, contain saponin and stimulate the flow.

In its fresh state, coconut sap has been found to contain between 12 and 18 per cent carbohydrates, with sucrose as the main constituent and small amounts of glucose. Apart from the sugars, the sap also contains about 25 per cent nonfermentable matter, of which 0.5 per cent is ash and 2 per cent organic solids, including proteins and fats. The sap of the coconut palm is widely used in the producing areas to make a number of products, notably toddy, spirits,

**Table 3.8 Composition of coconut husk and coir fibre (percentage dry weight basis)**

| | Composition | | | | | | | | | | | | | | | | | |
|---|---|---|---|---|---|---|---|---|---|---|---|---|---|---|---|---|---|---|
| | Organic matter | Mineral matter (ash) | Water-soluble substances (ws) | | | | | Water-insoluble substances (wi) | | | | | Mineral elements | | | | Nitrogen (N) | Protein (N × 6.25) |
| | | | Total | Pectins | | Hemicel-luloses | | Total | Lignin | | Cellulose | | K | P | Ca | Mg | | |
| | | | | % | % of total ws | % | % of total ws | | % | % of total wi | % | % of total wi | | | | | | |
| **Coconut husks** | | | | | | | | | | | | | | | | | | |
| Mature | 95.0 | 5.0 | 26.0 | 14.2 | 54.6 | 8.5 | 32.7 | 74.0 | 29.2 | 39.5 | 23.9 | 32.3 | 0.78 | 0.04 | 0.08 | 0.05 | 0.23 | 1.4 |
| Young | 95.4 | 4.6 | 29.0 | 14.8 | 51.0 | 8.1 | 27.9 | 71.0 | 31.6 | 44.5 | 19.2 | 27.0 | 0.64 | 0.07 | 0.16 | 0.17 | 0.54 | 3.4 |
| Very young | 96.2 | 3.8 | 38.5 | 15.2 | 39.5 | 9.0 | 23.4 | 61.5 | 20.1 | 32.7 | 14.4 | 23.4 | 0.62 | 0.08 | 0.18 | 0.17 | 0.57 | 3.6 |
| **Coir fibre** | | | | | | | | | | | | | | | | | | |
| Mature | 98.8 | 1.2 | 5.2 | 3.0 | 57.7 | 0.25 | 4.8 | 94.8 | 45.8 | 48.3 | 43.4 | 45.8 | 0.02 | 0.01 | 0.06 | 0.04 | 0.35 | 2.2 |
| Young | – | – | 16.0 | 2.7 | 16.9 | 0.15 | 9.4 | 84.0 | 40.5 | 48.2 | 32.9 | 39.2 | – | – | – | – | – | – |
| Very young | – | – | 15.5 | 4.0 | 25.8 | 0.25 | 16.1 | 84.5 | 41.0 | 48.5 | 36.1 | 42.7 | – | – | – | – | – | – |

Source: Nathaniel, 1964.

sugar and vinegar. Toddy is prepared by natural fermentation of coconut sap. Coconut sap ferments readily owing to the action of micro-organisms picked up from the surroundings; these include yeasts (*Saccharomyces* spp.) and other mould species resembling yeasts.

Fully fermented samples of toddy are reported to contain 6–10 per cent alcohol. Toddy is used without further processing or is distilled to yield a spirit called Arrack, which is produced commercially in Sri Lanka and the Philippines. Sweet toddy is boiled down, after straining, to crystallizing point to give a 12–15 per cent yield of jaggery or gur, the temperature of the 'boil' being about 115°C. When well prepared, this rough sugar is hard, semi-crystalline and golden brown in colour.

If coconut sap is allowed to ferment beyond the alcoholic beverage stage, within 24 hours it becomes unpalatable due to acetic acid production. In Sri Lanka and the Philippines the process is used commercially for the preparation of coconut vinegar. The collected sap is allowed to ferment by storage in acetifying vats; when the vats are full, they are covered lightly with wooden boards and jute hessian, or sometimes with fronds. Free access of air is necessary for acetification. Acetification usually takes 10–14 weeks, at the end of which the scum of yeast and mould growth sinks to the bottom, and the liquid appears clear. It is then matured in closed casks for up to six months. The vinegar usually contains 4–5 per cent acetic acids. Before bottling it is coloured with caramel.

## Coconut husk

Coconut husk is the fibrous mesocarp which lies between the exocarp (outer covering) and the endocarp. It constitutes about 35 per cent of the total weight of the mature fruit, and is for most part a by-product of copra production. The main value of coconut husk lies in its fibre content and its use as fuel for indirectly fired copra kilns. Husks are also a useful source of potash and a valuable mulch, both for the conservation of moisture in the soil and the maintenance of soil organic content. The main components of coconut husk are pectins, lignin, cellulose and hemicellulose (Table 3.8).

### *Coir fibre*

Coir is the fibre from the husk of coconuts. There are three main types of coir fibre: the longest and finest, known as mat or yarn fibre, is suitable for spinning into coir yarn for making mats and ropes; a coarser fibre, known as bristle fibre, is used for brushmaking; and a shorter staple fibre, known as mattress fibre, is used for stuffing mattresses and upholstery. Curled fibre, a partially processed mixture of bristle and mattress fibre in the form of rope, is made in particular for use in the preparation of rubberized coir.

The best quality coir fibre, yarn fibre, which shows a low degree of lignification, is obtained from the husks of unripe nuts. The harvesting of green, unripe nuts has, however, an adverse effect on the quality of copra produced from them; in this way, coir making and copra making compete with one another. The preparation of yarn is therefore confined to those areas where coconuts are grown mainly as part of a subsistence economy and coir, not copra, presents the main cash commodity. Coconut growers in southern India, southern Sri Lanka and Tanzania emphasize the production of coir, whereas those in the Philippines emphasize copra production.

### *Yarn fibre*

Yarn fibre is produced by traditional methods which include the natural bacteriological process of retting, whereby the husks are soaked in water until the pith decomposes. Soaking lasts for a period of up to ten months. Essentially, the retting process involves the hydrolysis of pectins by enzymes from *Micrococcus* spp., as also occurs in the retting of sisal (Kirby, 1963).

Pandalai *et al.* (1957) have established that the retting water should be saline, and that periodical changes of water – as occurs with tidal action in backwaters – is an important condition for the production of good quality fibre.

When retting is regarded as complete, the husks are removed, washed and squeezed in water to eliminate the mud and bad smell. After removing the tough exocarp, the remaining fibre is beaten with wooden mallets until all the pith is removed. It is then dried and the dry fibre is beaten again and put through a winnowing machine, consisting of a number of knives with saw-like teeth fixed to a shaft which is rotated by hand in a drum, thus removing any small particles of pith.

On average 1000 husks yield about 80 kg of coir. The fibre is spun into yarn, either by hand or on a *charka* or *spinning wheel.* Hand-spinning is done mainly by women. The fibre is rolled into short lengths of 15–23 cm. Two of these are twisted together until near the ends, when fresh short lengths are added, one after another, until the required length of yarn (up to 7 m) is obtained; this is then made into a hank and a knot is made at the end. Spinning on a wheel requires two units – one wheel is stationary and carries

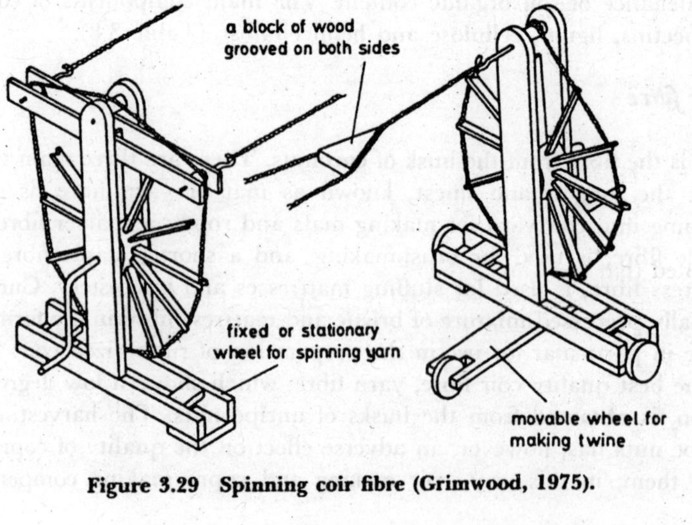


**Figure 3.29 Spinning coir fibre (Grimwood, 1975).**

two spindles set in motion by a central handle, the other is mounted on three castors and has one spindle (Figure 3.29).

Spinning is usually carried out by two experienced workers assisted by a boy or a girl. Slivers of fibre are taken by the two spinners, who start by attaching a small loop of fibre to each spindle of the stationary wheel. The spinners build up the fibre slivers to the required thickness on the starting loop, walking backwards until the required length of strand is obtained. During this operation, the stationary wheel is turned to give the necessary twist (TPI, 1970).

The two strands are passed by means of a triangular block of wood, which is rotated in the opposite direction to that of the fixed counter twist, preventing tangling, and binding the strands closely as the spindle on the movable wheel is turned. The twisted yarn is wound into hanks, usually 15–18 m in length. On average one pair of spinning wheels can produce, depending chiefly on the ability of the spinners, up to 15 kg of yarn daily. Spinning on the wheel gives a better quality yarn with a hard twist, which is suitable for the manufacture of matting and other such materials.

### *Bristle and mattress fibre*

The manufacture of bristle fibre involves soaking or retting the husks, extraction of fibre, washing and hanking, and hackling and grading.

Coconut husks are soaked or 'retted' for periods between a week and a month. They are then crushed by feeding through iron rollers before the actual separation of fibres takes place. This process, sometimes called milling or decortication, is carried out with two machines. The first one, called the *breaker*, is essentially a revolving wooden drum about 0.9 m in diameter. Projecting from its wide rim are iron nails, spaced 4–5 cm apart. As the drum revolves, the operator presses the husks, first one half then the other, against the rim; the nails tear away the pith and much of the short fibre from the long-stapled bristle fibres.

The hank of bristle fibre is then treated in a similar way on the *finisher* or *cleaner*, a machine similar to the breaker, but with closer spaced, thinner nails, which clean the bristle fibre from any short-stapled fibre not removed by the breaker. Finally the bristle is washed and laid out to dry, after which it is *hackled*, that is combed by drawing it through a row of steel spikes 10–15 cm long, spaced about 2.5 cm apart. Bristle fibre is tied into hanks by means of cords of twisted fibre, the number of cords indicating the quality. Long bristle fibre must be at least 20 cm in length for the brush trade. It is generally accepted that one pair of drums (breaker and finisher) can handle 2000 husks in eight hours and produce 100 kg of bristle and 200 kg of mattress fibre.

The crude wet mattress fibre from the breakers and finishers is treated in a sifting rotating wire-mesh drum, sometimes known as a *willowing machine*; the hard exocarp and the pith are thereby removed. The cleaned mattress fibre is then dried by spreading it on the ground up to a depth of about 45 cm. In good sunny weather, drying takes about eight days, during which the fibre is turned over a few times to ensure uniform drying.

### *Dry milling of coconut husk*

The traditional method of extracting fibre from coconut husk, as described above, includes prolonged soaking. In the dry milling process, soaking is either completely eliminated or is considerably reduced. Dry milling is commonly referred to as decortication. This process was first introduced with the intention of producing mattress type fibre in areas where soaking facilities are limited or not available. The *Downs decorticator* for dry or partially dry milling was first put onto the market in 1950. The basic principle on which the Downs decorticator works is that of disintegrating the husk segments by use of metal beater bars revolving at high speed, after which sifters separate the non fibrous matter from the fibre. The machine consists of the following main parts:

1. The husk-bursting mill for disintegrating and partially opening husk segments which have been soaked for at least three days. It has steel beater bars attached to an iron disc rotating at high speed in a cast iron chamber.
2. The sieve or sifter, fitted with horizontal rotating steel beaters, in which the opening of the husk segments is completed and most of the dust and the very short fibres are removed. It is connected to the bursting mill by a duct.
3. The turbo screen, where final cleaning takes place.

## Utilization of coconut shells

The most important product derived from coconut shells is charcoal. Coconut shell charcoal is widely used in the countries of origin as a fuel, occasionally to fire copra kilns but more commonly by goldsmiths, blacksmiths, laundries and bakeries. A second product derived from coconut shells is coconut shell flour, which is used as a filler in the thermoplastic industry and as an abrasive for cleaning machinery.

Coconut shells are similar in composition to hardwoods, though lower in cellulose and higher in lignin. Table 3.9 gives the chemical composition of coconut shells.

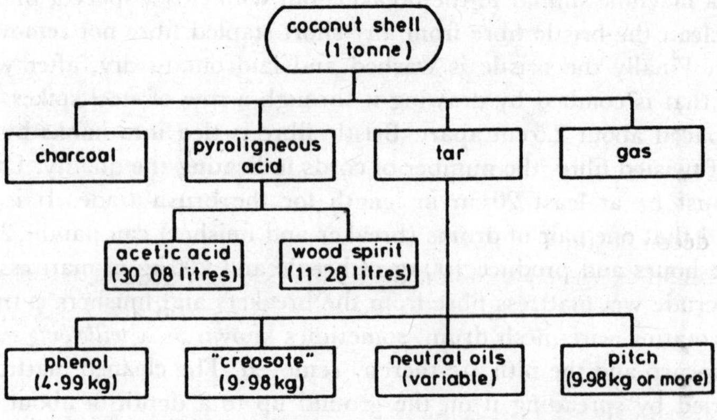


**Figure 3.30 Principal products from the destructive distillation of coconut shells (Purseglove, 1975).**

**Table 3.9 Composition of coconut shells**

| Constituent | Composition (%) |
|---|---|
| Moisture | 8.0 |
| Ash | 0.6 |
| Solvent extractives | 4.2 |
| Lignin | 29.4 |
| Cellulose | 26.6 |
| Pentosans | 27.7 |
| Uronic anhydrides | 3.5 |

Source: Child, 1947.

## *Manufacture of coconut-shell charcoal*

The manufacture of charcoal from coconut shells is a relatively simple process, consisting essentially of burning shells in a limited supply of air so that they are only carbonized and not burned to ash. In operating kilns a fire of shells is initially started at the bottom of the pit. More shells are added to this until these too catch fire. Regarding the shape of the pit, circular pits are preferred as the firing is said to be easier to control. The process is then continued until the pit is completely charged, making sure that before each new addition the fire is burning well. When the full charge of shells is flaring steadily, the fire may be damped down if necessary by sprinkling with water. This is deemed an appropriate stage for the exclusion of air from the kiln, which is carried out by covering the glowing mass with corrugated iron sheets and then lining them on top with green coconut fronds, damp turf and soil. However, sufficient space is allowed under the sheeting for the large volume of smoke and vapour to escape, as slow carbonization proceeds within. The pit is opened after the third day. The best charcoal is made from the shells of ripe nuts, because too much undesirable fibrous matter is associated with shells from unripe and green nuts. On average about 20 000 whole shells yield one tonne of charcoal. The unprocessed charcoal finds local use as a fuel and in metallurgical work.

### Destructive distillation

'Destructive' or 'dry distillation' are the terms for the changes that wood, palm-nut shells and similar vegetable products containing lignocellulose undergo when subjected to heat in the absence of air. When coconut shells are heated to a sufficiently high temperature in the absence of air, the highly complex substances of which they are composed break down, and the shells undergo decomposition. The reaction becomes exothermic at about 290°C. Factors influencing the formation of distillation products are shell maturity and moisture content, temperature, speed of distillation and pressure.

The primary products of the distillation of coconut shells are charcoal, pyroligneous acid, tar and gas (Figure 3.30). The proportion of these varies according to conditions of distillation. Child (1947) using laboratory and semi-commercial methods found that 45.4 kg of shells yield 15.4 kg of charcoal, 18.2 kg pyroligneous acid, 2.7 kg of tar and 9.1 kg of gas.

## *References*


**Aten, A., Manni, M.** and **Cooke, F.C.** (1958). *Copra processing in rural industries.* FAO Agric. Development Paper No. 63.

**Bezard, J., Bugat, M.** and **Clement, G.** (1971). Triglyceride composition of coconut oil. *Journal American Oil Chemist's Society,* **48**(3): 134–9.

**Burkill, I.H.** (1935) *A dictionary of the economic products of the Malay Peninsula,* London.

**Child, R.** (1940). The packing and keeping qualities of desiccated coconut. *J. C. I.,* **4**(1): 9–16.

**Child, R.** (1947). Coconut shell of charcoal; activated charcoal *Current Science,* **16**: 5–8.

**Child, R.** (1974). *Coconuts,* 2nd Edition, Tropical Agriculture Series, Longman, London.

**Clemente, A.,** and **Villacorte, M.** (1933). Some colloidal properties of coconut milk. *Philippine I.Sci.* **3**: 7–10.

**Cocoa, Chocolate and Confectionery Alliance** (1966). *Draft standard specifications for desiccated coconut,* London.

**Coconut Bulletin** (1964). *Home-made hot air driers of Tonga make worlds's best copra.* **18**:3.

**Coconut Industry Board,** Jamaica. *Second Report of the Research Department,* June 1961–June 1962.

**Coconut Research Institute, Ceylon,** (1956). *Smallholders' copra kiln.* Leaflet No. 23.

**Cooke, F.C.** (1956). Copra production by the smallholder. *World Crops,* **8**: 435–8; 489–91.

**De Vos, L.** (1956). *A quick drying method for copra.* Surinam Landb., **4**:119.

**FAO** (1970). *Coconut breeding.* Special issue of the Plant News (October 1970). Rome. Document AGP: NL/13.

**FAO** (1984). *FAO Production Yearbook,* Vol. 38. Rome.

**FAO** (1986). *FAO Mon. Bull. Statist.,* Vol. 9. No. 7/8.

**FAO/WHO** (1970). *Draft code of hygienic practice for desiccated coconut.* Rome. FAO/WHO Standards Programme, Codex Alimentarius Commission. Document ALINORM 71/13.

**Franklin B.** (1971). *Baker's coconut for the food industries.* Dover, Delaware, General Foods Co.

**Grimwood, B. E.** (1975). *Coconut Palm Products – Their Processing in Developing Countries.* FAO Rome.

**Jarman, C.G.** and **Jayasundera, D.S.** (1975). *The extraction and processing of coconut fibre.* Tropical Products Institute London.

**Kaufmann, H.P.** and **Thieme I.G.** (1956). Neuzeitliche Technologie der Fette und Fettprodukte Part. I. Münster, Aschendorff.

**Kirby, R.H.** (1963). *The vegetable fibres.* Leonard Hill, London, 355–372.

**Nathanael, W.R.N.** (1960). Chemical and technological investigations on coconut products. *Ceylon Coconut Quart.* **11**: 46.

**Nathanael, W.R.N.** (1964). Annual report of the Ceylon Coconut Research Institute for 1963. *Ceylon Coconut Quart.* **15**: 16–20.

**Palmer, I.A.** (1968). *Some aspects of copra drying. Cocoa and coconuts in Malaya.* Proceedings of a Symposium in Kuala Lumpur, September 1967.

**Pandalai, K.M.** (1958). Coconut water and its uses. *Coconut Bull.* **12**: No. 5, 167–173.

**Pandalai, K.M., Nair, U.K.** and **Menon, K.P.V.** (1957). A note on the quality of water in relation to the retting of coconut husks. *Coir,* **1**(3): 30–3.

**Popenoe, I.** (1969). *Coconut and Cashew. North American Nut Trees.* W.F. Humphrey Press, New York.

**Popper, K., Notter, G.G.** and **Nurry, F.S.** (1966). New products from coconuts: non-dairy chip dips and 'cream-cheese'. *Food Process. Market.* **27**: No. 12, 92, 96.

**Purseglove, I.W.** (1975). *Tropical Crops. Monocotyledons.* Longman Group Ltd. London.

**Rajasekharan, N., Pandalai, K.M., Bhatia, D.S.** and **Bhatia, B.S.** (1962). Processing of desiccated coconut. *Indian Coconut Journal,* **15**: 70–83.

**Salz, A.G.** (1971). *Advantages of controlled forced hot-air drying of copra.* Conference on Cocoa and Coconuts. Kuala Lumpur, Paper No. 16.

**Samson, A.S.** (1971). Heat treatment of coconut meats and coconut meal. *Journal Science Food and Agriculture,* **22** (6): 312–16.

**Schaffner, C.P., Mosbach, K., Bibit, V.C.** and **Watson, C.H.** (1967). Coconut and salmonella infection. *Appl. Microbiol.,* **15**.

**Tergast, G.C.W.C.** (1937). *Artificial drying of copra.* Landbouw, **13**:128.

**Thieme, J.G.** (1969). *Coconut Oil Processing.* FAO Rome.

**Tropical Products Institute Report G.52** (1970). *The Preparation of coir or coconut fibre by traditional methods.* Ministry of Overseas Development, London.

**Ullrich, H.** (1967). *Mechanische Verfahrenstechnik.* Springer Verlag Berlin/New York/Heidelberg.

**Woodroof, J.G.** (1979). *Coconuts – reproduction, processing, products,* AVI Westport, Connecticut.

# 4

# Coffee

## Production and economic species

Coffee is produced from the seeds of an evergreen tree or shrub belonging to the family Rubiaceae, which contains about 400 genera and 5000 species (Purseglove, 1976). The genus *Coffea* L. is the only member of the Rubiaceae used as a beverage. Several others are used as sources of drugs, such as quinine and ipecacuanha. The taxonomy of *Coffea* is still confused and there are discrepancies regarding the number of species that exist. The number of species given by various authorities ranges from 25 to over 100. Wellman (1961), for example, lists 64 species. The most important economic species are:-

1. *Coffea arabica* L. (arabica coffee), which provides about 70 per cent of the coffee consumed world-wide. It originated in Ethiopia.
2. *Coffea canephora* L. (robusta coffee), which provides about 30 per cent of the coffee consumed world-wide. It is indigenous to the West Coast of Africa, the Congo basin and Angola. Robusta coffee is cheaper to produce than arabica, and is used particularly in the manufacture of instant coffees.
3. *Coffea liberica* L. (liberica coffee) is found in Liberia and Sierra Leone.
4. *Coffea excelsa* L. (excelsa coffee) is indigenous to West Africa (the Ivory Coast, Guinea and Cameroon) and is only grown in small amounts.

Coffee is an important cash crop in over 70 countries in Central and South America, Africa, Asia and Oceania. Coffee is second only to petroleum in primary commodity trade and provides over 25 per cent of the foreign exchange earnings of 16 countries in Latin America and Africa. It also provides employment for at least 20 million people (Mwandha *et al.*, 1985). The area harvested for green coffee is over 100 million hectares and green coffee production in 1986 amounted to nearly 5.2 million tonnes (FAO Monthly Bulletin, 1986).

Productivity is directly related to the number of trees per unit area, and coffee is mostly planted on a square pattern with about 140–215 trees per hectare. The average yields in kilograms per hectare in 1984 of some selected countries are shown in Table 4.1 (FAO Production Yearbook, 1984).

**Table 4.1 Average yields of coffee for selected countries, 1984.**

| Country | Yield (kg/ha) |
|---|---|
| Brazil | 552 |
| Colombia | 825 |
| Mexico | 632 |
| Zimbabwe | 1250 |
| Kenya | 679 |
| Costa Rica | 1485 |
| Uganda | 927 |

## *Uses*

The dried coffee beans are roasted, ground and brewed to make a stimulating and refreshing beverage. Coffee and tea are at present the most important beverages in America and Europe. Coffee is consumed in almost every country in the world, and Europe alone imported over 2 million tonnes of green unroasted coffee beans in 1981. Although coffee is African in origin, relatively few indigenous people drink coffee. In Ethiopia dried coffee berries (cherries) have been used as a masticatory since ancient times; ground roasted coffee is also mixed with fat and eaten. A drink may be prepared from dried coffee pulp.

Coffee pulp and parchment are used as manures and mulches. Coffee pulp may be extracted with water or organic solvents to obtain caffeine and other compounds. The residue of such a process may be utilized as an animal feed (Braham and Bressani, 1979). Attempts have been made to extract the protein from coffee pulp as well as the pectic substances, including pectinolytic enzymes. Other approaches include hydrolysis to produce molasses, fermentation to yield alcohol, and a variety of extracts for soft drinks, jams, and similar types of food (Aguirre, 1966). Coffee oil has been used in specialty soaps, giving them good lathering and detergent properties.

## Chemical composition and physical properties of coffee beans

The coffee bean has a very complex chemical composition. When the ripe fruit is picked, the endosperm contains about 48–50 per cent moisture and 50–52 per cent dry matter. The crude fibre content of arabica coffee produced in Brazil as well as in Central America and Indonesia ranges from 22–27 per cent on a dry weight basis. Coffee fat consists of a non-drying oil and a wax, and contains large quantities of unsaponifiable substances (sterols) due to the fatty acid monoesters of cafestol or kahweol in coffee fat (Figure 4.1). The fatty acid monoesters of cafestol are also responsible for the fact that coffee fat is not suitable for use as an edible oil. Coffee fat occurs in the form of small droplets in the outer horny layer of the endosperm.

Kahn and Brown (1953) examined the fatty acids of the oil obtained from roasted arabica coffee in Brazil and found the composition shown in Table 4.2.

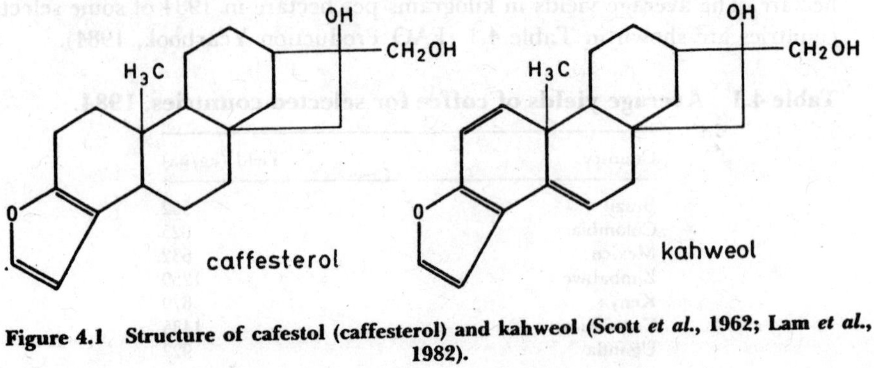


**Figure 4.1 Structure of cafestol (caffesterol) and kahweol (Scott *et al.*, 1962; Lam *et al.*, 1982).**

**Table 4.2 Fatty acid composition of the oil from roasted arabica coffee**

| Fatty acid | Composition |
|---|---|
| Palmitic acid | 32.0 |
| Palmitolic acid | 0.9 |
| Stearic acid | 7.6 |
| Oleic acid | 8.2 |
| Linoleic acid | 46.3 |
| Acids with $C_{20}$ $C_{22}$ and $C_{24}$ | 5.0 |

Coffee contains a high percentage of caffeine (0.9–2.4 per cent). Caffeine is 1:3:7 trimethylxanthine $C_8H_{10}O_2N_4$, crystallized with one molecule of water but becoming anhydrous when heated to 80–100°C. Its sublimation begins at 120°C and is completed at 178°C. This explains the losses in caffeine during roasting. Besides caffeine there is trigonelline (0.3–1.2 per cent) which has a rather weak physiological action. Trigonelline ($C_7H_7O_2N$) is an alkaloid derived from the pyridine nucleus (Figure 4.2).

The caffeine content of different types varies considerably; it amounts to 0.7–1.5 per cent in *C. arabica*, 1.7–2.4 per cent in *C. canephora* and 0.6–1.7 per cent in *C. liberica*. Contrary to popular belief it is now recognized that caffeine plays a very limited role in the bitter flavour of coffee. Caffeine is reported to account for not more than 10 per cent of the bitterness (Belitz, 1975; Panghorn, 1982). Belitz (1975) suggests that heterocyclic compounds may be responsible for the bitter flavour.

The composition of coffee beans varies according to species. These variations are difficult to interpret and are connected with the taste and aromatic qualities of the roasted coffee. The properties of coffee depend on the variety, soil, climate, altitude and cultivation practices. *C. robusta* yields coffees which have body but no fragrance or acidity. They are reported to be inferior to the coffees which can be obtained from well-processed *C. arabica* cherries, while *liberica* and the *excelsa* types produce harsher and more pungent coffees.

Table 4.3 gives the approximate chemical composition of green coffee beans. Carbohydrates represent the largest proportion of the green coffee beans, about 50–60 per cent (Sivetz and Desrosier, 1979).

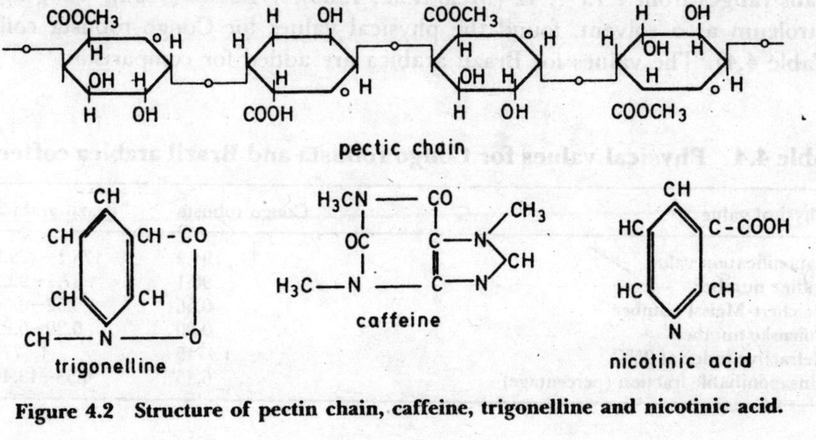


**Figure 4.2 Structure of pectin chain, caffeine, trigonelline and nicotinic acid.**

### Table 4.3 Chemical composition of green coffee beans

| Constituent | Composition on a dry weight basis (approximate percentage) |
|---|---|
| **Carbohydrates** | |
| Reducing sugars | 4.0 |
| Sucrose | 3.8 |
| Glucose, fructose, galactose | 2.0 |
| Starch | 10.0 |
| Pentosan | 5.0 |
| Hemicellulose | 15.0 |
| Holocellulose | 18.0 |
| Lignin | 2.0 |
| **Oils** | 13.0 |
| **Protein** (N x 6.25) | 13.0 |
| **Ash** as oxide | 4.0 |
| **Non-volatile acids** | |
| Chlorogenic | 7.0 |
| Oxalic | 0.2 |
| Malic | 0.3 |
| Citric | 0.3 |
| Tartaric | 0.4 |
| **Trigonelline** | 1.0 |
| **Caffeine** (arabica 1 per cent, robusta 2 per cent) | 1.0 |

The beans of robusta and arabica coffees and certain other varieties can be distinguished by their macroscopic characters. The beans of *C. arabica* are larger than those of *C. robusta* and the length/width ratio is higher. The length varies from 8–12 mm and the width from 6–8 mm, the length/width ratio usually being 1.3–1.5. *C. robusta* beans are smaller and rounder; they are usually 6–8 mm long and 6–7 mm in width; their length/width ratio being 1.0–1.15. *C. liberica* beans are very large and elongated, 14–17 mm long, 7–9 mm in width, and with a length/width ratio of about 1.9–2.0.

The beans of *C. excelsa* are similar in shape to those of *C. liberica* but they are much smaller, although they are larger than those of *C. arabica*.

Individual green beans vary in mass from below 100 mg to over 200 mg, with an overlap between the arabicas and robustas. The relative density of the beans ranges from 1.15–1.42 (Melo *et al.*, 1980). Wilbaux (1956), using light petroleum as a solvent, found the physical values for Congo robusta coffee (Table 4.4). The values for Brazil arabica are added for comparison.

### Table 4.4 Physical values for Congo robusta and Brazil arabica coffees

| Physical value | Congo robusta | Brazil arabica |
|---|---|---|
| Saponification value | 194.9 | 176.1–179.2 |
| Iodine number | 96.1 | 87.1–92.5 |
| Reichert-Meissl number | 0.56 | 0.52–0.66 |
| Polenske number | 0.20 | 0.20–0.25 |
| Refractive index at 25°C | 1.4715 | 1.4773 |
| Unsaponifiable fraction (percentage) | 6.35 | 6.53–13.49 |

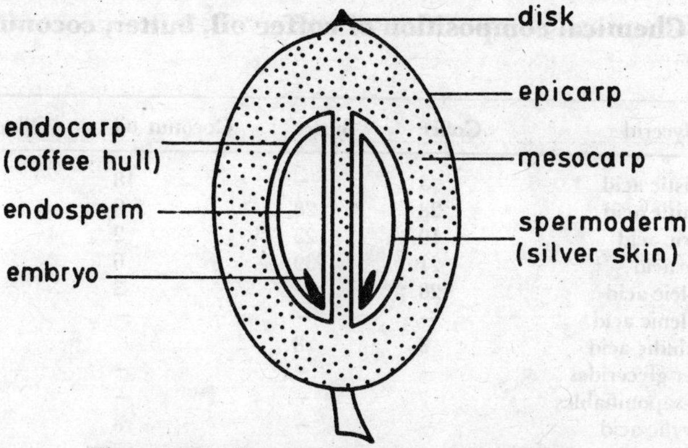


**Figure 4.3 Cross-section of a coffee cherry.**

The coffee fruit is a globular to oblong drupe which is commonly called a cherry or a berry. It is composed of the following parts (Figure 4.3).

1. The coffee bean proper or endosperm, which consists of a cuticle, an outer hard layer made up of small cells containing droplets of oil, and inner layers made up of larger soft cells. The endosperm is folded and contains the embryo.
2. The spermoderm or testa, often called the silverskin, which is a thin skin covering the bean, consisting of parenchyma cells. The testa is silvery and thin in *C. arabica* and copper-coloured and thicker in *C. robusta*; it contains fats and a little chlorogenic acid.
3. The endocarp or parchment, the main constituent of which is fibrous. The parchment is often used for fuel. It can be made into briquettes without any binder. The maximum heat value of parchment with 8 per cent moisture is reported to be about 17976 joules.
4. The mesocarp or mucilage fraction, which adhers to the parchment after pulping, consists of hyaline, juicy tissues (except in *C. liberica*); the tissues are rich in sugars and pectins but do not contain caffeine or tannins.
5. The epicarp, or thin red skin, which is usually called the pulp. It is thick in *C. liberica* and supple and thinner in *C. arabica* and *C. robusta*.

The epicarp (pulp) together with the mesocarp (mucilage) form the pericarp which represents 45–65 per cent of the weight of the fruit, depending on variety, species, size and degree of ripeness. The drupe, cherry or berry normally contains two beans which lie with their flat faces towards each other. Because of the viscous nature of the mucilage, slight pressure applied to the cherry is enough to expel the two beans from the fruit.

## *Coffee oil – its composition and physical properties*

Green or roast coffee bean oil has a chemical composition similar to that of many edible vegetable oils, Table 4.5. It is liquid at room temperature and at about 7°C. Roast coffee has more fatty acids than green coffee. Roast coffee oil may contain up to 5 per cent free fatty acids. Some of the physical properties of coffee oil are shown in Table 4.6.

**Table 4.5 Chemical composition of coffee oil, butter, coconut oil and olive oil***

| Triglyceride | Coffee | Butter | Coconut oil | Olive oil |
|---|---|---|---|---|
| Myristic acid | 3 | – | 18 | – |
| Palmitic acid | 28 | 28 | 9 | 15 |
| Stearic acid | 10 | 25 | 2 | 10 |
| Oleic acid | 21 | 39 | 6 | 70 |
| Linoleic acid | 28 | – | 3 | 5 |
| Linolenic acid | – | – | – | – |
| Arachidic acid | 3 | 8 | – | – |
| Other glycerides | – | – | – | – |
| Non-saponifiables | 7 | – | – | – |
| Caprylic acid | – | – | 8 | – |
| Capric acid | – | – | 7 | – |
| Lauric acid | – | – | 47 | – |

* Values extracted from 'Chemical composition of fats and oils (approximate)' in Coffee Technology, Sivetz and Desrosier (1979), AVI Publishing Co. Westport, Connecticut.

**Table 4.6 Physical properties of coffee oil**

| Physical property | Value |
|---|---|
| Acid number | 1–2% |
| Saponification number | 170–199 |
| Iodine number | 95–100 |
| Specific gravity | 0.944–0.945 |
| Refractive index | 1.468–1.469 |

## Green coffee processing

After harvesting, the cherries or berries of the coffee tree undergo a series of unit operations with the aim of obtaining storable and easily transportable green coffee (Figure 4.4). Technically, the basic problem to be solved is the removal of the coffee seeds from within the cherry by removing the various enveloping layers in the most effective manner, and the production of marketable green coffee beans with a moisture content of less than 12–13 per cent. Two main processes have been used for many years, traditionally distinguished as the dry process, common in Brazil and the robusta producing countries of Africa, and the wet process. Green coffee processed by these two methods is known as either dry process (natural) or wet process (washed) coffee. Detailed accounts of these methods or processes are numerous in the literature, notably by Harer (1962), Foote and Sivetz (1963), Wilbaux (1952) and Sivetz and Desrosier (1979).

The unit operations involved in the wet process include grading or classification (unripe and damaged cherries are removed), pulping, fermenting, washing, drying and storage. The dry process involves classification, drying, hulling or dehusking and storage. The following sections describe the general principles involved in the wet and dry processes. Variations in these processes exist in the various coffee producing countries. Roasting and grinding of the

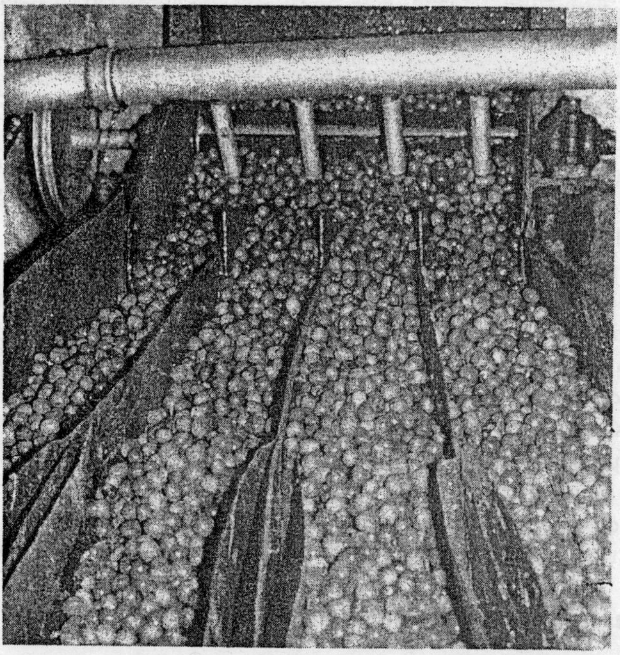

**Figure 4.4 Harvested coffee cherries ready for processing.**

dried beans prior to preparing the beverage in the cup is treated under the section 'Manufacture of instant coffee'.

## *The wet processing method*

In the wet processing method the harvested ripe cherries are hand-pulped by pressing them between the thumb and forefinger to force out the two beans, or the ripe cherries are squeezed in a pulping machine which removes most of the soft outer pulp or fibrous fruit flesh, leaving a slippery exposed layer of

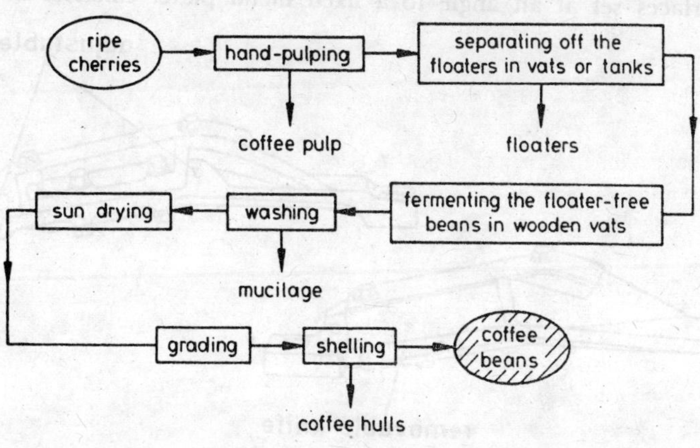


**Figure 4.5 Flowchart for wet processing of coffee.**

mucilage. After pulping, the beans are put in tanks or vats to separate the floaters. The 'sound' beans are then fermented in suitable vats for 48–60 hours. After fermentation the beans are washed and dried either in the sun or in mechanical driers. Sun drying on wire mesh trays is the commonest method with not more than 12 kg of parchment coffee to the square metre. Depending on the weather conditions and the available drying area, drying time ranges from 6–15 days. Drying is followed by shelling/hulling and grading. The wet process is illustrated in the flowchart in Figure 4.5.

The wet processing method is more expensive and more sophisticated than the dry processing method. It produces a cleaner, better quality bean, which usually commands a higher price. The method requires substantial quantities of water, and is used everywhere except in Brazil for the production of mild (arabica) coffee. It is also used occasionally for other coffees. The method is used extensively in Kenya and Colombia.

The dry process is the older and cheaper method, which is preferred in drier cultivating regions with a high harvest (Brazil). The cherries are spread out thinly and dried in the sun, taking about 1–3 weeks. Dried coffee cherries are dark brown and brittle. After drying, the cherries are hulled and polished.

## *Operations and equipment for the wet method*

### Pulping and pulpers

Freshly picked cherries are put into a tank, where the badly blemished and light fruits float off naturally, while at the bottom, heavy materials such as stones and dirt are removed. The good cherries are put into the pulper to be pulped. Pulping is the part of the wet method used to remove the epicarp (red skin) and part of the adhering mesocarp.

Arabica cherries pulp more easily than robusta cherries but the latter have a stronger parchment layer, less easily broken by pulpers; liberica is particularly difficult to pulp because the pericarp (epicarp and mesocarp) is harder. Machines used for pulping are called pulpers. They consist of rollers with rough surfaces set at an angle to a fixed metal plate. Cherries are forced

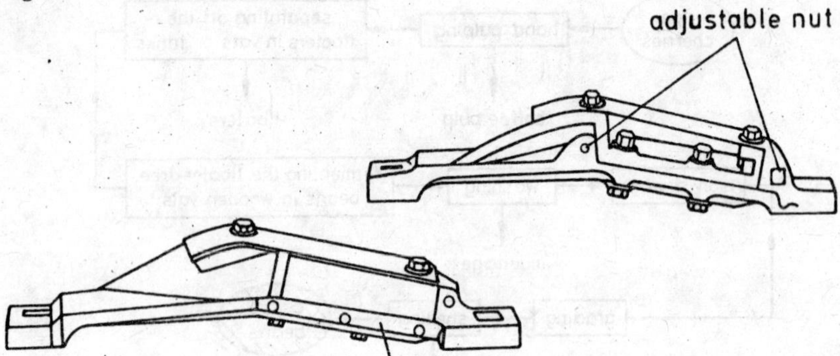


**Figure 4.6 Pulping bar for disc pulper.**

through the rollers, pushed between the rollers and the plate and the beans are squeezed out.

## Disc pulpers

Disc pulpers consist mainly of one or more usually 457.2 mm discs fixed to a horizontal shaft. The discs are made of cast iron with bulb-shaped projections on the surface, or from a punched copper plate. The pulping chops and bars are set side-by-side and the removable knife is adjustable (Figure 4.6).

The cherry is conveyed to the discs, forced along by the bulb-shaped projections and torn by the pulping bars so that the beans are separated and the empty fruit skins can be removed. The gap between the bars and the discs should be such that no beans can pass through with their skins still attached. The normal revolving speed is about 120 rpm. The broken skins are discharged through an inclined chute.

## Cylinder pulpers (drum or barrel type)

A cylinder pulper is a rotating drum or barrel covered with a copper plate punched over its whole surface, and an adjustable breastplate with pulping channels tapering down to outlets through which the pulped coffee is discharged. The pulping knives are adjustable and correspond to the pulping bars of disc pulpers. The depth of the pulping channels can be regulated by hand wheels. The breastplate is fixed to the cylinder by two screws (Figure 4.7).

Cherries are fed into the space between the rotating cylinder and the breastplate, and are drawn into the pulping channels where they are squeezed to free the beans. The pulped beans are retained by adjusting gates and they escape through the apertures at the bottom of the pulping channels. The skins are dragged down to the back of the machine by the punched projections on the cylinder. These punched projections vary in size to suit the variety of coffee to be pulped (Figure 4.8). Some pulpers have rubber breastplates which

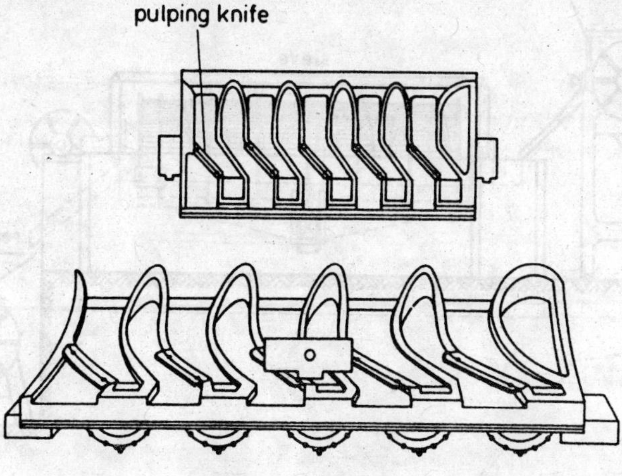


**Figure 4.7 Breastplate for cylindrical pulper.**

have a more gentle action. These cannot, however, pulp black cherries, which have partially dried on the tree, without damaging them.

Good pulping results were reported to have been obtained in Kivu with the following adjustments for disc pulpers (Wilbaux, 1963):

(a) Distance between disc and chop : 2 mm.
(b) Distance between disc and pulping bar : 4.8 mm.

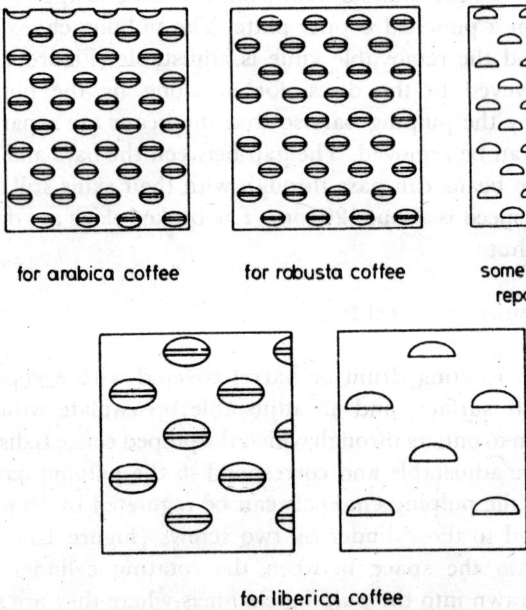


**Figure 4.8 . Specimens of punchings in pulpers.**

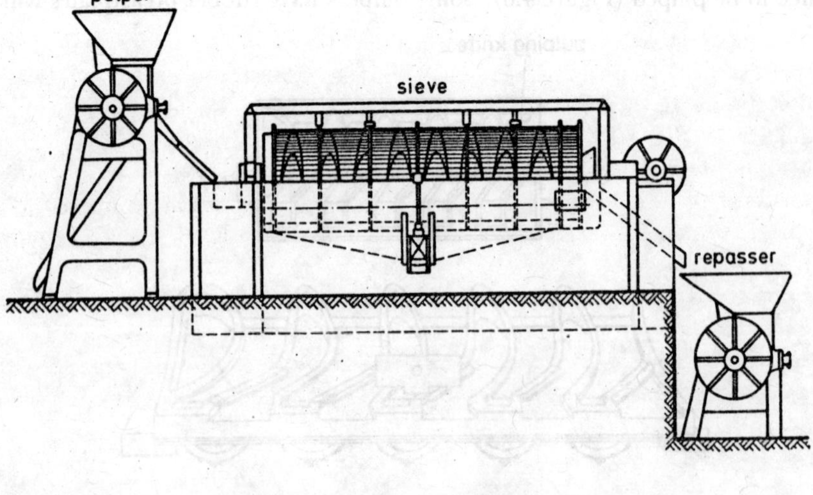


**Figure 4.9 Pulper-sieve-repasser unit.**

Disc pulpers are less powerful and Wilbaux (1963) considers them better for processing arabica coffee than the cylinder type which is more suitable for robusta coffee. The pulping bars of the disc pulper are readily adjusted because they have ample clearance and the copper plate covers are easier to replace than on a cylinder pulper. Some of the pulpers now used are listed in Table 4.7.

**Table 4.7 Classification of some of the pulpers used for processing coffee**

| Pulper | Maximum theoretical hourly capacity in kg of cherries | Rated speed in rpm | Power required in kW |
|---|---|---|---|
| * One-disc pulper – diam. 457.2 mm | 1000 | 120 | 0.73 |
| * Two-disc pulper – diam. 457.2 mm | 2000 | 120 | 0.95 |
| Three-disc pulper – diam. 457.2 mm | 3000 | 120 | 1.4 |
| Four-disc pulper – diam. 457.2 mm | 4000 | 120 | 1.8 |
| Cylinder pulper – drum diam. 381 mm | | | |
| –size 304.8 mm | 1000 | 120 | 0.73 |
| –size 609.6 mm, 4-hole breast | 2000 | 120 | 1.1 |
| –size 609 mm, 5-hole breast | 2500 | 120 | 1.3 |
| Cylinder pulper – drum diam. 381 mm | | | |
| –size 152.4 mm, 1-hole breast | 500 | 150 | 0.36 |
| –size 304.8 mm, 2-hole breast | 1500 | 150 | 0.73 |
| –size 609.6 mm, 4-hole breast | 2000 | 150 | 1.3 |
| VOGTS pulper – drum diam. 457.2 mm | | | |
| –size 419.1 mm | 2200 | 200 | 2.40 |
| VOGTS pulper – drum diam. 609.6 mm | | | |
| size 419.1 mm | 3300 | 200 | 2.77 |

Pulpers marked (*) can be operated by hand; in this case the actual capacity will not exceed 30–40 per cent of the rated maximum output.
Source: Wilbaux, 1963.

## Pulper-sieve-repasser unit

In a simple pulper, small cherries are only partially pulped or may remain unpulped while larger ones are only nipped. To circumvent this drawback, pulpers may be combined with repassers. For this purpose a separator is attached to the pulper, either a flat oscillating sieve or a rotary screen, horizontal or inclined, with spaced rods or oval perforations to let the pulped beans through. The unpulped or partially pulped cherries are conveyed to the repasser, which is in effect an additional pulper, where they are pulped (Figure 4.9).

Combined pulper-sieve-repasser units can handle about 25 per cent more cherries than a pulper alone, though power consumption is increased by about 50–60 per cent. However, they pulp so efficiently that they are recommended for use on medium and small-scale plantations (Wilbaux, 1963).

## Roller pulpers

Roller pulpers consist essentially of two rollers, one grooved obliquely so as to direct the cherries towards the centre, and the other smooth; these rollers rotate in opposite directions, the cherries being pulped between them. The pulped beans pass between the rollers while the pulp is held back and discharged separately. Roller pulpers are used mainly for processing liberica coffee.

## Raoeng pulpers

The Raoeng pulper removes not only the red skin and the adhering portion of the mesocarp but also the mesocarp attached to the endocarp or parchment. The pulper consists of:

1. A solid steel cylinder with ring channels on the feeder side, along about 12 per cent of its length, to bring forward the cherries. Further along there are circular ribs with intervening vertical divisions parallel to the shaft of the cylinder; at least 4 $m^3$ of water at a minimum pressure of 0.5 $kp/cm^2$ are required per tonne of cherries. The cylinder turns at 400–500 rpm, depending on the pulper model.
2. A circular perforated plate cover, which lets the pulp, mesocarp and water through but holds back the pulped beans. The perforations vary according to the variety of coffee:
(a) 3.75 x 22 mm for liberica.
(b) 3.2 x 22 mm for arabica.
(c) 2.8 x 22 mm for robusta.
3. A discharge chute for the pulped coffee, regulated by a weight-loaded discharge gate.

The Raoeng pulper is well suited to large plantations producing robusta coffee, which has a tough parchment. If the pulper is not properly adjusted the beans may be nipped or the parchment may be torn, especially with arabica.

The advantages of this machine are that on large estates the machine makes processing more efficient, the processing chain is shortened through the elimination of fermentation and the beans usually leave the pulper after the removal of all the mucilage that visibly adhers to the parchment. The disadvantages are that more parchment is torn as compared with standard pulpers and a large supply of water is needed for pulping alone. To avoid nipping of the beans, the discharge gate is set less tightly so that the beans come out with the parchment still covered with a thin layer of mesocarp palisade cells. The beans are therefore left to ferment for a period of 5 hours for robusta coffee. The main disadvantage of the Raoeng pulper is its high power consumption (3–6 times higher than in the standard method).

The specifications of some German models are shown in Table 4.8.

**Table 4.8 Specifications of three German Raoeng pulpers**

| Specification | Type 1 | Type 2 | Type 3 |
|---|---|---|---|
| Rated capacity (kg cherry per hour) | 750 | 1500 | 3000 |
| Power required (kW) | 5.9 | 10.4 | 18.5 |
| Revolving speed (rpm) | 500 | 475 | 400 |

### *Fermentation of coffee*

After pulping, the light beans are separated from the heavy ones before fermentation of the 'sound' beans takes place. Separation is carried out in reception tanks. Parallel channels can be used instead of tanks for the separation of floaters from sinkers in water. The light and empty beans, together with the waste pulp, float on the surface while the heavy beans fall to the bottom and are graded by density. The channels are usually 60 cm wide, 40 cm high, 40 m long and inclined at a gradient of 0.6–0.75 per cent. This method of rough-grading is widely practised in Indonesia for robusta coffee and in the Congo and Kenya for arabica coffee.

Fermentation is carried out to liquefy and dissolve the mucilage (mesocarp) adhering to the parchment. According to Groenewege quoted by Wilbaux (1963) the mucilage is dissolved during fermentation mainly by bacteria of the *Bacterium lactic aerogenes* group which form lactic acid at the expense of the sugars and at the same time hydrolize the pectins which bind the cell walls.

Berkley (1931) investigated enzyme fermentation in coffee, other than microbial, and demonstrated the enzymatic activity of broken tissues under aseptic conditions. The naturally occurring enzymes were diastatic in type, attacking starches and pectic compounds. Processes that occur during fermentation of coffee may be summarized as follows:

1. The first micro-organisms to flourish in the pulped beans are the yeasts. On multiplication they cause a rapid rise in the alcoholic content of the pulped mass. It is believed that during this period some character called 'body' is developed in the bean. When the yeast population starts to subside during wet fermentation, naturally occurring bacteria begin to multiply.
2. Alcohol is no longer formed with the cessation of yeast activity, and typical bacterially induced lactic acid fermentation begins, as occurs in any fruit or vegetable acid fermentation. During this time the acidic flavour develops and is enhanced in the bean.
3. The lactic acid organisms stop multiplying after reaching a certain intensity, and other types of organism start to develop. These contaminants are the kinds that decay oils and similar components, and are considered to be butyric ferments. They can be identified when there is evidence of an ill-smelling scum and froth on the surface of the mass in the fermenting vat.

The time of fermentation depends on various factors:

(a) The environmental temperature; the hotter it is the more rapid the process will be.

(b) The degree of cherry ripeness. Overripe fruits ferment more rapidly; unripe cherries, with the protopectin not demethoxylated as completely as in the ripe fruit, require a longer fermentation period.

(c) The intentional inoculation (with cultures of bacteria and yeasts) or unintentional (introduction of pectolysing bacteria by polluted water), which hastens the rate of mucilage digestion.

In hot areas, robusta coffee is fermented for 12–24 hours, or even 36 hours; in cooler areas, arabica coffee is left to ferment for 36–60 hours. These figures do not apply in all cases but, in general, fermentation is considered to be finished when the beans appear to be ready for washing. To ascertain when the beans are ready for washing, a handful are taken out and rubbed between the hands; a grating noise due to the rubbing of the rough parchment should be clearly heard. Also, on shaking some of the beans in water, the remaining mucilage should easily detach from the parchment.

## Fermentation methods

There are two fermentation methods in the processing of coffee. These are dry fermentation (draining off and removing the ferment fluid) and wet fermentation (fermenting the beans under water).

Dry fermentation is nearly always used for robusta coffee as well as for arabica coffee in Costa Rica and Columbia. The lower sluice gate of the fermenting tank is usually left open to let the ferment fluid out. To avoid the top layer drying it is covered by wet sacks, banana leaves or a wooden cover. When the mucilage is sufficiently loosened from the beans to be washed off easily, fermentation is considered to be completed. The lower sluice gate is then closed and water is let in to cover the mass of beans completely. After a period of time the beans are taken to the washing unit.

Wet fermentation is used to a large extent in Kivu and in the East African countries which produce very fine arabica coffees. After filling the tank and draining off the surplus water the lower sluice gate is closed and clean water is allowed in to cover the beans, which are then left to ferment.

## Other methods of breaking down mucilage

For removing the mucilage adhering to the parchment, mechanical methods may be applied, for example, by using Raoeng or Aquapulpa machines. The mucilage is then removed by friction as the beans pass between a revolving perforated drum and an inner perforated tube, with a countercurrent of water.

On the one hand, these machines make it possible to circumvent faulty fermentations, while on the other hand, beans produced in this way are somewhat neutral, with relatively little character unless they are put in clean running water afterwards for 12–20 hours. These so-called demucilaging machines are suitable for large estates growing robusta coffee, which has a strong parchment. The major disadvantage with mechanical removal is that traces of mucilage always remain adhered to the parchment and these may ferment during predrying so that the coffee may acquire a fruity or tainted flavour.

The chemical removal of mucilage by NaOH or lime and the use of pectic enzymes to accelerate mucilage degradation have been studied by a number of researchers for the past 50 years (Roelofsen, 1939; Wootton, 1965, 1967a; Ehlers, 1980). By treating beans with dilute potassium salts to remove the slimy mucilage, Roelofsen (1939) is reported to have obtained successful results. On the whole, however, alkali-treated coffee is believed to 'lack character' and may be 'less acidic' than coffee made from standard fermented beans.

### *Fermentation equipment*

The African small-scale producers usually ferment arabica coffee in wooden vats, baskets or piled up in heaps. The major drawbacks of this simple method are that fermenting mucilage promotes the proliferation of harmful microbial species and that fermentation is often allowed to continue far longer than is necessary. Detrimental effects, such as unfavourable flavours and aromas, resulting from this sort of practice are not unusual.

Robusta and arabica beans are nearly always fermented in stone or brick-work tanks coated with smooth cement plaster (Figure 4.10). The capacity of the tanks varies from about $0.5-15\,m^3$. The fermenting tanks are usually rectangular with a 5−8 per cent inclined floor sloping towards the outlet. They are fitted with grates located below, to remove the water and ferment fluid into a draining pit. They also have a sluice gate placed above the grate which allows the beans be flumed out to the washer (Figure 4.11).

### *Washing*

The objective of washing is to remove all traces of mucilage from the parchment. The importance of this washing operation lies in the fact that, during draining, predrying and even the final drying, any remaining mucilage may undergo further fermentation, which can be detrimental to the flavour of the final product.

Washing may be carried out in the fermenting tank by adding water to the tank and stirring the beans by hand. The turbid water produced in this way is subsequently drawn off and the operation is repeated until the water comes out clean. This method can be used on very small plantations. When the bean layer is more than 60−70 cm deep, stirring by hand is difficult and tiring. An amount of 1300 litres of water is required to wash one tonne of wet parchment coffee. Washing may be done mechanically by means of equipment known as

**Figure 4.10 Fermention of coffee after pulping in cement lined tanks.**

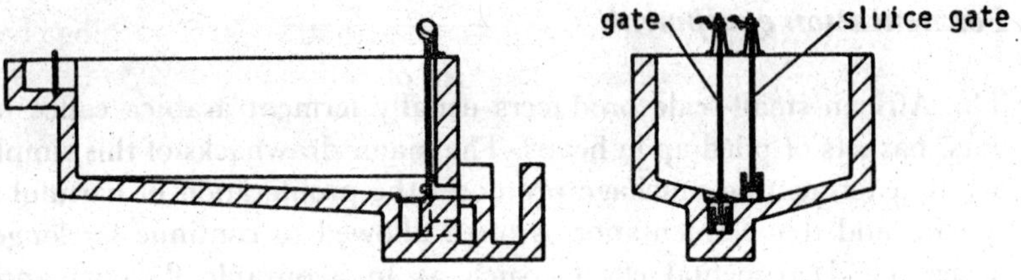


**Figure 4.11 Coffee fermenting tank.**

washers. These work either batchwise or continuously.

### Batch washers

These pieces of equipment consist of a wire gauze horizontal cylinder 65 cm in diameter and 1.25–2 m long, reinforced on the inside by metal semicircular sections, and set in a tank filled with water. A central shaft, on which beaters are fixed, revolves inside the cylinder at 90 rpm to mix the coffee thoroughly in the course of washing. A shutter closes the discharge outlet to prolong the operation. Batch washers have a small capacity.

### Horizontal, cylindrical, continuous-flow washers

These washers consist mainly of a galvanized sheet iron cylinder in which a central shaft with beaters turns at 100–180 rpm to mix the coffee and to promote a spiral, fowards movement towards the discharge end. The lower part of the cylinder unit is often corrugated lengthwise to increase mechanical action.

To operate the washers, they are filled with a mixture of equal volumes of fermented beans and water. The main drawback of these machines is that the beans are not thoroughly washed and the coffee comes out with mucilage-laden water. The purpose of washing therefore is not achieved so that the

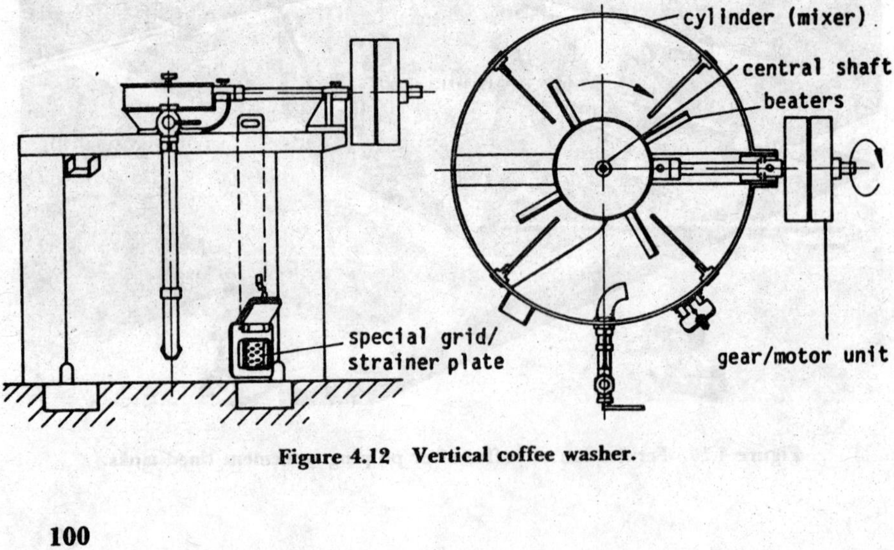


**Figure 4.12 Vertical coffee washer.**

process must be completed in another machine or in a washing channel with a fresh supply of clean water. The use of continuous-flow washers is therefore limited.

### Vertical washers

The vertical washer is designed as a mixer into which water is fed and allowed to overflow. The wash water, along with the mucilage, is drained away through the discharge door, the beans being retained in the washer by a special grid or strainer plate (Figure 4.12). Vertical washers are reported to be more efficient than other washers. Combined washing and flotation in washing channels is a method commonly used in processing arabica coffee in Africa and Central America. The channels usually have a cross-section of 0.6 x 0.4 m and a slope of 0.5–1 per cent. A channel 60 m long is large enough to wash a batch of fermented beans from 1300–1600 kg of cherries in about 25 minutes. One tonne of cured coffee (about 6250 kg of cherries) requires 20–25 $m^3$ of water.

## Processing coffee by the dry method

While the best quality mild arabica coffee is prepared by the wet method, robusta and Brazilian coffee are processed by the dry method. One of the variations of the dry system of processing coffee is popularly known as the East Indies method. The coffee is picked, brought in as whole fruits, and spread on mats in the sun. It is dried to glass-hardness and then hulled by pounding with long handled wooden pestles in hollowed wood mortars.

Dry processing is economical in tropical regions where the cherries are picked in the dry season. In the dry method the amount of moisture to be removed from cherries containing 65 per cent moisture is roughly 3.25 kg of water per kg of cured coffee. On the other hand, when parchment coffee with 50–52 per cent moisture is prepared by the wet method, the figure amounts to 1.5 kg per kg of the finished product. The dried fruit, called 'em coco' in Brazil (coffee in the dry pod), should contain less than 20 per cent moisture for satisfactory hulling.

During drying, the coffee bean undergoes many changes; the colour of the red pericarp turns black, fermentation easily takes place and usually has a detrimental effect on the finished product and the beans turn yellowish-brown, instead of blue-green as they do when properly prepared by the wet method. Further, the beans as well as the husks shrink and harden. Drying is known to be complete when, on shaking the dry cherries, the beans rattle. The rate of shrinkage for red-ripe cherries is about 39 per cent for robusta and 40 per cent for liberica coffee (Llewelyn, 1955).

### *Sun drying*

Cherries can be dried in the open on a floor of beaten earth. However, better results are achieved on cement platforms, slightly inclined to allow easy drainage, or on trays. In Brazil the cherries are spread out on exposed drying platforms known as terreiros. Under favourable weather conditions drying is completed in 10–14 days.

With layers 8 cm deep the load should be about 50 kg of ripe robusta cherries, equal to 10 kg of the finished product, to the square metre. For liberica coffee, Nosti Nava (1953) recommends not more than 25 kg per square metre. Table 4.9 summarizes the weight loss for robusta coffee sun-dried in Indonesia.

**Table 4.9 Weight loss for robusta coffee during sun drying in Indonesia**

| Time (days) | Weight of cherries as related to initial wet weight (%) | Moisture content of cherry (%) | Moisture content of bean (%) |
|---|---|---|---|
| 0 | 100 | 61.8 | – |
| 1 | 77.1 | 50.5 | 50.9 |
| 2 | 66.6 | 42.6 | 46.3 |
| 3 | 60.2 | 36.6 | 39.0 |
| 4 | 55.2 | 30.8 | 33.5 |
| 5 | 52.0 | 26.6 | 28.0 |
| 6 | 46.9 | 18.6 | 18.5 |
| 7 | 45.9 | 16.8 | 17.2 |
| 8 | 45.9 | 16.8 | 16.3 |
| 9 | 45.3 | 15.7 | 14.6 |
| 10 | 44.5 | 14.2 | 14.1 |
| 11 | 43.9 | 13.0 | 13.7 |
| 12 | 38.3 | 0 | 0 |

Source: Roelofsen, 1939.

## *Mechanical drying of cherries*

Cherries can be dried with the same driers as those used for parchment coffee. The static drier, for example, is extensively used in the Congo. In Brazil and Angola, to avoid the drawbacks of rapid drying, mechanical drying is carried out slowly, alternating drying with storage of the heated cherries. One drier of this type is the so-called 'Torres' drier.

### The Torres drier for ripe cherries

The Torres drier consists essentially of an outer wire-mesh cylinder, turning on rollers at 2 rpm for models with a capacity (net volume) of 900–3600 litres and at 1.5 rpm for the 7600 litre-model; an axial helical blower forces the heated air coming from the furnace into an inner concentric cylinder also made of wire-mesh.

The drying unit therefore comprises a central air distribution cylinder and an outer cylinder. The coffee is introduced into the space between the two cylinders which must be well filled otherwise there will be a serious waste of heated air. The air is heated to a temperature of 50°C. The Torres drier is suitable for regions where the cherries are picked during a pronounced dry season or when they are partially dried before they are put into the drier, as in Brazil and Angola. It is considered in Brazil to be the most efficient drier of its type in use.

### *Hulling and peeling*

The coffee bean must be hulled or shelled, depending on the method used in its preparation. Coffee prepared by the dry method (dry cherries) are hulled while wet prepared coffee (parchment coffee) is shelled or peeled. Before hulling or peeling, stones, dust and other extraneous materials are removed from the beans by means of, for example, simple winnowers, an airfloat stoner, a vibrating screen or pneumatic separators. Machines for hulling or peeling are accordingly called hullers or peelers. Some of these machines will take only one type of coffee, parchment or dry cherries. Others, on the other hand, with minor adjustments, will handle both types of coffee.

#### The Africa huller

The main feature of this huller is a steel hulling roller with a feed-screw end about two-ninths of the total length, fitted with spiral or oblique bars; the spiral action of these bars moves the coffee onto the hulling section of the roller, fitted with slightly oblique hulling bars. The bars are detachable. A stationary adjustable and removable hulling knife is fitted to the side of the machine, parallel to the roller. The coffee is carried forwards by the hulling bars to the discharge spout and hulling takes place between the bars on the roller and the inside edge of the knife (Figure 4.13). Africa hullers are extensively used and will hull dry cherries or shell parchment coffee. Whatever the type of coffee being hulled it must be quite dry, especially for hulling cherries, otherwise clogging and overheating will occur. Africa hullers are built to the sizes shown in Table 4.10.

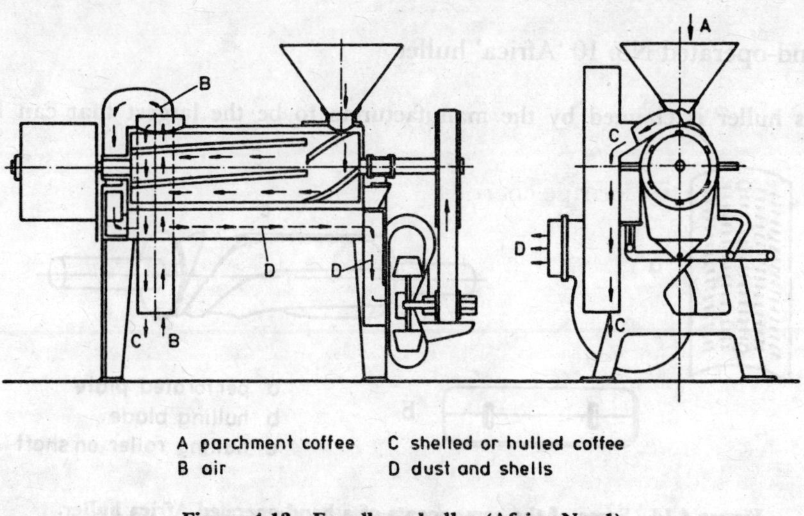


**Figure 4.13 Engelberg huller (Africa No. 1).**

**Table 4.10 Specifications of various Africa hullers**

| | Rated hourly output | | Power required (kW) | Speed of driving pulley (rpm) |
|---|---|---|---|---|
| | parchment coffee | dry cherry coffee | | |
| No. 1 | 650 | 440 | 7.3–8.8 | 400 |
| No. 1½ | 410 | 280 | 5.8–7.3 | 400 |
| No. 2 | 320 | 220 | 4.4–5.8 | 450 |
| No. 5 | 170 | 110 | 3 | 550 |

The rated outputs in Table 4.10 are the maximum outputs for arabica coffee. For robusta coffee they can be increased by about 25 per cent and reduced by 30 per cent for liberica coffee, which requires special screens.

## Okrassa peeler and polisher

As the name suggests this machine consists mainly of two compartments; the coffee is peeled in the first compartment and polished in the second. An air current is provided to draw off the shells as the beans move onto the polishing compartment, where any beans still coated are shelled and the beans are polished.

The Okrassa peeler and polisher is used for shelling both arabica and robusta coffee. It usually comes in 3 sizes with the specifications shown in Table 4.11

**Table 4.11 Specifications of the Okrassa peeler and polisher**

| Specification | No. 1 | No. 2 | No. 3 |
|---|---|---|---|
| Capacity of parchment coffee, (kg/h) | 1000 | 1500 | 2000 |
| Speed of pulley, (rpm) | 200 | 125 | 110 |
| Speed of pulley on fan, (rpm) | 1900 | 1500 | 1200 |
| Power required, (kW) | 6.6 | 8.8 | 14.6 |

## Hand-operated No. 10 'Africa' huller

This huller is claimed by the manufacturers to be the largest that can be

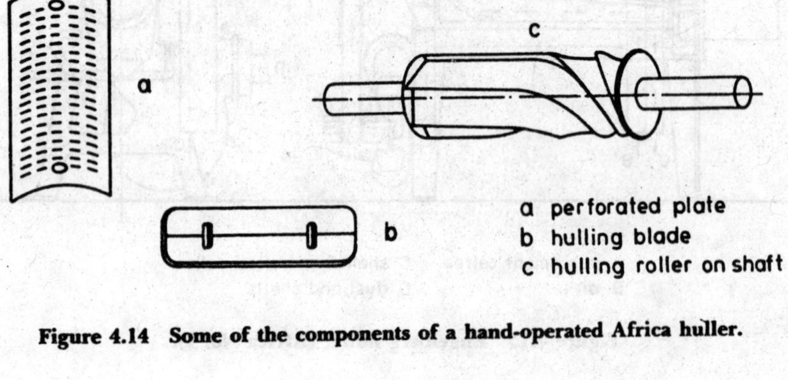


**Figure 4.14 Some of the components of a hand-operated Africa huller.**

hand-operated. It is fitted with two hulling blades and these must be adjusted so that there is minimum breakage of the beans. A perforated plate, which is easily detachable for cleaning purposes, is fitted to the bottom of the machine, and any dust and small flakes of shell pass through it (Figure 4.14).

The huller is normally fitted with two handles, but it can also be provided with a pulley 228.6 mm in diameter × 76.2 mm wide and should then run at 250 rpm, requiring 1.0 kW. It can shell dry coffee cherries and parchment coffee and can be used for arabica, robusta and liberica coffees. The output of clean arabica coffee, using parchment coffee, is 36 kg/h, and 22 kg/h for dry cherry coffee. For robusta the outputs indicated are about 30 per cent more and for liberica about 30 per cent less.

## Descascador, a Brazilian huller

A representative of this machine is the Blasi huller, which consists of a feed hopper, a perforated cylinder, a hulling rod, wooden trunking, a ciranda screen and a fan (Figure 4.15). It is designed for dry cherry coffee prepared by the dry method. The coffee is moved from one end of the machine to the other in two perforated plate cylinders within which the hulling rods rotate. These, each fitted with seven arms, are topped by bevelled steel knives at an adjustable distance 14–17 mm from the side of the perforated cylinder.

The hulled coffee, together with the husks, passes through the perforations and falls into the wooden trunking, fixed at the bottom with an inclined woven-wire sheet with 3 mm mesh, which lets through small stones. The wire sheet directs the coffee to a separator screen (sururuca or ciranda). A current of air drawn up through the screen into the cylinders carries off the husks to the outside. The output is 450 kg of clean coffee per hour, and the power requirement is 8.8 kW. The main drawback of this machine is the fairly rapid wear on the perforated plates, the vertical shafts and the gudgeons.

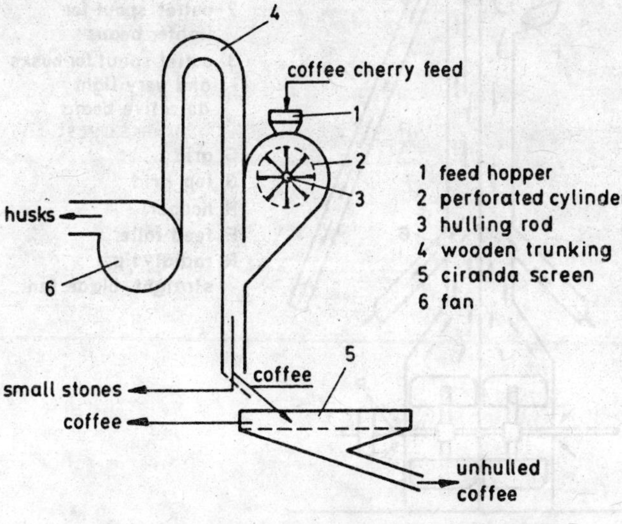


**Figure 4.15 Schematic layout of the Blasi huller.**

This huller, however, is reported to be superior to all European makes for dry cherry coffee. A trial run with 200 tonnes of robusta coffee is said to have given only 1.98 per cent broken beans by number, that is less than 1 per cent by weight. There is very little or no overheating during hulling.

## Coffee grading

Coffee is graded by means of catadors and air-floaters. Pneumatic coffee

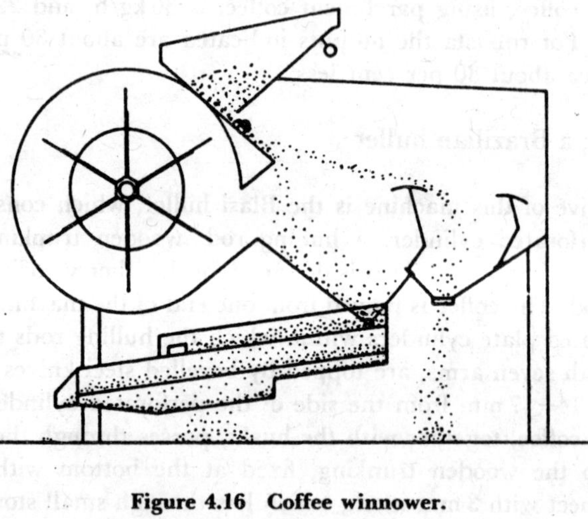

**Figure 4.16 Coffee winnower.**

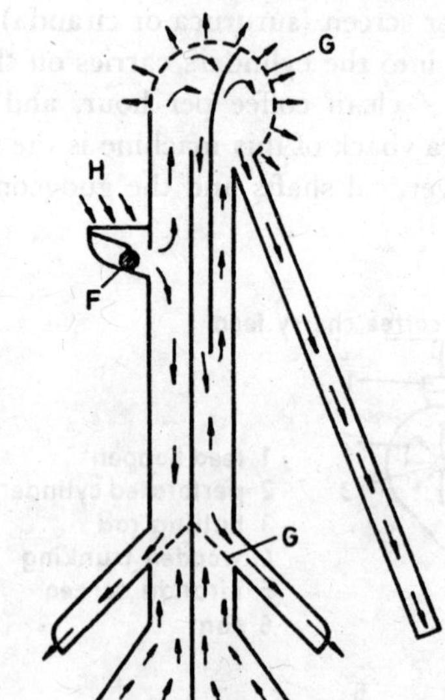


**Figure 4.17 Double-column catador.**

graders, usually called catadors, work on the principle of a strong upward current of air which passes through the coffee as it enters the trunking. The heavy beans fall downwards to an outlet spout while the lighter beans and refuse are carried upwards. Winnowers are often used by small-scale coffee processors for cleaning and separating coffee (Figure 4.16).

Two different catadors may be distinguished; double-column catadors (duplex) separate the sound beans into heavy and medium heavy, and single-column catadors (simplex) are designed for only one class of heavy beans. Figure 4.17 shows the basic arrangement of a double-column catador. A radial-tip fan induces a strong upward current of air through the two columns of the machine, separated by a partition. The air flow can be adjusted separately for each fan inlet. Radial-tip straight-blade fans turning at 550 rpm are generally used. The air is forced upwards through the two wire grids and escapes through a grid at the top. The beans freed from the husks and other refuse are fed into hoppers and spread out uniformly by a feed roller. The heaviest beans fall downwards to an outlet spout in the first column. The lighter beans and the husks are carried upwards and transferred to the second column or air passage with a less powerful air current, where beans fall into the second spout. Husks and very light defective beans are blown upwards so that they fall through the first spout.

The net power input for these catadors is usually 1.8−2.2 kW. The velocity of the air in the first air passage when the catador is empty is about 13.3 m/s.

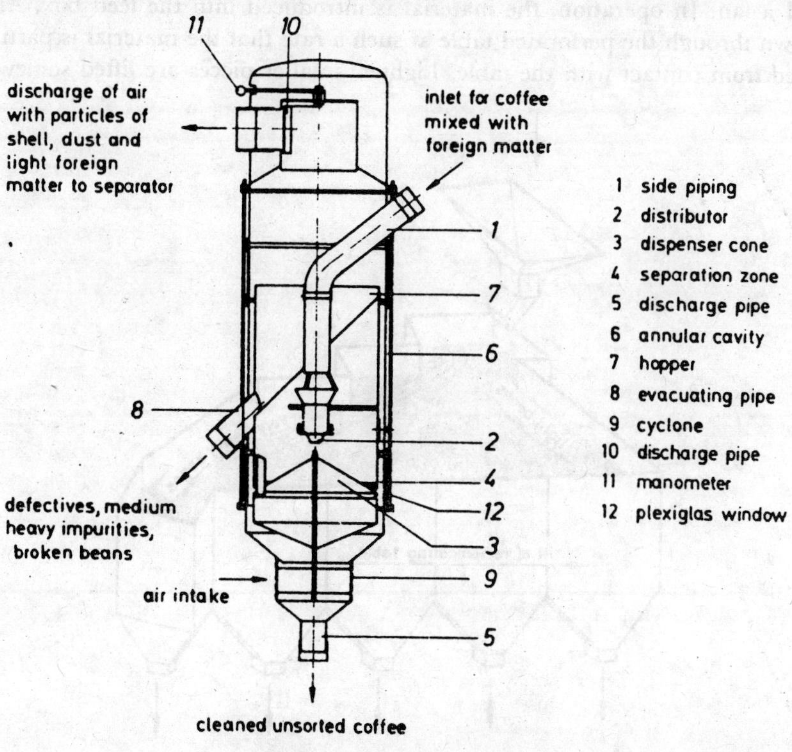


**Figure 4.18 Zulauf catador.**

The capacity is between 600 and 800 kg/h.

The single-column catadors have no central partition and there are two discharge outlets instead of three. The beans exit at the lower outlet while the refuse is collected at the top. For the same coffee intake this type of catador requires only 70 per cent of the power used in the double-column catadors. Single-column catadors are widely used in Brazil for arabica coffee and in Angola for robusta coffee. In Brazil catadors are generally shorter and are made almost entirely of wood.

In Germany the so-called 'Zulauf' catador operating on the vacuum principle has been designed by MIAG. The main feature of this design is the cylindrical trunking into which the coffee is introduced; a vacuum is created by an exhaust fan over a cyclone, and air flow and vacuum are adjusted by a regulator fixed at the top of the trunking. The heavy beans fall through a discharge pipe at the bottom of the trunking. The lighter beans and brokens rise through the annular cavity, fall into the hopper and are evacuated through a piping connected to the lock-chamber, which is designed to prevent air from entering. The very light impurities are carried off through another piping to a cyclone fitted to the lock-chamber (Figure 4.18).

### *Air-float separators*

Air-float separators consist mainly of a perforated inclined plane, a hopper and a fan. In operation, the material is introduced into the feed box. Air is blown through the perforated table at such a rate that the material is partially lifted from contact with the table. Lighter, smaller pieces are lifted somewhat

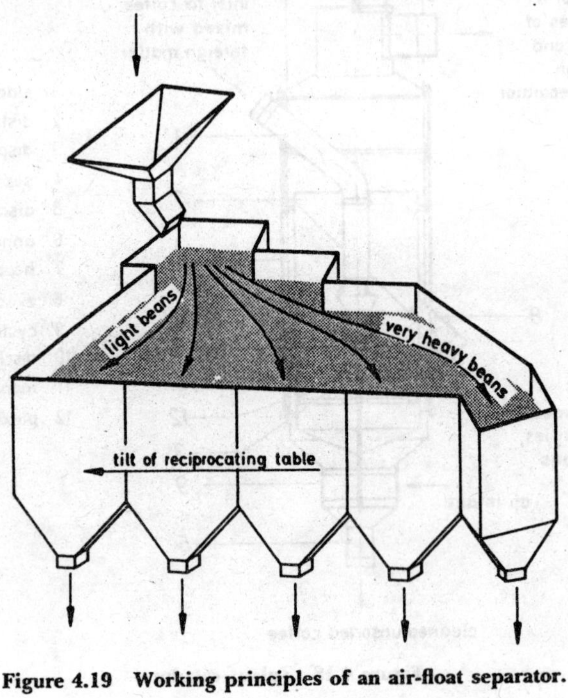


**Figure 4.19 Working principles of an air-float separator.**

higher and 'float' down the table towards the discharge edge. The large, heavy particles are not lifted by the air; they are moved by the oscillating motion of the table in the direction of conveyance and are discharged at the right edge of the table (Figure 4.19). The optimum hourly capacity for the regular models with a 4.4 kW fan is 700–780 kg. Power consumption is 5 kW/tonne of coffee processed.

## Drying of parchment coffee

Parchment coffee contains about 45–50 per cent moisture after pre-drying. The beans have to be dried until the moisture content is reduced to 8–12 per cent; 8 per cent for robusta and 10 per cent for arabica coffee. During drying, the volume of the beans shrinks by 12–15 per cent. Drying methods vary according to whether the species is robusta, arabica or liberica. Arabica coffee is dried in different ways depending on whether the final product is to be fair average quality (f.a.q.) or fine coffee.

The drying of coffee has many similarities with the drying of other heat-sensitive substances in which resistance to the diffusion of moisture from the inside to the surface is the main factor limiting the rate of evaporation. Coffee, and especially arabica coffee, is particularly sensitive to drying operations. Llewelyn (1955) distinguishes 3 stages in the drying of arabica coffee:

1. Drying from fully wet to skin dry, when the moisture content of the beans is reduced from 60–65 per cent to 45–50 per cent. This corresponds to pre-drying when about 25 per cent of the total moisture to be evaporated is removed.
2. Drying from skin dry to just past the so-called black stage, that is from 45–50 per cent moisture to 15–20 per cent. The beans are dark blue.
3. Drying from just past the black stage to the correct moisture of about 11 per cent.

Heating the air to a high temperature at the beginning of drying loosens the

**Figure 4.20 Parchment coffee drying in the sun.**

silverskin. Artificial drying of robusta coffee with air at 100°C, for 1–2 hours gives the beans a more pronounced bluish colour than drying at an initial temperature of 80°C, which makes them yellowish green. This is possibly attributed to the fact that the moisture diffusing from the centre to the surface when the beans are very wet facilitates the migration of chlorogenic acid. The use of high temperatures to dry arabica coffee spoils its quality.

### *Sun drying parchment coffee*

Sun drying is suitable in areas where there is plenty of sunshine, and is carried out in a number of different ways (Figure 4.20); for example, the parchment coffee may be placed on a permanent foot-wide staging of wire-mesh covered with loose strips of strong hessian, which may be held at the corners and carried under cover as necessary, or have loose rainproof covers that can be placed over the coffee (Haarer, 1962).

In favourable weather conditions drying on trays in the sun takes 10–15 days. The temperature of the layer of beans, usually 4 cm deep, does not exceed 42°C for quantities corresponding roughly to 28 kg of pre-dried coffee and 10–15 kg of dried parchment coffee. Knaus (1935) recommends not more than 15 kg of washed robusta beans (approximately 55 per cent moisture) to the square metre, while Ringoet suggests 10 kg, and during peak harvesting period as much as 20 kg. Ten kilograms of washed beans are equivalent to about 5 kg of parchment coffee.

Removable trays 2 m × 1 m are preferable. They are raised off the ground to allow the air to circulate freely over the drying surface. For best quality coffee, the beans must be turned regularly and protected from intense sunshine. McCloy (1959) investigated the effect of sunlight on the quality of fine arabica coffee, and found that yellowing of the parchment was due mainly to short wave ultraviolet radiation below about $32 \times 10^{-7}$ m wavelength. Moreover, the bright blue-green colour was probably caused by visible frequencies $40 \times 10^{-7}$ m – $70 \times 10^{-7}$ m and infra-red radiation (above $70 \times 10^{-7}$ m) could cause severe damage to liquor quality. It is considered that sunlight bleaches the colour of the silverskin because of the destruction of plastids assumed to contain chlorophyll.

For top quality arabica coffees any machine drying is bound to be harmful as it reduces the time of exposure to sunlight. Lack of light treatment causes incomplete bleaching of the silverskin and the presence of the green-yellow colour in the bean instead of blue. Excessive sourness in machine-dried coffee is due to too high a drying temperature. The most efficient light frequency in promoting bean quality is reported to be about $46 \times 10^{-7}$ m, that is blue light.

### *Artificial drying*

In humid and rainy areas and on large estates, artificial driers are largely used because of the many operations involved in sun drying and the space taken up by the apparently simple and cheap process of sun drying. Some form of heat with or without forced ventilation is required for artificial drying of parchment coffee. The driers used in coffee drying are static driers, rotary driers and batch driers adapted from grain driers.

## Static driers

These consist mainly of a perforated galvanized sheet-iron flooring lying at least 4 m above a bank of fire tubes and roofed over to keep off rain. Figure 4.21 illustrates the layout of this type of drier. A furnace, usually fired with wood, connects to a header conveying combustion gas into tubes. Air inlets below the bank of fire tubes bring in cold air which is heated by the fire tubes, rises to the sheet-iron floor, circulates through the layer of beans and exits through the skylight.

The main advantages of this drier are:

1. No mechanically driven equipment (fan or other units) or power supply needed.
2. Simple construction and low maintenance, limited to cleaning the fire tubes and headers.

The drawbacks are:

1. Low thermal efficiency, especially towards the end of the drying operation.
2. Inefficient utilization of the heated air in the case of small loads spread out in thin layers (4 cm).
3. The necessity of relying on the skill of labourers who have to turn the bean layer continually to ensure uniform drying.
4. The difficulty of adjusting to different heating temperatures the flow of hot flue gases in the various inlet and outlet tubes connected to the headers.

This type of drier, despite its defects, is reported to be popular and widely used in Java and in the Congo, particularly for drying robusta coffee, which is spread out in layers 8–10 cm deep (about 25–30 kg/m$^2$). The initial temperature may be set at 80°C and later reduced to 60° or 50°C towards the end of

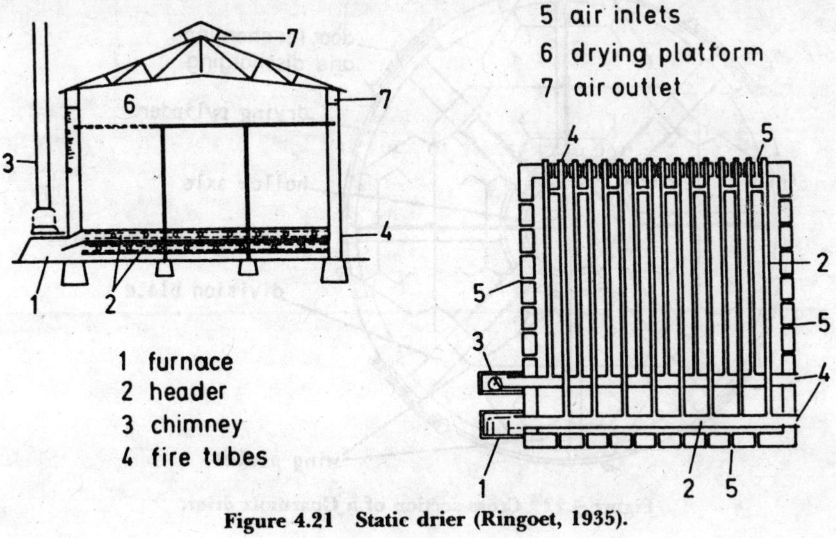


**Figure 4.21 Static drier (Ringoet, 1935).**

the drying operation. At these temperatures a batch of coffee is dried in about 20 hours. To obtain a bluish tinge the air must be heated to 95–98°C. and reduced to 80°C after about 2 hours. Fuelwood consumption is reported to be between 1.5 and 1.8 kg/1 kg clean coffee.

When drying average quality arabica coffee the temperature of hot air should not be higher than 50°C and should be reduced to 40°C towards the end of the drying operation. The beans must be spread in a thin layer, not more than 15 kg to the square metre, and the drying time must be extended to over 40 hours. Fine and very fine arabica coffees cannot be obtained with the static drier, because of the difficulty of adjusting the temperature.

## Rotary driers

The *Guardiola drier* is a representative of the rotary driers. It consists essentially of a 1.8 m diameter cylinder made in various lengths which is built around a hollow axle for conveying the heated air (Figure 4.22). The drying cylinder is divided into four equal compartments by division plates fixed to the axle in the form of a star. Perforated shell plates enclose the cylinder to contain the coffee and let the air escape, and are fitted with one or more doors for loading or unloading. Hot air is introduced through one end of the hollow axle and circulated through the batch by perforated radial tubes connected to the hollow axle. Wing plates to ensure mixing of the coffee are attached to the longitudinal partitions. The correct charge for a 1.8 m diameter cylinder is 1.860 litres/m run, for a gross volume of 2630 litres. A high-pressure, multi-bladed suction fan (rpm range from 1100 to 2500 according to size) draws in the air from outside and forces it into the heating system, which may be either direct or steam heating. The Guardiola is a non-conducting batch drier in which the heated air circulates rapidly.

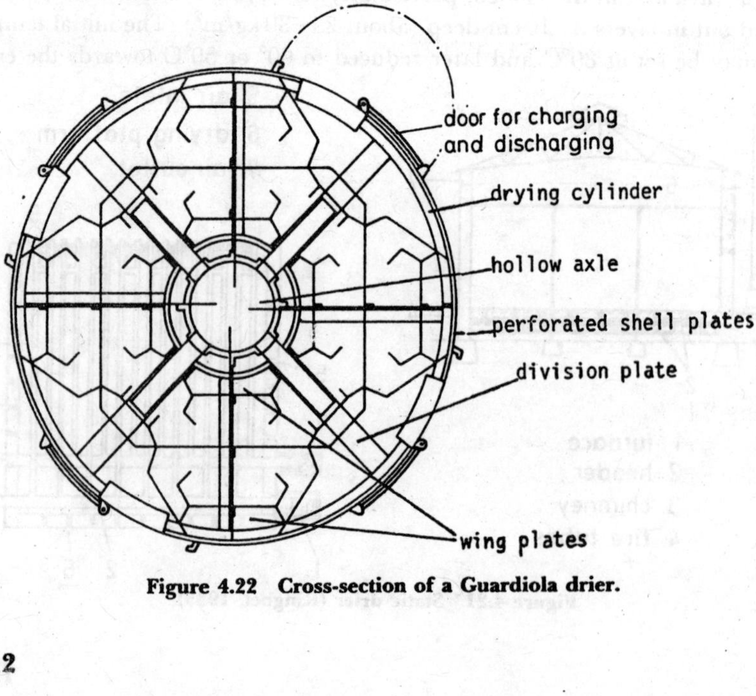


**Figure 4.22 Cross-section of a Guardiola drier.**

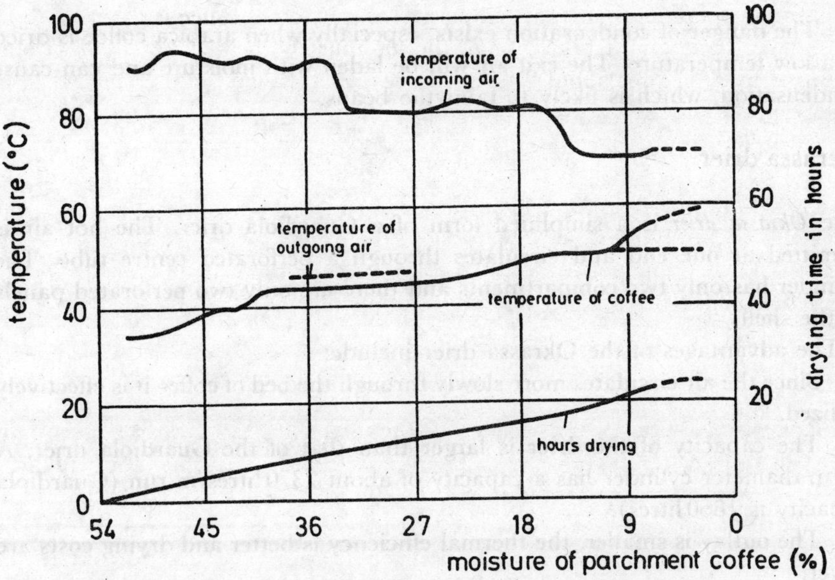


**Figure 4.23 Drying rate of robusta coffee (Wilbaux, 1963).**

Robusta coffee can be dried in 24 hours at temperatures starting at 90°C and afterwards reducing to 80°C and finally 60°C. Figure 4.23 shows the rate of drying for 54 per cent wet robusta coffee exposed to hot air at 90°C for 10 hours, then at 80°C for another 10 hours and lastly, at 70°C for 4 hours. At this drying rate 1.5–1.7 kg of fuelwood and 11.7 kW per hour are required to produce 1 kg of clean coffee. Arabica coffee cannot be dried in less than 40 hours with an initial air temperature of 60°C, and reducing the temperature afterwards to 50°C. Combined sun and machine drying is used for arabica coffee. Specifications of some Guardiola driers are shown in Table 4.12.

**Table 4.12 Specifications of Guardiola driers**

| Model | 1.8 × 0.9 m | 1.8 × 1.8 m | 1.8 × 3 m | 1.8 × 3.6 m | 1.8 × 4.8 m |
|---|---|---|---|---|---|
| Charge in kg of dry parchment coffee | 540 | 1080 | 1820 | 2180 | 2900 |
| Drier kW | 0.6 | 0.73 | 0.9 | 1.1 | 1.5 |
| Fan kW | 2.2 | 3.1 | 4.4 | 5.5 | 7.7 |
| Total kW | 2.8 | 3.8 | 5.3 | 6.6 | 9.2 |
| Fan Speed (rpm) | 2600 | 2600 | 1650 | 1400 | 1400 |

Guardiola driers are reported to be good machines, giving satisfactory results. However, the following disadvantages are associated with them:

1. Their power requirements are high, about 12 kW per hour/kg of clean dry coffee. Like all non-conducting batch driers their thermal efficiency is low.
2. The outlay is high considering their drying capacity especially in the case of arabica coffee which has to be dried at a lower temperature for about 48 hours as against 24 hours for robusta coffee.

3. The danger of condensation exists, especially when arabica coffee is dried at a low temperature. The exit air will be laden with moisture and can cause condensation, which is likely to taint the beans.

## Okrassa drier

The *Okrassa drier* is a simplified form of a Guardiola drier. The hot air is admitted at one end and circulates through a perforated centre tube. The cylinder has only two compartments and there are only two perforated panels in the shell.

The advantages of the Okrassa drier include:

1. Since the air circulates more slowly through the bed of coffee it is effectively utilized.
2. The capacity of the drier is larger than that of the Guardiola drier. A 1.8 m diameter cylinder has a capacity of about 2370 litres/m run (Guardiola capacity is 1860 litres).
3. The outlay is smaller, the thermal efficiency is better and drying costs are lower.

The main drawback is that the drier must always have a full load otherwise the cylinder is out of balance.

Also based on the Guardiola design, the *Mason drier* has a special arrangement of the division plates, which keeps the beans moving and prevents the cylinder from getting out of balance. The drum runs on rollers instead of on trunnions.

## Wilken drier

The *Wilken drier* is essentially a stirred bin drier consisting of a circular vat 6 m in diameter and 0.45 m deep with a perforated false floor. A central shaft is provided with radial arms each of which has four stirrers to keep the beans moving in an even layer and to convey them to the discharge outlet at the end of the drying operation (Figure 4.24). During drying, the central shaft is set in motion for 5 minutes every 3 hours.

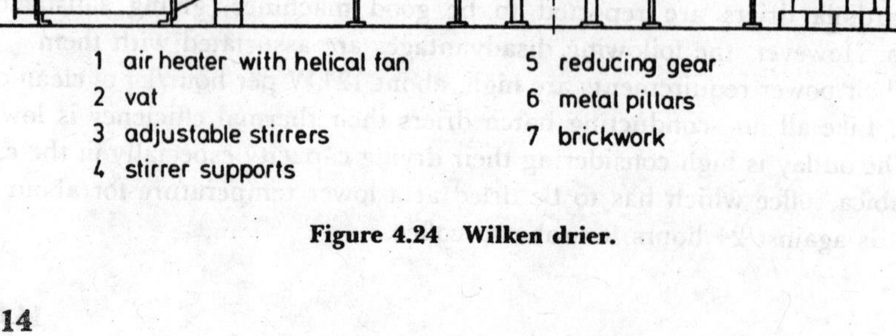


**Figure 4.24 Wilken drier.**

The capacity of the drier is 2 tonnes of parchment coffee and drying is completed in 23 hours. The drier is equipped with a gas-oil air heater fitted with a 2.2 kW fan unit. The load of the motor driving the central shaft may not exceed 3.5–3.7 kW. The Wilken drier, which was designed in Kenya, is widely used for drying arabica parchment coffee.

### *Combined mechanical and sun drying of coffee beans*

Rapid drying at a relatively high temperature is good for robusta coffee, especially in the initial stages, while arabica coffee needs to be dried much more slowly at a lower temperature. Very fine arabica coffee needs to be exposed to sunlight; if for some reason the beans cannot be completely sun-dried, then mechanical drying is employed to complete the drying process.

For very fine arabica coffees the beans should be dried in the sun until the moisture content is reduced to 20 per cent. Mechanical drying is reported to have no negative effect on the quality of the coffee if the temperature of the drying air does not exceed 49°C. Coffee can first be mechanically dried at an inlet temperature of 49°C for about 3 days to reduce the moisture content to 25 per cent and then finally sun-dried. McCloy (1959) recommends making mechanical drying as brief as possible. With an inlet temperature of 49°C he obtained the results shown in Table 4.13

**Table 4.13 Drying of arabica coffee**

| Moisture content after mechanical drying | Time for final sun drying (days) | Colour of raw coffee | Taste of liquour |
|---|---|---|---|
| 54% not mechanically dried, 'fully wet' | 5 | Grey–green, good | Fair–good |
| 45% skin dry | 4½ | Grey–green, fair-good | Fair–good |
| 25% soft black | 3 | Grey–green, fair | Fair–good |
| 15% hard black | 1 | Grey–green, poor | Sourish |
| 11% fully dry | 0 | Yellowish-green with brown tinge, very poor | Very sour and unpleasant |

### *Storage of coffee beans*

Green coffee is usually sold in jute sacks or bags, generally at 60 kg net weight, and may need to be stored either in warehouses in the producing country or at its overseas destination. Green coffee is susceptible to changes affecting appearance and flavour, but especially at elevated temperatures (above 25°C) and in conditions of high relative humidity where the moisture content can rise above 12–13 per cent. During storage of the green bean, particularly at elevated temperatures, it has been reported that there are changes due in part to proteolysis, for example, increases in alanine, isoleucine and tyrosine, and losses of free amino-acids in non-enzymic browning reactions. Such losses are accompanied by losses in reducing sugars, the production of a brown discolouration and a reduction in beverage quality (Pokorny *et al.*, 1975).

Figure 4.25 illustrates the equilibrium moisture of green beans, roast beans

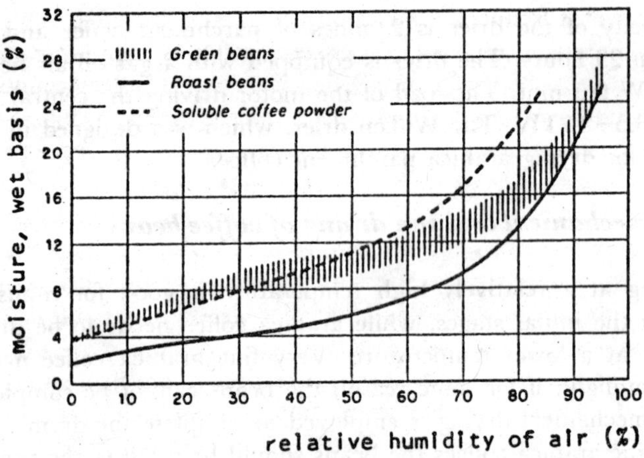


**Figure 4.25 Equilibrium curve for green beans, roast beans and soluble coffee powder (Sivetz and Foote, 1963).**

and of soluble coffee powder, at various temperatures and relative humidities. It is important to keep the bean moisture low so that respiration is relatively dormant, and conditions so dry that the beans tolerate higher temperatures without damage.

Staling is the loss and change in composition of the volatile coffee aroma constituents. This takes place when coffee and coffee products are stored prior to the preparation of the beverage by the consumer. The phenomenon of staling has been studied extensively as it determines the commercial shelf life (Clifford, 1985). During storage many volatiles decline in concentration; certain others increase in concentration, possibly as a result of oxidation. For example, dimethyl sulfide (DMS) is not only volatile but also readily oxidizes to methyl sulphoxide. The loss of this potent aromatic component will downgrade the aroma and flavour of most coffee. Contact with oxygen is particularly detrimental to the quality of coffee. Radtke (1979) reports that roast and ground coffee has a commercial storage life of six months at 23°C provided the residual oxygen content does not exceed 1 per cent and the water content does not exceed 4 per cent. An increase in variables like temperature, residual oxygen content and water content reduces the storage life. Staling of coffee is retarded by storage of the beans or grounds in air-tight containers at temperatures below 0°C.

## Manufacture of instant coffee

The dried coffee beans take one of two paths. They are purchased by the consumer in the roasted and ground form or their soluble constituents are extracted and they reach the consumer as a dry soluble powder – popularly known as instant coffee. Instant coffee is a convenience food compared with brewed roast coffee; it eliminates all brew processing and equipment. One advantage of instant coffee processing is that it is able to use lower grade green coffee beans, yet it frequently makes an acceptably flavoured beverage. However, the manufacture of soluble or instant coffee is highly capital intensive and so its production is concentrated mainly in industrialized countries. In the last 20 years, however, raw coffee producing countries such as Brazil, Columbia, Mexico and the Ivory Coast have themselves manufactured substantial amounts of soluble or instant coffee.

The processing of dried coffee beans into instant coffee essentially involves roasting, grinding, percolating, drying, cooling and packaging (Figure 4.26). Different types of green coffee roast and percolate differently, and also give varying results when processed as a blend. Robusta coffees easily give higher yields of solubles but have the disadvantage of giving more tars. Brazilian coffees behave similarly to robustas with respect to physical strength, but do not give the high yields of solubles or tars. Arabica mild coffees have more body and flavour, which carries through into the instant coffee cup. Robusta coffees constitute a large proportion, over 50 per cent, of commercial instant coffee blends in the USA and Europe. Some instant blends are even 100 per cent robusta. By 1967 about 15 per cent of American green coffee imports went into the manufacture of soluble coffee. It was estimated by the International Coffee Organization that 19 per cent of raw coffee was used in the manufacture of instant coffee in the world in 1980 (Quarterly Statistical Bulletin, Nos. 17/18, ICO 1980).

### *Roasting of coffee*

Roasted coffee is obtained by dry heating coffee beans at a temperature above 100°C until the desired changes in colour, aroma and strength are achieved. In the initial phase of roasting, energy is used up to evaporate the water contained in the beans. With the setting in of the browning reaction (above 100°C) the process becomes exothermic. Up to a temperature of 150°C, an

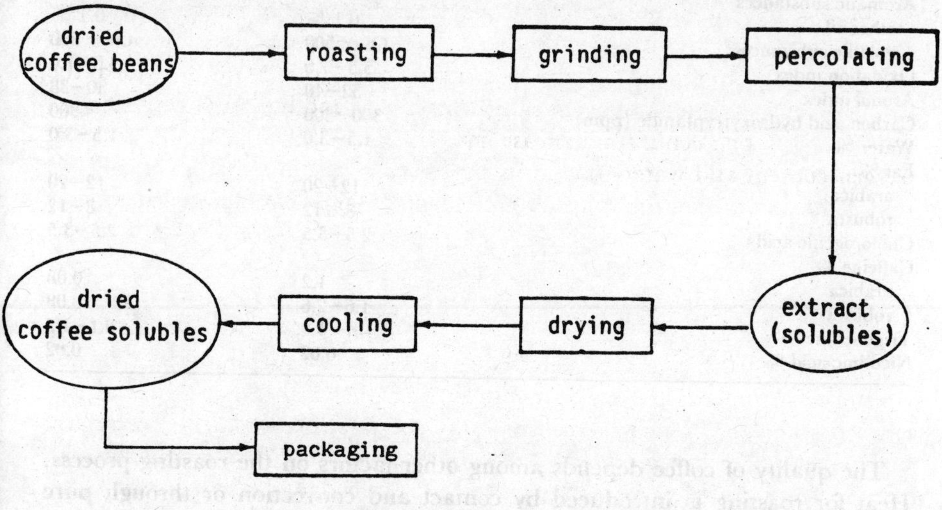


**Figure 4.26 Flowchart (simplified) for the processing of instant coffee.**

intensive gas development takes place (carbon dioxide, carbon monoxide, vapour). The beans become brittle, swell to 40–100 per cent of their volume and become millable.

Between temperatures of 180–200°C the beans begin to crackle and the coffee aroma is formed. Generally, coffee is roasted at 180–240°C for 15–20 minutes. The moisture content of the beans after roasting amounts to 2–3 per cent and is dependent on the roasting grade and roasting process. A light roast entails a 14–16 per cent loss in weight, a medium roast 16–18 per cent and a full roast (espresso-roasting) over 18 per cent. The following changes take place during roasting:

1. Apart from causing loss of water, roasting brings about formation of gases – mainly carbon dioxide – and the development of aroma and flavour, which are water-soluble on infusion.
2. The volatile substances are lost to some extent.
3. Sugars are partially caramelized and the carbohydrates and crude fibre partially carbonized.
4. The protein may be partially hydrolyzed or decomposed by the heat.
5. The caffeine content does not undergo discernible changes.

The composition of roasted coffee depends on the type of coffee, the method of processing the raw coffee and the roasting grade. Table 4.14 shows an overview on the colour and composition of roasted coffee.

**Table 4.14 Composition of roasted coffee**

| Constituent | With caffeine | Caffeine-free |
|---|---|---|
| Colour | Brown (light to dark) | Brown (light to dark) |
| Cut | | |
| wet processed | White, yellowish-white | Brownish |
| dry processed | Brown | Brown |
| Aromatic substances | | |
| ether oil | 0.1 % | 0.1 % |
| volatile substances | Over 500 | Over 500 |
| Oxidation index | 5.5–7.0 | 5.5–7.0 |
| Aroma index | 32–40 | 30–38 |
| Carbon acid hydroxytryptanide (ppm) | 300–800 | <300 |
| Water % | 1.5–3.0 | 1.5–3.0 |
| Fat % | | |
| arabica | 12–20 | 12–20 |
| robusta | 8.–12 | 8–12 |
| Cholorogenic acid | 2.5–3.5 | 2.5–3.5 |
| Caffeine % | | |
| arabica | 1.2 | 0.08 |
| robusta | 1.6–2.6 | 0.08 |
| Trigonelline | 0.1–1.2 | 0.1–1.2 |
| Nicotinic acid % | 0.02 | 0.02 |

The quality of coffee depends among other factors on the roasting process. Heat for roasting is introduced by contact and convection or through pure convection.

### *Contact-convection roasting*

#### Drum roasting

The roasting container may be a cylinder or a sphere. For batch operations, roasters with stirrers or rotating drum roasters are used. They are heated by gas or oil. Heat, directly heated air and combustion gases or indirectly heated air, is introduced through the double wall of the drum. Roasting time amounts to 12–30 minutes at full charge of 120–180 kg, quick roasting lasts 5–12 minutes at a finishing temperature of 210–240°C. At the end of the first roasting phase (170–180°C), heat is minimized and the coffee is roasted further until the desired colour is obtained. A simplified arrangement of a typical drum roaster is shown in Figure 4.27.

Roasting temperature and roasting time depend on the area of the roasting container. For a roasting duration of 10 minutes an area of 850 $cm^2$ is required for 1 kg coffee beans, and for 15 minutes, 600 $cm^2$ is required.

Non-perforated roasting drums, arranged one above the other and with divisions into many single chambers, make continuous roasting possible. Batchwise roasters usually have a capacity of 1800–3000 kg/hour of raw coffee beans, while continuous roasters have a capacity of 1500–3000 kg/hour.

#### Drum-pressure roasting

Closing the valves at a temperature of 200°C prevents the escape of volatile substances. Pressure of up to 5 bars is developed in the drum; this pressure is further increased through the injection of water in the end phase of the roasting. The water content of coffee roasted under pressure is about 1–2 per cent higher than that of coffee roasted normally.

For continuous working pressure roasters, perforated horizontal cylinders

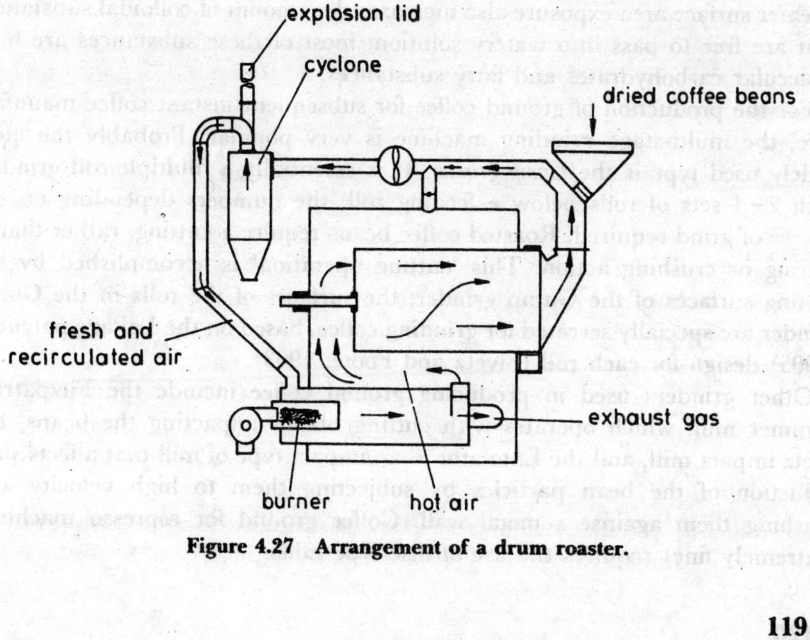


**Figure 4.27 Arrangement of a drum roaster.**

are used, through which the coffee beans are spirally moved (torque-tube principle). After roasting, the beans are passed through a cold zone.

### Convection roasting

*Fluidized bed method*

Coffee beans, floating and without contact with metal, are roasted in turbulent hot air currents in two phases:

1. Extraction of moisture and pre-roasting.
2. Aroma-roasting. At temperatures around 200°C the roasting times amount to 2.5 minutes.

*Centrifugal and turbine systems*

On a rotating horizontal bowl, coffee beans flowing centrally are carried to the periphery of the bowl by centrifugal motion with hot air support. The beans pass to a fixed lamella ring (turbine-blade arrangement) through which they are redirected to the centre. After a roasting period of 4–6 minutes the beans are brought to a cooling sieve within 2 seconds and are cooled for 2.5 minutes. The capacity amounts to 1500–4000 kg/hour.

## *Grinding roast coffee*

After developing the coffee flavour through roasting, the roast coffee is ground to allow sufficiently efficient and rapid infusion or percolation of its soluble content by boiling or hot water. Grinding exposes a much greater surface area for the liberation of $CO_2$ gas and the absorption of hot water, while at the same time it shortens the distance from the centre of each particle to the surface, thereby greatly reducing the diffusion distance for coffee solubles. Greater surface area exposure also increases the amount of colloidal substances that are free to pass into watery solution; most of these substances are high molecular carbohydrates and fatty substances.

For the production of ground coffee for subsequent instant coffee manufacture, the multi-stage grinding machine is very popular. Probably the most widely used type is the *Gump grinder*. It is essentially a multiple-roll grinder, with 2–4 sets of rolls below a feeding roll, the numbers depending on the degree of grind required. Roasted coffee beans require a cutting, rather than a tearing or crushing action. This 'cutting operation' is accomplished by the cutting surfaces of the Gump grinder; the surfaces of the rolls in the Gump grinder are specially serrated for grinding coffee, based on the LePage patented (1905) design for each roll (Sivetz and Foote, 1963).

Other grinders used in producing ground coffee include the Fitzpatrick hammer mill, which operates with cutting blades impacting the beans, the Rietz impact mill, and the Entolator – an impact type of mill that affects size-reduction of the bean particles by subjecting them to high velocity and crushing them against a metal wall. Coffee ground for espresso machines (extremely fine) requires the use of disc-type mills.

### *Extraction/percolation and percolation equipment*

After grinding, the extraction of roast and ground coffee is the next stage in the manufacture of instant coffee. Extraction or percolation consists of three distinctly different processes as described by Sivetz and Foote (1963) and Sivetz and Desrosier (1979):

1. Wetting. Gas is evolved from the coffee particles and interstitial voids as they are wetted with hot extract. Dessication by the coffee of the rising extract raises its solubles concentration. Selective water and solubles absorption by the coffee particles prepares the particles for solubles extraction.
2. Extraction of coffee solubles occurs rapidly after the particles are wetted with selective water absorption and absorbed solubles. The solubles content of the coffee is higher at this time than ever before or after.
3. Hydrolysis in coffee grounds is mostly the breakdown of large molecules of water-insoluble carbohydrates into smaller molecules which are water soluble. These are mostly reducing sugars, but there are also larger sugar molecules. Proteins also undergo hydrolysis under these conditions.

Percolators or extractors are used in the extraction process for roast and ground coffee. A detailed description of typical large-scale percolation battery plants is given by Sivetz and Foote (1963). Percolators consist of a battery of 5–7 vessels with an elaborate piping manifold; they range in diameter from 305–762 mm, and in height from 1.8–6 m. Pressurized hot water at about 163°C enters the column with the coffee grounds that are most spent and progressively moves upwards through each column of less spent grounds. The flow is in series through each column. The extract moving forwards continuously fills the interstitial spaces between the grounds, displacing gases upwards and outwards. The coffee extract leaves the freshest coffee column with a concentration of coffee solubles ranging from 25–35 per cent. The hot water feed flows continuously into the column with the spent coffee grounds. The extract from the freshest roast and ground coffee column is withdrawn intermittently and has to be cooled through heat exchangers, and kept cool before proceeding to subsequent operations.

### *Spray drying of soluble coffee*

After percolation and subsequent cooling the coffee extract is subjected to drying. Spray drying or freeze drying may be used; whereas spray drying produces powder, freeze-drying produces granules. In general, the capital and running costs of freeze drying are higher than those of spray drying. On the other hand freeze drying offers better quality and a more appealing appearance than spray drying.

Spray drying of soluble coffee means the atomization of liquid coffee extracts into minute droplets which, upon mixing with hot air, give up most of their moisture. There is marked cooling in the instantaneous alteration of liquid, with 65–75 per cent water, to powder containing 3 per cent moisture. The coffee spray drier vessel is 305–508 mm in diameter and has a 70° slope to minimize powder build-up. It usually projects down from a bustle that is about twice the size of the upper vessel in diameter. The effective particle drying zone is usually only about half to two thirds the overall drier height. Air, heated to about 296°C is passed through the drier by one or two fans, and is then uniformly guided around the atomizing nozzle.

The lower part (bustle) of the spray drier is essentially a particle collector and air separator. Secondary collectors capture fine soluble coffee particles before the air is discharged to the atmosphere. The resulting powder is cooled, coarse lumps are scalped, and the powder is conveyed to collection vessels.

After drying, the coffee is packed into appropriate containers, impermeable to moisture. With instant coffee processed to retain a high proportion of the aromatic volatiles, it is necessary to pack the coffee in an atmosphere of inert gas, either carbon dioxide or nitrogen, or mixtures of both, so that the residual oxygen content is as low as possible (less than 4 per cent).

## References


**Aguirre, B.E.** (1966). *La utilicaiòn industrial del graoo de café y de sus productos.* Guatemala, Instituto Centroamericano de Investigación y Technologia Industrial No. 1, 43 pp.

**Beckley, V.A.** (1931). Fermentation of coffee – Dept. of Agric. of Kenya Bull., 8.

**Belitz, H.D.** (1975). *Geschmacksaktive Substanzen in Kaffee,* 7th International Colloquiem on the Chemistry of Coffee, ASIC, Paris, 243–52.

**Braham, J.E.** and **R. Bressani** (1979). *Coffee Pulp, Composition, Technology and Utilization.* IDRC – 108 e, Ottawa, Ont. Canada.

**Clifford, M.N.** (1985). In *Coffee Botany, Biochemistry and Production of Beans and Beverage* (Clifford M.N. and K.C. Wilson, eds.), Grown Helm, London and Sydney.

**Ehlers, G.M.** (1980). *Possible applications of enzymes in coffee processing,* 9th International Coloquium on the Chemistry of Coffee, ASIC, Paris, 267–72.

**FAO** (1984). *FAO Production Yearbook,* Vol. 38.

**FAO** (1986). *FAO Monthly Bulletin of Statistics,* Vol. 9.

**Haarer, A.E.** (1962), *Modern Coffee Production,* 2nd edition. Leonard Hill. London.

**ICO** (1980). *Quarterly Statistical Bulletin,* Nos. A/18, International Coffee Organisation.

**Kahn, N.A.** and **Brown J.B.,** (1953). The composition of coffee oil and its component fatty acids. *Amer. Oil Chem. Soc.* **12**: 606–609.

**Knaus, C.** (1933). *Handbook voor de Koffie cultuur. Ze part. De Koffie Bereiding.* Soerabaja, H. van Ingen.

**Lam, L.K.T., Sparnis, V.L.** and **Watternberg, L.W.** (1982). Isolation and identification of kahweol palmitate and cafestol palmitate as active constituents of green coffee beans that enhance glutathione S-transferase activity in the mouse. *Cancer Research,* **42**: 1193–8.

**Llewilyn, D.A.B.** (1955). Coffee drying. A summary and preliminary report to the Coffee Board of Kenya. *Coffee Board of Kenya,* **6**: 153–159.

**McCloy, J.F.** (1959). Mechanical drying of arabica coffee, *Kenya Coffee,* **280**: 117–33

**Melo, M., Farudi, L.S., Texeira, A.A.** and **Amorim, H.V.** (1980). Chemical, physical and organoleptic alterations on storage of coffee beans. *Ciencia e Cultura* (Sao Paulo), **32**: 468–71.

**Mwandha, J., Nichols, J.** and **Sargent M.** (1985). *Coffee: The International Commodity Agreements.* Gower Publishing Company Limited. England.

**Nosti Nava, J.** (1953). *Cacao, café y te Barcelona,* Salvat Ed.

**Panghorn, R.M.** (1982). Influence of water composition, extraction procedures, holding time and temperature on quality of coffee beverages. *Lebensmittel Wissenschaft und Technologie,* **15**: 161–8.

**Pokorny, J., Con, N.H., Smidrkalova, E.** and **Janice, G.** (1975). Non-enzymic browning XII. Maillard reaction in green coffee beans on storage. *Zeitschnft fur Lebensmittel Untersuchung und Forshumg,* **158**: 87–92.

**Purseglove, J.W.** (1976). *Tropical Crops; Dicotyledons,* Longmans, London.

**Radtke, R.** (1979). Information about oxygen consumption of roast coffee and its influence on the sensorial evaluated quality of coffee beverage, *Chemie, Mikrobiologie, Technologie der Lebensmittel,* **6**: 36–42.

**Ringoet, A.** (1935). *Note sur la préparation du café.* Public INEAC Bruxelles Série Techn. **1**: 1–52.

**Scott, A.I., Sim, G.A., Ferguson, G., Young, P.W.** and **McCapra, F.** (1962). Stereochemistry of the diterpenoids. Absolute configuration of Cafestol. *Journal of the American Chemical Society*, **84**: 3197–9.

**Sivetz, M.** (1963). *Coffee Processing Technology*, Vol. 2. AVI Publishing, Connecticut, USA.

**Sivetz, M.** and **Desrosier, N.W.** (1979). *Coffee Technology*. AVI Publishing Company, INC. Westport, Connecticut.

**Sivetz, M.** and **Foote, H.E.** (1963). *Coffee Processing Technology*, Volume I. AVI Publishing Company, Westport, Connecticut.

**Wellman, F.L.** (1961). *Coffee: Botany, Cultivation and Utilization*, Leonard Hill Ltd. World Crops Books, London.

**Wilbaux, René** (1963). *Coffee processing*, Informal Working Bulletin No. 20, FAO, Rome.

**Wootton, A.E.** (1965). *The importance of field processing to the quality of East African coffee*, 2nd International Colloquium on the Chemistry of Coffee. ASIC, Paris, 247–58.

**Wootton, A.E.** (1967b). *A report by the working group on the wet processing of coffee*, 3rd International colloquium on the Chemistry of Coffee ASIC, Paris, 398–404.

**Wurziger, J.** (1977). *Kaffee in Ullmanns Enzyklopädie der technischen Chemie*, Bd. 13, Chemie, Weinheim, New York.

# 5

# Groundnut

## Production and yield

The groundnut or peanut, *Arachis hypogaea* L., belongs to the family Leguminosae. It is grown as an annual crop on about 19 million hectares in tropical and subtropical regions and the warmer areas of temperate regions of the world, principally for its edible oil and protein rich kernels or seeds, borne in pods, which develop and mature below the soil surface (Figure 5.1). The groundnut is a herbaceous plant of which there are two major varieties, bunch and runner.

Bunch varieties, common in the United States, grow 30–46 cm high and do not spread. Runner varieties, the most common in West Africa, are shorter and run along the ground for 30–60 cm. Apart from the bunch and runner varieties, many intermediate forms or hybrids exist (Woodroof, 1973; Irvine, 1974). After soya beans, *Glycine max*, groundnut is the second most important

**Figure 5.1 Groundnut plant bearing pods which mature below the soil surface.**

source of vegetable oil in the world. In 1985, the recorded world-wide output of soya beans (about 101 million tonnes) contained about 21.2 million tonnes of oil. The 14.4 million tonnes of groundnut kernels, corresponding to the reported production of about 21 million tonnes in their shells (at a shelling percentage of 70 per cent), contained 7.2 million tonnes of oil. In that year, the other main sources (sunflower, brassicas, oil palm and copra from coconut) each contributed 3.5–5.1 million tonnes of oil (FAO, 1986).

Yields of kernels generally range from 0.5–4.0 tonnes/hectare. In the developing countries, where 80 per cent of the crop is produced, the average yields are around 1 tonne/hectare. In India, the world's largest producer, the average yield is about 900 kg/hectare. The other major world producers of groundnut are China, USA, Nigeria, Senegal and Brazil. The USA is the largest exporter, producing kernels, cake, meal and oil equivalent to about 360 000 tonnes in shell (Miller, 1983).

## *Main products and uses*

Groundnut crops are grown for their kernels, the oil and meal derived from them, and the vegetative residue (haulms). As human food the kernels are eaten raw, lightly roasted or boiled, sometimes salted or made into a paste, which is known as peanut butter. In Senegal, the leaves of the plant are used as a vegetable in soups. Groundnut oil is the most important product of the crop. At present about 40 per cent of the world crop is crushed for oil, which is used mainly for cooking. Many of the people in West and Equatorial Africa use a traditional method of extracting the oil contained in the kernels. The kernels are roasted, ground, made into a paste and boiled; oil rises to the surface and is skimmed off. Groundnut oil forms an important part of the diet in West Africa. On average the oil content of the kernel is 40–50 per cent. The best oil is obtained by simple pressure and is called 'cold pressed oil', whereas oil of inferior quality is obtained by further pressure and heat and is mainly used for soap-making. The oil may also be used for margarines and vegetable ghee, for shortening in pastries and bread, for pharmaceutical and cosmetic products, as a lubricant and emulsion for insecticides, and as a fuel for diesel engines (Duke, 1981).

The residue (press cake), after the oil has been extracted, contains 40–50 per cent protein and is a valuable feed mainly for poultry. Groundnut flour, produced from the cake, can be used for enriching tuber flours such as cassava flour, which is low in protein. The so-called Mysore flour, which is a mixture of 25 parts of groundnut flour and 75 parts cassava flour and has a protein content of 12 per cent, is accepted as a partial substitute for cereals in certain parts of India. Paushtik atta, a blend of wheat, groundnut flour and cassava flour is also used in India for the preparation of chapattis – unleavened bread. The dry pericarp of the mature fruit or pods, known as shells or husks, may be used for fuel, as a soil conditioner, as a filler in feeds and as a source of furfural, processed as a substitute for cork or for hardboard, or composted with the aid of lignin-decomposing bacteria (Adams and Hartzog, 1980). In the USA, about half the groundnut harvest is used to make peanut butter;

salted groundnuts and confectionary products are also important. Groundnut haulms are nutritionally comparable with grass hay and are used in many countries (Gibbons, 1980).

## Composition and nutritive value of groundnut kernels

The pod of a groundnut is about 12.5–75 mm in length and roughly cylindrical in shape. The shell of the pod comprises 20–30 per cent of the whole nut and may be separated from the kernels. The kernel consists of two cotyledons (halves) and the germ (heart) enveloped in a thin red-brown, purple or white skin called the testa (Figure 5.2). The kernels are made up of about 72.4 per cent cotyledons, 4.1 per cent skins and 3.3 per cent germs.

The groundnut kernel is composed of approximately equal weights of fatty and non fatty constituents, the relative amounts of each depending upon variety and maturity. Most of the fatty constituents are contained in the cotyledons, some are found in the germs or hearts and small amounts are located in the testa or skins.

As in many of the seeds of other grain legumes, the protein is nutritionally inferior to that of the Standard Reference Protein (SRP, which approximates the average amino-acid profile of human proteins) because it contains relatively small proportions of lysine, methionine and threonine, and sometimes isoleucine and valine (Pancholy *et al.*, 1978; Young *et al.*, 1973). Aspartic and glutamic acids and arginine constitute about 45 per cent of the total amino-acids, and their proportion is greater than in the SRP (Young *et al.*, 1973). 100 g of raw groundnut kernels provide about 570 k calories (2.39 mJ) of dietary energy (Ahmed and Young, 1982). The proteins of groundnuts have certain unique functional properties, such as low viscosity at relatively high concentrations (5–10 per cent), good compatability with bread dough systems, white colour and bland flavour. On the other hand, they are lower in nutritional value than soya proteins because of their low levels of lysine, threonine and methionine. However, when blended with other protein sources or fortified with other necessary amino-acids, groundnut protein can make a valuable contribution to the human diet. Table 5.1 shows the essential amino-acid composition of groundnuts.

The mature kernel contains 20–25 per cent carbohydrate, of which about

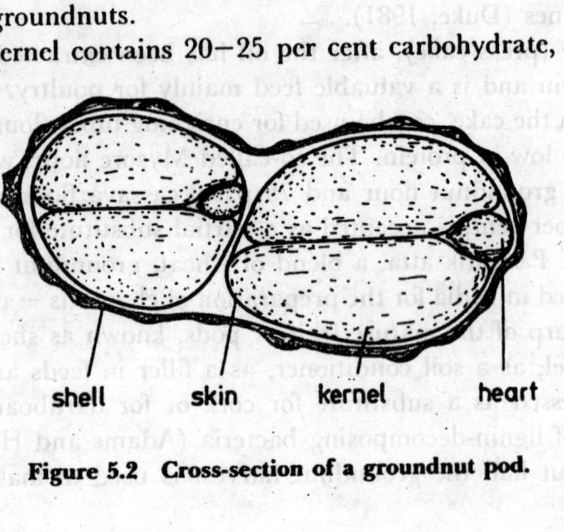


**Figure 5.2 Cross-section of a groundnut pod.**

8–10 per cent is cellulose and hemicellulose, 4 per cent is starch, and 10–12 per cent is sugar (Patee *et al.*, 1974). Sucrose is the principal sugar, varying from 2.86–6.35 per cent depending on the cultivar. When the kernels are roasted, the sucrose is hydrolysed to fructose and glucose, which then reacts with free amino-acids to form the numerous substances which give the characteristic flavour to the roasted product (Mason *et al.*, 1969). Groundnuts are useful sources of thiamine, niacin, tocopherol (vitamin E) and folic acid in the human diet (Ahmed and Young, 1982). Table 5.2 shows the nutrient, vitamin and mineral composition of raw groundnuts and various groundnut products.

**Table 5.1 Essential amino-acid composition of groundnuts**

| Amino-acid | Groundnut whole (mg/gN) | Groundnut protein (arachin) (mg/gN) |
|---|---|---|
| Arginine | 775 | 763 |
| Histidine | 150 | 119 |
| Isoleucine | 250 | 413 |
| Leucine | 438 | 425 |
| Lysine | 319 | 250 |
| Phenylalanine | 325 | 344 |
| Tyrosine | 220 | 300 |
| Cysteine | 81 | 81 |
| Methionine | 88 | 38 |
| Threonine | 244 | 144 |
| Tryptophan | 70 | 56 |
| Valine | 313 | 244 |

Source: Aykroyd, W.R. and Doughty, J. In *Legumes in Human Nutrition*, FAO, Rome, 1982.

## Groundnuts and aflatoxins

It is now generally recognized that the toxic products from moulds, or mycotoxins as they are called, constitute a potential and ever present environmental hazard to animals and man alike. Since the outbreak of 'Turkey X' disease in 1960 in the UK (about 100 000 young turkeys and thousands of ducklings died), much research has been devoted to fungal toxins, in particular those which are specifically associated with certain strains of the mould *Aspergillus flavus* and related species on groundnuts and other crops. As a result of intensive investigations, the toxic metabolites responsible for the disease were isolated and called *aflatoxins*.

Aflatoxins occur naturally when a toxin-producing strain of *Aspergillus flavus* or *Aspergillus parasiticus* grows on a substrate in environmental conditions conducive to the development of the mould. The presence of mouldiness is not, in itself, indicative of toxin production. In view of the fact that the *A. flavus* group of moulds is capable of development over a wide range of temperatures on substrates of high carbohydrate content, agricultural commodities and their products are most vulnerable to contamination (Heathcote and Hibbert, 1978). The first known natural occurrence of aflatoxins was associated with peanuts, as it was from Brazilian peanut meal that Sergeant *et al.* (1961)

## Table 5.2 Composition of edible groundnut products (per 100 g)

| Constituent | Raw | | Boiled | Roasted | | Peanut butter | | Spread | Flour |
|---|---|---|---|---|---|---|---|---|---|
| | with skins | without skins | | with skins | without skins, salted | added fat, salt | added fat, salt, sugar | | |
| Water % | 5.6 | 5.4 | 36.4 | 1.8 | 1.6 | 1.8 | 1.7 | 2.2 | 7.3 |
| Energy (joules) | 2361 | 2378 | 1574 | 2436 | 2449 | 2432 | 2465 | 2516 | 1553 |
| Protein (g) | 26.0 | 26.3 | 15.5 | 26.2 | 26.0 | 27.8 | 25.2 | 20.3 | 47.9 |
| Fat (g) | 47.5 | 48.4 | 31.5 | 48.7 | 49.8 | 49.4 | 50.6 | 52.1 | 9.2 |
| Carbohydrate (g) | 18.6 | 17.6 | 14.5 | 20.6 | 18.8 | 17.2 | 18.8 | 22.0 | 31.5 |
| Starch (g) | 4.0 | – | – | – | – | – | – | – | – |
| Reducing sugars (g) | 0.2 | – | – | – | – | – | – | – | – |
| Disaccharide sugars (g) | 4.5 | – | – | – | – | – | – | – | – |
| Pentosans (g) | 2.5 | – | – | – | – | – | – | – | – |
| Fibre (g) | 2.4 | 1.9 | 1.8 | 2.7 | 2.4 | 1.9 | 1.8 | 1.5 | 2.7 |
| Ash (g) | 2.3 | 2.3 | 2.1 | 2.7 | 3.8 | 3.8 | 3.7 | 3.4 | 4.1 |
| Calcium (mg) | 69 | 59 | 43 | 72 | 74 | 63 | 59 | 50 | 104 |
| Phosphorus (mg) | 401 | 409 | 181 | 407 | 401 | 407 | 380 | 322 | 720 |
| Iron (mg) | 2.1 | 2.0 | 1.3 | 2.2 | 2.1 | 2.0 | 1.9 | 1.5 | 3.5 |
| Sodium (mg) | 5 | 5 | 4 | 5 | 418 | 607 | 605 | 597 | 9 |
| Potassium (mg) | 674 | 674 | 462 | 701 | 674 | 670 | 627 | 530 | 1186 |
| Vitamin A (IU) | 26 | 26 | – | – | – | 300 | 300 | – | – |
| Carotene (provitamin A) (µg) | 1.0 | – | – | – | – | – | – | – | – |
| Vitamin E (mg) | 22.9 | 22.9 | – | – | – | – | – | – | – |
| Thiamine (mg) | 1.14 | 0.99 | 0.48 | 0.32 | 0.32 | 0.13 | 0.12 | 0.10 | 0.75 |
| Riboflavin (mg) | 0.13 | 0.13 | 0.08 | 0.13 | 0.13 | 0.13 | 0.12 | 0.10 | 0.22 |
| Pyridoxin (mg) | 0.3 | – | – | – | – | – | – | – | – |
| Niacin (mg) | 17.2 | 15.8 | 10.0 | 17.1 | 17.2 | 15.7 | 14.7 | 12.4 | 27.8 |
| Ascorbic acid (mg) | 5.8 | 5.8 | – | – | – | 13.0 | 13.0 | – | – |
| Choline (mg) | 170 | – | – | – | – | – | – | – | – |
| Folic acid (mg) | 0.28 | – | – | – | – | – | – | – | – |
| Inositol (mg) | 180 | – | – | – | – | – | – | – | – |
| Biotin (mg) | 0.034 | – | – | – | – | – | – | – | – |
| Pantothenic acid (mg) | 2.715 | – | – | – | – | – | – | – | – |

Sources: Watt and Merrill, 1963; Cobb and Johnson 1973

first isolated a toxic substance. After this, other natural occurrences of aflatoxin were reported. Groundnut crops in northern Nigeria (McDonald and Harkness, 1963), groundnuts, groundnut meal and cake from tropical areas in Asia, Africa and Latin America (Hiscocks, 1965), and peanuts and peanut meal in India were all found to be contaminated with aflatoxins (Screenivasamurthy *et al.*, 1965). In the USA, aflatoxins were detected in peanuts and peanut meal (Prickett and Salmon, 1965).

Many researchers have examined the biological effects of aflatoxins over the past two decades and have established that aflatoxins are lethal to various animals, and that aflatoxin $B_1$ is the most acutely toxic of the first four aflatoxins to be identified, followed by $G_1$, $B_2$, and $G_2$, in order of decreasing potency.

Some information about the effects of aflatoxins on human beings has been reported in the literature. From the numerous cases reported, five have been selected and described here. The death of a 15 year old boy (abdominal pain and swelling) was reported to have been caused by cassava consumption. The cassava was found to be mouldy and contained aflatoxins at a concentration of 1.7 mg/kg (Serck-Hanssen, 1970). Aflatoxins were also implicated in an outbreak of hepatitis in India in 1974 (Krishnamachari *et al.*, 1975). Keen and Martin (1971) reported on a situation in Swaziland where hepatoma is a serious problem, and found that the distribution of liver cancer correlated well with the geographical distribution of aflatoxin contamination of groundnuts. Van Rensburg *et al.* (1974) reported on the high incidence of primary liver cancer in the Inhambane district of Mozambique. The death rate was reported to be over 25 per 100 000 of the population per year. The consumption of aflatoxins by each adult was estimated to be, on average, 15.6 mg/day. A 68 year-old man died from a tumour – bronchiolar carcinoma – just over one year after he had spent three months working on a method of sterilizing Brazilian peanut meal contaminated with aflatoxins. An extract of lung contained a compound which had the same chromatographic properties as aflatoxin $B_1$ (Dvorackova, 1976).

## *Production of aflatoxins*

The effects of various levels of atmospheric gases (carbon dioxide, oxygen and nitrogen) on aflatoxin production under conditions of varying temperature and humidity were reported by Diener and Davis (1972). Aflatoxin production in sound, mature, peanut kernels decreased with increasing concentrations of carbon dioxide from 0.03–100 per cent. A reduction in oxygen concentration generally resulted in decreased aflatoxin production.

Diener and Davis (1966) found that the optimum temperature for toxin production by *A. flavus* was 25°C on both peanuts and liquid media. The incubation period for maximum production of toxin was from 7–15 days. Moisture is reported to be a crucial factor for the growth of *A. flavus* and aflatoxin production. The optimum relative humidity for fungal growth is 85 per cent or more. Diener and Davis (1968) reported that the lowest relative humidity for aflatoxin production on peanuts was 83 per cent, with production increasing up to 99 per cent relative humidity. Maximum production of aflatoxins on peanuts was obtained with a kernel moisture content of 25 per cent at a temperature of 30°C (Ashworth *et al.*, 1965). It has been found that groundnuts are more likely to become visibly infected with *A. flavus* and other fungi when they are:

(a) Damaged by certain of the small fungus grower termites.

(b) Punctured and gnawed by the bollworm.

(c) Seriously affected by drought.

(d) When the pods burst in the soil as the result of alternating humid and drought conditions (Sellschop, 1965).

## *Measures to control the development of aflatoxins*

Contamination of plants by aflatoxins may take place while plants are growing in the field, during harvesting, or during storage and processing. It is important that during harvesting of groundnuts every effort is made to avoid physical damage to the fruit since damaged groundnuts are more susceptible to mould invasion under favourable conditions than their sound counterparts. Another important factor is the soil remaining on the groundnut pods. Since *A. flavus* appears to be indigenous to the soils in many groundnut fields, any soil remaining on the pods will considerably increase the risk of post-harvest moulding.

Reports on the development of *A. flavus* on stored agricultural products have shown that levels of moisture and temperature are the most important factors in the protection of stored produce against aflatoxin contamination. Fungi such as *A. flavus* will not invade grain and oil-seeds when their moisture contents are in equilibrium with a relative humidity of 70 per cent or less. Diener and Davis (1970) report that a relative humidity of 83 per cent is the upper limit for broken, mature and immature groundnuts stored at 30°C. Another means of controlling aflatoxin production in groundnuts is by regulation of the atmospheric gases in the store. Aflatoxin production in sound, mature groundnuts decreases with increasing concentrations of $CO_2$ gas. Particularly in areas of high ambient temperature and of high relative humidity, this method would seem to be the most practicable of all the possible control measures for groundnuts in storage.

**Table 5.3 Effect of $CO_2$ on aflatoxin production in groundnuts**

| Percentage of gas ($N_2$ added to 100%) | | Total aflatoxin (mg/g) |
|---|---|---|
| $CO_2$ | $O_2$ | |
| 0.03 | 21.0 | 299.38 |
| 20.0 | 20.0 | 74.55 |
| 40.0 | 20.0 | 35.42 |
| 60.0 | 20.0 | 19.82 |
| 80.0 | 20.0 | 0.10 |
| 100.0 | 0.0 | 0.01 |

Source: Diener and Davis, 1972.

**Table 5.4 Effect of $O_2$ on aflatoxin production in groundnuts**

| Percentage of gas ($N_2$ added to 100%) | | Total aflatoxin (mg/g) |
|---|---|---|
| $CO_2$ | $O_2$ | |
| 0.03 | 21.0 | 511.89 |
| 0.0 | 15.0 | 519.29 |
| 0.0 | 10.0 | 316.12 |
| 0.0 | 5.0 | 154.12 |
| 0.0 | 1.0 | 5.93 |
| 0.0 | 0.1 | 0.07 |

Source: Diener and Davis, 1972.

The effect of carbon dioxide and oxygen on aflatoxin production in stored groundnuts is shown in Tables 5.3 and 5.4.

## Storage of groundnuts

Groundnuts are semi-perishable, that is, they may be held for a number of years under optimum conditions. Under unfavourable conditions, however, they become inedible within a month due to mould, insects, absorption of foreign flavours or rancidity.

Undesirable effects resulting from improper storage are cumulative and irreversible. For successful storage of groundnuts certain requirements should be met, and some of these are:

1. The temperature should be low. In general, the lower the temperature, the longer the expected storage life of groundnuts, though there is no critical temperature at which groundnuts must be held. At 21°C, unshelled groundnuts may be expected to retain edible quality for six months; when shelled the time is reduced to about four months. At this temperature groundnuts are prone to insect infestation, development of an 'amber' colour, staling and rancidity (Figure 5.3). At 8°C, unshelled groundnuts may be held for nine months, and

**Figure 5.3 Treatment of stored groundnuts with insecticide to prevent the spread of groundnut seed beetles.**

shelled nuts for six months. Insects development is 'arrested' at this temperature. Between 0–2°C the storage life of shelled groundnuts may be extended to two years, at about −4°C it may be five years, and at −12°C it may be as long as ten years. Storage life of in-shell nuts should be 50 per cent longer (Woodroof, 1973). The effect of temperature on absorption of foreign flavour as well as on loss of natural flavour of shelled Spanish-type groundnuts is illustrated in Figures 5.4 and 5.5.

2. The relative humidity should be low. High moisture in groundnuts is reported to possibly be the cause of more deterioration than any other single factor. At a relative humidity of 65–70 per cent, groundnuts equalize at a moisture of about 7 per cent. Above 70 per cent humidity groundnuts are likely to become mouldy.

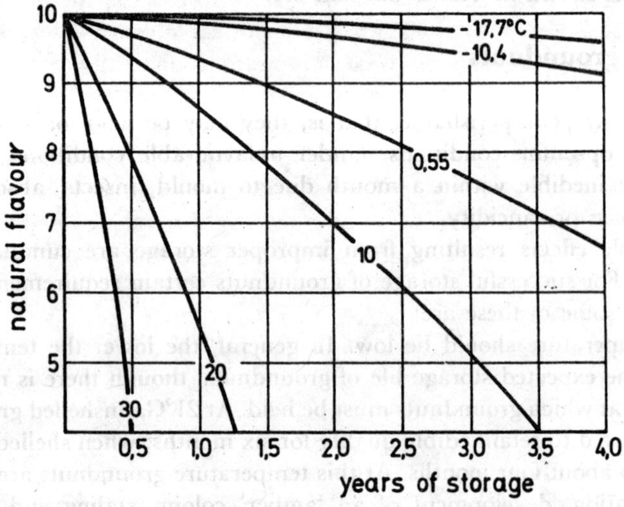


**Figure 5.4 Effect of temperature on loss of flavour (Woodroof, 1973).**

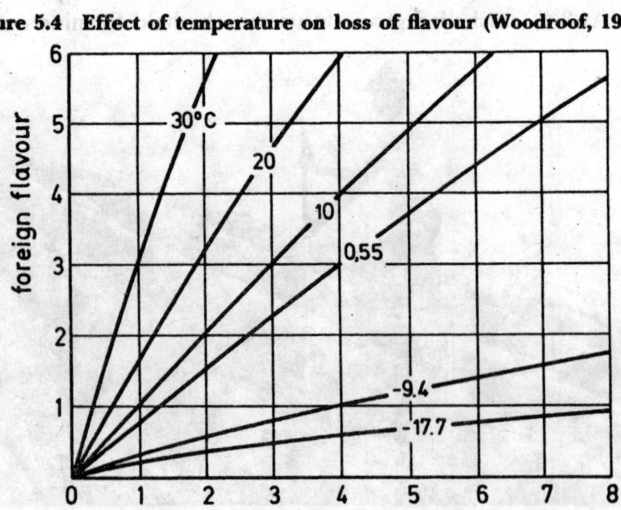


**Figure 5.5 Effect of temperature on flavour absorption (Woodroof, 1973).**

3. The storage atmosphere should be free of odours and should be well circulated. Ammonia, being the most effective cause of damage may be eliminated by burning sulphur sticks or releasing sulphur dioxide gas into the air. Also, exposure for some days should neutralize the ammonia.

## Harvesting and shelling

Groundnuts are best harvested when the soil is dry enough to fall freely from the stems and pods. If, however, the soil is very dry and crusted many of the pods may break off and be lost. On small-scale farms, harvesting consists of removing the entire groundnut plant from the soil, allowing the plant to wilt, picking the pods from the vines, and then drying the pods. In some instances, a groundnut digger is used in harvesting. The tap-roots of the plants are cut and the soil is loosened by the sharp blades of the digger, which slide along below the surface. At digging time, groundnuts generally contain 35–55 per cent moisture, and until the moisture is reduced below 10 per cent the nuts are susceptible to contamination by moulds, especially at warm temperatures and high humidity. The common practice is to leave groundnuts in wind-rows for several days after digging, until their moisture content drops to about 10 per cent. Contamination of nuts by toxin-producing moulds, particularly *Aspergillus flavus*, can be circumvented by 'green harvesting' and drying immediately.

### *Stripping of groundnuts*

The process of separating groundnut pods from the plant or haulm is known

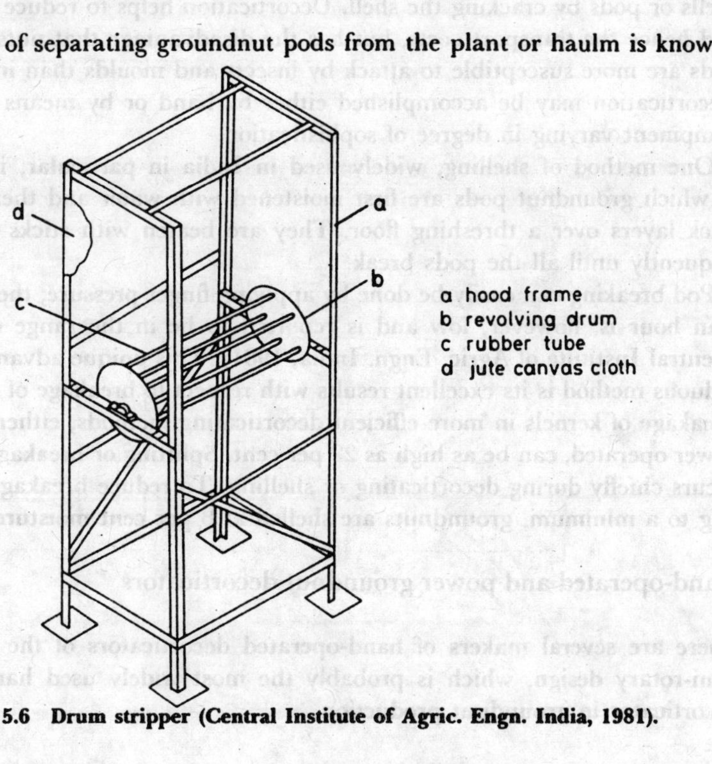


**Figure 5.6 Drum stripper (Central Institute of Agric. Engn. India, 1981).**

as stripping. Among subsistence farmers stripping is done manually by pulling the pods from the plants. The average stripping rate is 7 kg/hour/worker. The hourly output can be increased by using the so-called Comb type stripper, developed by the Coimbatore Centre (1981). The Comb stripper is a very simple unit, consisting of a four vertical legged frame with a strip of expanded metal fixed to each of its sides in the form of a comb. The pods are separated or stripped from the plants by drawing a handful of groundnut vines across the comb with slight force. Up to four persons can conveniently work on the stripper. Stripping rates of 10–12 kg/hour/worker can be achieved.

### Drum stripper

The drum stripper consists of a hollow drum formed by two metal discs, one at each end, connected by steel rods, which are covered by thick, soft rubber tubes (Figure 5.6). The drum is mounted on pedestal bearings and is free to rotate. Stripping is done by beating the handful of groundnut plants over the rubber-covered rods of the revolving drum. A deflecting jute or canvas cloth is wrapped around the stripper to avoid scattering of pods and to allow pods to be collected under the stripper. The output is reported to be about 16 kg/hour/worker.

## *Decorticating of groundnuts*

Decorticating or shelling is the process of freeing groundnut kernels from the shells or pods by cracking the shell. Decortication helps to reduce the volume and hence the transport costs, but has the disadvantage that nuts freed from pods are more susceptible to attack by insects and moulds than in-shell nuts. Decortication may be accomplished either by hand or by means of different equipment varying in degree of sophistication.

One method of shelling, widely used in India in particular, is a process in which groundnut pods are first moistened with water and then spread in thick layers over a threshing floor. They are beaten with sticks and stirred frequently until all the pods break.

Pod breaking can easily be done by applying finger pressure; the output per man hour is, however, low and is reported to be in the range of 1–2.5 kg (Central Institute of Agric. Engn. India, 1981). This unique advantage of this arduous method is its excellent results with respect to breakage of the kernels. Breakage of kernels in more efficient decorticating methods, either manual or power operated, can be as high as 22 per cent. Splitting or breakage of kernels occurs chiefly during decorticating or shelling. To reduce breakage and splitting to a minimum, groundnuts are shelled at 8 per cent moisture or higher.

### Hand-operated and power groundnut decorticators

There are several makers of hand-operated decorticators of the concave or semi-rotary design, which is probably the most widely used hand-operated decorticator in groundnut production.

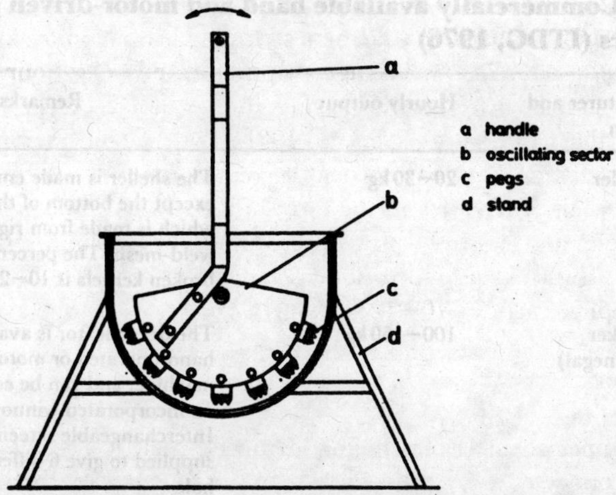


**Figure 5.7 Hand-operated groundnut decorticator.**

The basic design for this equipment is common to all manufacturers with some minor variations or modifications. A simple hand-operated groundnut decorticating machine has a semi-cylindrial screen closed on both sides. Across its centre and towards the top is fixed a shaft carrying a lever. On the lever swings a pair of plates with shoes or beater bars, which have blunt spikes on their undersides (Figure 5.7). The semi-rotary action of the shoes is effected by swinging them backwards and forwards, using the lever or handle manually. In this way the nuts are shelled or decorticated against the screen. The mixture of kernels and shells may be subjected to air suction for separation of the two components.

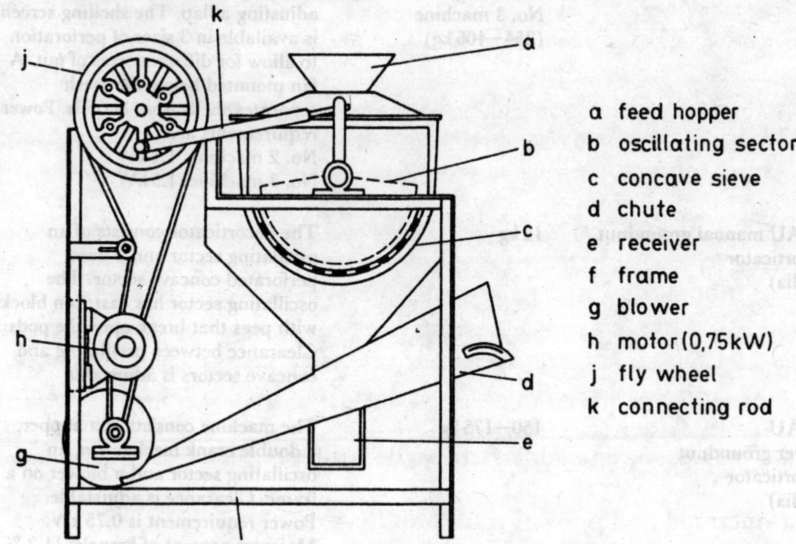


**Figure 5.8 Motor-driven groundnut decorticator (Central Institute of Agric. Engn. India, 1981).**

**Table 5.5 Commercially available hand and motor-driven groundnut decorticators (ITDG, 1976)**

| Name, manufacturer and country of origin | Hourly output | Remarks |
|---|---|---|
| Groundnut sheller (Kenya) | 20–30 kg | The sheller is made entirely of wood, except the bottom of the trough which is made from rigid 1.25 $cm^2$ weld-mesh. The percentage of broken kernels is 10–20 % |
| Alternating husker (SISCOMA; Senegal) | 100–150 kg | The decorticator is available as a hand-operated or motor-driven machine, and can be equipped with an incorporated winnower. Interchangeable screens are supplied to give 6 different sizes of hole |
| Groundnut decorticator (Atlas Works; UK) | | The decorticator uses galvanized, heavy gauge wire-mesh for a screen, and rubbing bars can be adjusted to suit the type of groundnut being decorticated. Four different sizes of screen can be supplied |
| Foot operated groundnut sheller (India) | 25 kg | The sheller is fitted with a flywheel for easier operation and with a blower to separate shells from kernels |
| Automatic groundnut decorticating machine (UK) | 150–400 kg according to model. No. 2 machine (152–304 kg), No. 3 machine (254–406 kg) | A ribbed feed roller feeds the nuts into a beater chamber below at a rate which can be varied by adjusting a flap. The shelling screen is available in 3 sizes of perforation to allow for different sizes of nut. A fan mounted below the mesh separates shells from kernels. Power requirements are:<br>No. 2 machine: 1.1 kW<br>No. 3 machine: 1.5 kW |
| TNAU manual groundnut decorticator (India) | 12 kg | The decorticator consists of an oscillating sector and a fixed perforated concave sector. The oscillating sector has cast iron blocks with pegs that break open the pods. Clearance between oscillating and concave sectors is adjustable |
| TNAU Power groundnut decorticator (India) | 150–175 kg | The machine consists of a hopper, a double crank mechanism, an oscillating sector and a blower on a frame. Clearance is adjustable. Power requirement is 0.75 kW.<br>Moisture content of kernels: 11.2 %<br>Moisture content of shell: 4.0 %<br>Breakage: 4.3 % |

Power groundnut decorticators consist essentially of a hopper, a beater chamber where the shelling takes place, a chute, a fan, and an electric or diesel engine. The fan, usually mounted below the wire-mesh chute, blows the shells upwards, and the kernels are collected through a spout at the bottom of the machine (Figure 5.8). The principle of the shelling operation in power decorticators is the same as that in hand-operated decorticators.

Table 5.5 lists some of the commercially available hand and motor-driven groundnut decorticators.

## Processing groundnuts for oil

All cells of the groundnut embryo contain oil in an extremely fine emulsion according to Woodroof and Leahy (1940). When kernels are broken or bruised, a sufficient number of cells are injured to cause tiny drops of free oil to ooze out and collect on the surface of the kernels.

Oil may be extracted from groundnut kernels by mechanical pressing or by the use of a solvent. Solvent extraction is capable of removing nearly all the available oil from oilseed meal or flakes. In addition to the high yield of oil, the method produces oil with better qualities, and a higher protein meal (Khan and Hanna, 1983). This method generally requires more capital expenditure than a mechanical process. The efficiency of a mechanical expression process seldom exceeds 90 per cent. It has, however, the advantage of yielding end-products free of dissolved chemicals and is a comparatively safer and less expensive process.

Research on expression of oil from oilseeds indicates that pretreatment of material, prior to pressing or extraction, operating temperature, pressing time and moisture content of material during expression are the factors which affect oil yield and oil quality.

### *Traditional West African method of obtaining groundnut oil*

In many areas of the semi-arid tropics, groundnut oil is the only affordable edible oil for cooking. The importance of this oil as a local food item warrants a detailed description of the method used for its extraction. Although oil presses have been introduced into most areas, the traditional method still survives.

In West Africa, groundnut oil is traditionally obtained in the following way. Groundnuts are shelled and cleaned and roasted lightly. Next the red skins are removed by rubbing individual kernels against each other by hand. The deskinned kernels may be pounded with a mortar and pestle or ground using grinding stones (Figure 5.9), to a smooth paste. The oil-rich paste is kneaded and pressed by hand to remove the oil. A small quantity of warm water is added following each pressing operation until as much of the oil-water mixture as possible has been extracted. The oil-water mixture is fried or 'fire-dried' over a low fire to remove most of the water. Finally, the oil is decanted into bottles or cans. Alternatively, depending upon the desired flavour of the oil, the paste is boiled and the oil rising to the surface is skimmed off. This

**Figure 5.9 Pounding groundnuts using grinding stones.**

separation operation continues until no sign of oil can be distinguished on the surface. The traditional method usually has a low extraction rate. However, the resulting press cake is generally used to prepare groundnut soup.

## *Hydraulic pressing of groundnuts*

Shelled and crushed groundnuts are subjected to heating with steam and spread on press cloths, with edges folded to prevent running. The optimum moisture content is about 6 per cent. The containers holding the heated mass are about 900 $cm^2$ in cross-section and 50.8 cm deep, and are arranged one above the other. The racks are placed in tiers about 1 m high, and a pressure of about 138 000 $kN/m^2$ is applied. The crude oil is collected beneath the tiers. As a press aid, a small amount of unshelled groundnut meal is mixed with the ground mass before pressing takes place. The cake from the hydraulic press usually contains around 7 per cent oil.

## *Expeller pressing*

For expeller pressing the groundnuts are crushed and fed continuously into an expeller, consisting of a screw which rotates within a sturdily built cylinder. The crushed groundnut mass is fed into the larger end of the expeller chamber and pressure is exerted as the screw turns, forcing the mass towards the smaller end (discharge end). Friction and pressure cause the mass to heat, which facilitates oil extraction. The groundnut oil passes through the perforated screen walls, and is collected beneath the expeller chamber, while the press cake is extruded from the discharge end. The optimum moisture content of the groundnut mass for oil recovery by expeller pressing is 6 per cent. The resulting press cake normally contains 5 per cent oil.

### *Solvent extraction of groundnut oil*

The process for solvent extraction of groundnut oil is similar to solvent extraction of soya bean oil. Generally, the groundnuts destined for oil are shelled and winnowed to remove the fibre-rich shells, and whitened by removing the tannin-containing red skins. Next, the nuts are cracked into pieces and conditioned to 10–11 per cent moisture at 70°C or more, and then flaked by passing through rolls. Sometimes the groundnut flakes are cooked before they are conveyed to the extractor. In the extractor, the oil is removed by means of a solvent. The solvent-laden flakes are then passed through a desolventizer which recovers the solvent. The defatted and desolventized cake may undergo further treatment before it is used as feed.

The crude oil is clarified by passing it through a filter press. After clarification, the oil may be dehydrated and sent directly to the oil refinery.

### *Refining*

The refining process is basically the same for most vegetable oils. The procedure can also be applied to crude oil extracted using a hydraulic press or screw expeller. Refining consists of alkali-refining, to neutralize the free fatty acids, using sodium hydroxide; bleaching, to improve flavour stability, using natural bleaching earths; and deodorizing, to remove odour, using steam distillation under vacuum.

If groundnut oil is to be used as a salad oil, it undergoes the process of winterization prior to deodorization. The alkali-refined and bleached oil, if necessary, is hydrogenated to an iodine value of about 105 before it is winterized to remove fats that crystallize out at about 0°C. Winterization is usually effected by slowly cooking the oil to about 13°C over a period of 12 hours or until the first crystals appear. The cooking brine is usually kept at about 4°C below the temperature of the oil until crystals start to form. Then over the next 18–20 hours the cooking rate is slowed to allow the crystals to grow and the temperature to drop slowly. When the temperature drops to about 5–6°C, the oil is maintained at that level for a holding period of at least 12 hours. The oil is then filtered through plate and frame filter presses to give salad oil and solid residue, sometimes called vegetable stearine. The oil after winterization usually undergoes deodorization.

Investigations on extraction rates of groundnut grits using three solvents (hexane, 95 per cent ethanol and absolute ethanol) have been made (Anon., 1962). Conclusions were:

(a) Absolute ethanol is a far better solvent for extracting groundnut oil from grits than 95 per cent ethanol.

(b) At a temperature of 43°C, hexane is superior to 95 per cent ethanol and slightly inferior to absolute ethanol as a solvent.

(c) More nonlipid solids are extracted at 14.5°C by 95 per cent ethanol than by absolute ethanol.
(d) Ethanol-extracted oils are slightly higher in colour and free fatty acids than hexane-extracted oils.

### *Composition and properties of groundnut oil*

As indicated in Table 5.1, the groundnut kernel contains up to 50 per cent or more oil, and when extracted from sound kernels, the oil is easily refined to give a pale yellow oil with a pleasant, mild or bland flavour. Fats and oils can be classified according to 'group characteristics'; the groundnut oil belongs to the so-called 'oleic-linoleic' fat group. The unsaturated acids oleic and linoleic comprise about 80 per cent of the total, and the fully saturated palmitic acid contributes up to 10 per cent. The amount of linoleic acid (essential fatty acid) in groundnut oil can be as high as 30 per cent. The flavour of groundnut oil does not readily deteriorate like other vegetable oils, due to the absence of linolenic acid; the linolenic acid component in other oils is often implicated in the formation of reversion flavours.

A correlation exists between the linoleic acid content and the stability of oil from different types of groundnuts (Ahmed and Young, 1982). Stability is highest for the oils containing the smallest amounts of linoleic acid. Oil from runner types of groundnut and Virginia cultivars contain the least linoleic acid and are the most stable. On the other hand, oils from bunch types and Spanish and Valencia cultivars, have larger amounts of linoleic acid and are less stable, and have a correspondingly short shelf-life.

The characteristic odour and flavour of groundnut oil is attributed to the presence of very minute amounts (1.8 g per tonne) of hydrocarbons ($C_{15}H_{30}$

**Table 5.6 Properties of groundnut oil**

| Property | Value |
|---|---|
| Melting point | 0–3°C |
| Iodine value | 82–106 |
| Thiocyanogen value | 58–75.5 |
| Saponification value | 188–195 |
| Acetyl value | 8.5–9.5 |
| Reichert-Meissl value | <0.5 |
| Polenske value | <0.5 |
| Free fatty acids | 0.02–0.6% |
| Unsaponifiable matter | 0.3–0.7% |
| Refractive index ($ND^{20}$) | 1.4697–1.4719 |
| Density at 15°C | 0.917–0.921 |
| Density at 25°C | 0.910–0.915 |
| Mean viscosity at 20°C | 71.07–86.15 centipoise |
| Titre | 26–32°C |
| Heat of fusion | 21.7 cal/g (unhydrogenated)<br>24.7 cal/g (hydrogenated) |
| Colour | |
| visual | Light yellow |
| lovibond, 1 in. | Yellow: 16–25; red: 1–2 |
| Taste and odour | Slightly nut-like |

Source: Cobb and Johnson, 1973.

and $C_{19}H_{38}$); both hydrocarbons are in a concentrated form, and have an acrid taste. The unsaponifiable material (about 248 ml per 100 g) is mostly phytosterol.

Antioxidants – tocopherols and related compounds – are also present in the oil. Maximum stability is exhibited at a concentration of about 0.05 per cent tocopherol. The iodine value (82–106), a measure of unsaturation, and the refractive index (1.47), are relatively high (Table 5.6).

## Blanching of groundnuts

Blanching or whitening in groundnut processing is the procedure for removing the red skins and germs from the kernels. Blanching also cleans the kernels of dust, mould and other foreign material.

There are at least four methods of blanching: dry blanching, water blanching, spin blanching, and alkali blanching. Blanching groundnuts for roasting differs from blanching for peanut butter, in that for roasting the germs are usually not removed, while for peanut butter the germs are always removed.

Dry blanching is accomplished by heating the shelled groundnuts to about 138°C for up to 25 minutes, depending upon the variety and degree of separation desired, to loosen and crack the skins. After a period of cooling, the groundnuts are subjected to thorough but gentle rubbing between brushes or ribbed rubber belts. As a result the skins are rubbed off and blown into a container, and the germs are separated from the cotyledons by screening.

In the case of water blanching the first step is cutting the skins of individual kernels longitudinally by rolling the kernels between sharp stationary blades and slitting the skins on opposite sides. The kernels are then run through hot water sprays to loosen the skins, and passed on a knobbed conveyor under an oscillating canvas-covered pad, which rubs off the skins. Finally the kernels are air dried to a moisture content of 12–16 per cent at a temperature of about 49°C over a period of at least 6 hours (Lawler, 1961). Water-blanched groundnuts are reported to be more attractive in appearance and more stable in storage than are heat-blanched groundnuts.

Water blanching has the advantage that the kernels are not heated to destroy natural antioxidants, but has the drawback that drying is necessary and the germs are retained.

In spin blanching, the slit kernels are placed in driers, where they are rapidly dehydrated at a temperature lower than that of roasting, to loosen the skins. They are next fed through a skin blanching machine in which, by cross-feeding onto belts, the kernels are made to spin to unwrap the skins (Reeve, 1962).

In alkali blanching, shelled groundnuts are immersed in a 1 per cent solution of sodium hydroxide for about 8 seconds, then dipped in a 1 per cent solution of hydrochloric acid, to prevent the red colour from 'running' and staining the kernels. The skins are then rinsed off with water, and the kernels are dried. Alkali blanching is appropriate for use in the home or by small-scale processors.

Hydrogen peroxide blanching, as described by Takeuchi and Mazumoto

(1970), is based on a biochemical reaction with the catalase existing in the nuts themselves. The groundnuts to be blanched are put in hydrogen peroxide for 30–60 seconds, the skins swell and the hydrogen peroxide is decomposed into water and oxygen by the action of the catalase which occurs in the skin and on the endosperm of the nut. The oxygen generated forms a jacket between the skin and the endosperm. The skin is removed easily by a nut blancher in the wet state or after the nut dries. The whole operation is reported to be complete within 10–15 minutes. The hydrogen peroxide sterlizes and bleaches the nuts. Very few solubles are removed from the nuts, and there are no residues.

### *Colour sorting*

Blanched groundnuts are generally passed through pneumatically operated colour sorters, 'electric eyes', to reject unblanched, partially blanched or discoloured kernels and those suspected of containing aflatoxins. The kernels to be sorted are passed, in single-file, through a battery of colour sorters or photocells, which view the kernels from several angles. The chamber where the scanning occurs provides a background of the colour or colour intensity that approximately equals the desired colour of the kernel. Acceptable kernels fall into the good product chute but kernels that are too dark or too light affect the photocells which in turn activate an electromagnetically controlled pneumatic rejection system which deflects the undesirable kernel into the 'reject' chute. The single-file movement of the kernels is made possible by feeding them into a constant-speed grooved belt, which carries them towards the optical or 'discrimination' box in a constant-rate single-file flow.

## Processing groundnuts into peanut butter

The making of peanut butter on a commercial basis is largely an American art, and peanut butter is by far the most important product made from groundnuts in the United States. In principle, the manufacture of peanut butter consists of shelling, dry-roasting and blanching the nuts, followed by fine grinding. Ingredients such as salt, hydrogenated fat and dextrose, glycerin and lecithin or antioxidants are usually incorporated to improve flavour, prevent oil separation and to control rancidity.

Shelled groundnuts are dry roasted by batch or continuous methods. In the batch method, groundnuts are roasted batchwise, with about 200 kg in a revolving oven heated to approximately 430°C. The nuts are heated to 160°C and held at this temperature for 40–60 minutes to obtain the desired level of roast (Beattie, 1936). High roasting temperatures are detrimental to the nuts as they break down the oils, scorch the surface of the nuts and char the loose skin. The first effect of roasting is rapid drying, in which the moisture content is reduced to about 0.5 per cent (Woodroof and Leahy, 1940). This is followed by the development of oily transculent spots on the surface of cotyledons, called 'steam blisters', caused by oozing of oil from the cytoplasm as free oil. Changes in colour are reported to be due to the cell walls becoming wet with oil. This stage is referred to as 'white roast'. The skins also become wet with oil and darker in colour. The final stage of roasting is the development of a brown colour, at which time the groundnuts are 'brown roasted' and roasting is completed (Woodroof and Leahy, 1940). After roasting, the nuts are cooled as quickly as possible to stop cooking at a definite point and to produce a uniform product. In large plants, the hot nuts are passed from the roaster directly to a perforated cylinder or cooler box where a large volume of air is blown through the mass by fans.

Cooling is followed by dry blanching. Next, the blanched nuts are inspected and screened to remove scorched and rotten nuts. The nuts are ground into butter by means of attrition mills, homogenizers, hammer mills, or colloidal mills. Peanut butter is usually made using two grinding operations. In the first operation the nuts are reduced to a medium grind and in the second to a fine, smooth texture. For fine grinding, clearance between plates is 0.8–0.07 mm. Salt and other ingredients, if any, are usually fed into the grinder with the groundnuts. Finally the peanut butter is packed in airtight containers to prevent oxidative rancidity.

Peanut butter is a very popular product, especially in the United States of America, where a large section of the population eat peanut butter at least once a week. Its popularity is attributed to its pleasant flavour, convenience in application and stability to bacterial and fungal growth.

## Preparation of salted peanuts

Groundnuts are consumed in large quantities in the salted form; the term salted peanut refers to those groundnuts that are shelled, roasted and salted to taste. Groundnuts for salting may be dry roasted as described for peanut butter. Dry-roasted nuts are sometimes glazed with 2 per cent of a special oil and mixed with 2.5 per cent of salt without blanching. The most common practice, however, is to blanch the nuts, then oil roast them or fry them in oil. The roasting or frying temperature is 140–150°C, and the time 3–10 minutes, depending on the variety and the degree of roasting required.

When the nuts are fried in solid shortening, the salt is applied before the nuts have cooled to room temperature. At this point the frying fat is still in a melted condition on the surface of the nuts. Upon cooling to room temperature, the shortening on the surface sets to solid fat. The salt is thus held firmly to the surface of the nuts and will stay there in spite of rigorous handling. This is the case with coconut oil in particular, since coconut oil sets at about 25°C. Below 25°C, coconut oil is a fairly firm fat.

When the nuts are fried in oils that are liquid at room temperature, as is the case with groundnut oil, olive oil or cotton-seed oil, there is very little binding action to hold the salt on the surface of the nuts. The type of salt added to groundnuts is important, because its components play a considerable role with respect to rancidity. Since groundnuts are rich in oils, and since copper and iron accelerate rancidity of oils and fats, it is essential that a salt with minimal copper and iron impurities is used. On average the best salt products contain less than 1.5 ppm copper or iron. Nuts are salted at a level of about 2 per cent.

### *Nutritive value of groundnuts after processing*

Heat processing improves the aroma, flavour and texture of groundnuts. However, heat processing has undesirable effects on some of the nutritive properties of groundnuts. The shelf-life of the oil component is reduced by destroying natural antioxidants. The proteins, vitamins (excluding thiamin) and minerals are reported to be stable during processing. Blanching or mechanical removal of the skins further reduces thiamin content as this vitamin is concentrated in the skins. The red skins of unroasted groundnuts contain 25 per cent of the total thiamin. Water-soluble nutrients are not removed in the course of processing, since groundnuts need not be washed before they are processed into various products.

The thiamin content decreases with increased roasting. Pickett (1944) found that a high proportion of thiamin in groundnuts was destroyed very rapidly during roasting, especially at temperatures over 150°C. Other vitamins, such as niacin, choline and riboflavin, are little affected during processing.

The red skins, representing 2–3.5 per cent of the kernels, contain tannins and related pigments which give an undesirable colour to the protein unless removed during processing.

Pickett and Holley (1952) found that carbon dioxide accounted for 98 per cent of the volatile substances produced during roasting of groundnuts. Other volatiles produced were aldehydes, vanillin, ammonia, and hydrogen sulphide.

Cellulose substances react with amino-acids at high temperatures to produce carbon dioxide and colours.

The browning reaction is responsible for the main changes occurring in colour and flavour during groundnut roasting, and sucrose is the leading carbohydrate involved; crude fibre is also involved, but to a lesser extent.

The bitter flavour in groundnuts is reported to be attributed to at least four saponins, which are about 20 times as concentrated in the germs of groundnuts as in the cotyledons (Fischer, 1959).

### *References*


**Adams, F.** and **Hartzog, D.** (1980). *Peanut Science*, **7**: 120–123.

**Ahmed, E.M.** and **Young C.T.** (1982). In (Pattee, H. and Young, C.T., eds.) *Peanut Science and Technology*. Yoakum, Texas: American Peanut Research and Education Society, 635–88.

**Anon** (1962). Peanut oil solvents rated. *Food Eng.* **34**: No. 8, 91–92.

**Ashworth, L.J., Schroeder, H.W.** and **Langley, B.C.** (1965) *Science*, **148**: 1228.

**Beattie, W.R.** (1936). *Making and using peanut butter.* USDA Circ. 384.

**Central Institute of Agricultural Engineering India** (1981). *Development and Evaluation of Processing Equipment.* Technical Bulletin No. CIAE/PHTS/81/24 (A. Alam, ed., compiled by J. Singh and Y. Singh).

**Cobb, W.Y.** and **Johnson, B.R.** (1973). Physicochemical Properties of Peanuts, in *Peanuts – Culture and Uses*, Am. Peanut Res. and Education Assoc., Stone Printing Co., Roanoke, Va.

**Diener, U.L.** and **Davis, N.D.** (1966). *Phytopathology*, **56**: 1390.

**Diener, U.L.** and **Davis, N.D.** (1968). *Trop. Sci.*, **110**: 22.

**Diener, U.L.** and **Davis, N.D.** (1970). *J. Am. Oil Chem. Soc.*, **47**: 347.

**Diener, U.L.** and **Davis, N.D.** (1972). *Highlights Agric. Res.*, **19**: No. 3.

**Duke, J.A.** (1981). *Handbook of Legumes of World Economic Importance.* New York: Plenum Press.

**Dvorackova, I** (1976). *Br. Med. J.*, **1**: 691.

**FAO** (1982). *Legumes in Human Nutrition*, FAO, Rome.

**FAO** (1986). *FAO Mon. Bull. Statis.* Vol. 9 July/August, Rome.
**Fisher, C.H.** (1959). Highlights of peanut utilization research. *Peanut P. Nut World,* **38**: No. 6. 14, 16, 31–33.
**Freeman, A.F., Morris, H.J.** and **Willich, R.K.** (1954). *Peanut Butter.* U.S. Dept. Agr. AIC – 370.
**Gibbons, R.W.** (1980). In *Proceedings of the Workshop on Groundnut.* Patancheru, India, ICRISAT, 12–16.
**Heathcote, J.G.** and **Hibbert** (1978). *Aflatoxins: Chemical and biological aspects,* Elsevier Scientific Publ. Co. Chapter 2.
**Hiscocks, E.S.** and **Wogan, G.N.** (eds.) (1965). *Mycotoxins in Foodstuffs,* M.I.T. Press, Cambridge, Mass. 291.
**Irvine, F.R.** (1974) *West African Crops,* Oxford University Press, London.
**Keen, P.** and **Martin, P.** (1971). *Trop. Geogr. Med.* **23**: 44.
**Khan, L.M.** and **Hanna, N.A.** (1983). Expression of oil from oilseeds, *J. Agric. Engn. Res.* **28**: 495–503.
**Krishnamachari, K.A.V.R., Bhat, R.V., Nagarajan, V.** and **Tilak, T.B.G.** (1975). *Lancet,* **1**: 1061.
**Lawler, F.K.** (1961). Ingenuity creates new process. *Food Eng.* **33**: No. 2, 38–40.
**Mason, M.E., Newell, J.A., Johnson, B.R., Koehler, P.E.** and **Waller, G.R.** (1969). *Journal of Agric. and Fd. Chemistry,* **17**: 782–92.
**McDonald, D.** and **Harkness, C.** (1963). *Trop. Sci.,* **5**: 143.
**Miller, M.L.** (ed.) (1983). *Peanut Industry Guide.* Washington, D.C.: National Peanut Council.
**Pancholy, S.K., Deshpande, A.S.** and **Krall, S.** (1978). In *Proceedings of the American Peanut Research and Education Association,* **10**: 30–37.
**Pattee, H.E., John, E.B., Singleton, J.A.** and **Sanders, T.A.** (1974). *Peanut Science,* **1**: 57–62.
**Picket, T.A.** (1944). Thiamin content of peanut butter. *Georgia Agr. Expt. Sta. Circ.,* **146**.
**Picket, T.A.** (1947). Some effects of heat treatment on peanuts – peanut oil. *Georgia. Agr. Expt. Sta.* JS 168.
**Picket, T.A.** and **Holley, K.T.** (1952). Peanut roasting studies. *Georgia Agr. Expt. Sta. Tech. Bull.,* **1**.
**Prickett, C.O.** and **Salmon, W.D.** (1965). *Fed. Proc.,* **24**: 627.
**Reeve, K.J.** (1962). Split skins spun off in new nut blancher. *Food Eng.* **34**: No. 8, 51.
**Sargeant, K., Sheridan, A., O'Kelly, J.** and **Carnagan, R.B.A.** (1961). *Nature,* London, 1095.
**Screenivasamurthy, V., Jayarman, A.** and **Parpia, H.A.B.** (1965). In (G.N. Wogan ed.), *Mycotoxins in Foodstuffs,* M.I.T. Press, Cambridge, Mass. 251.
**Sellschop, J.P.F.** (1965). *Symposium on mycotoxins.* Heer Printing Co., Pretoria, South Africa.
**Serck-Hanssen, A.** (1970). *Arch. Environ. Health,* **20**: 729.
**Takeuchi, A.K.** and **Mazumoto, Y.** (1970). New skinning process. **Peanut J. Nut. World, 50**: No. 1, 12.
**Tools for Agriculture** (1976). *A buyer's guide to low cost Agricultural Implements* (compiled by John Boyd), Intermediate Technology Publications Ltd. London, England.
**Van Rensburg, S.J.,** and **van der Markham, R.** (1974). *S. African Med. J.* **48**, 2508a.
**Watt, B.K.** and **Merrill, A.L.** (1963). *Composition of Foods,* USDA Agricultural Handbook 8.
**Woodroof, J.G.** (1973). *Peanuts – Production, Processing, Products.* AVI Publishing Company, Connecticut, USA.
**Woodroof, J.G.,** and **Leahy, J.F.** (1940). *Microscopic studies of peanuts.* Georgia Agr. Expt. Sta. Bull. 205.
**Young, C.T., Walker, G.R.** and **Hammons, R.O.** (1970). *Crop Science,* **50**: 521–23.

# 6

# Maize

## Production and varieties

Maize, also referred to as corn or Indian corn, belongs to the family Gramineae (Figure 6.1). Maize was the dominant staple food for the early civilizations of the Western Hemisphere, and today it still plays an important role in the diet of millions of people because of its capacity to produce a large amount of dry matter per hectare, its ease of cultivation, versatile food uses and storage characteristics. Maize is grown throughout the world and is a staple food crop, particularly in Latin America, and the southern, western and eastern parts of Africa. The average maize consumption in Guatemala, Zimbabwe and Malawi is 104, 145 and 169 kg per head per year, respectively (FAO, 1979).

Most of the maize produced in Western countries is fed to animals, while in Asia, Africa and Latin America almost the entire production is used for human consumption. In 1984, the total area under maize cultivation was

**Figure 6.1 Maize plant with mature cob.**

about 130 million hectares, with a world production of about 449 million tonnes. In that same year the USA alone produced over 194 million tonnes (FAO, 1984). The USA, China and India are the major maize growing countries of the world.

About 2000 cultivars, based on the vegetative characteristics of the plants, the tassels, the ears and physiological and cytological characteristics, have been reported (Wellhausen *et al.*, 1957). However, all the cultivated species grown for food, feed or industrial purposes have been classified as *Zea mays*. The most important maize varieties are flint corn, dent corn, flour or soft corn and pop corn. Other varieties of minor importance include sweet corn, waxy corn and starchy sweet corn.

## Utilization

### *Human food*

Maize is a staple food in the diets of people in many countries (Figure 6.2). Hoffman (1965) estimated that some 100 million people in the world consume maize in the form of thin, round, unleavened cakes or as porridge. In Africa, south of the Sahara, porridge is the most widespread form in which maize is consumed. In western countries maize for human food is generally industrially processed, whereas food products from maize are still often prepared at home or at the village level in developing countries. In West Africa one of the most popular foods prepared from maize is kenkey. Kenkey is made by soaking maize kernels, followed by grinding. After thorough mixing, the maize meal is left to ferment to produce an acidic dough which, after steaming or cooking, is eaten with fish or vegetable soup. Throughout Central and South America, a maize product called tortilla appears with practically every meal and is a basic food for the majority of the people.

**Figure 6.2 Maize harvest in Tanzania.**

Maize may be parched, soaked, or steeped before grinding. Maize meal or dough is used in the preparation of cakes and other foods, or the meal is simply boiled in water and eaten as a porridge or gruel. Another maize product consists of roasted ground maize mixed with peanuts and red peppers before rolling into balls or making into sticks and frying. Koga is a popular food in Cameroon. It is prepared by mixing ground maize kernels, palm oil, salt and ground capsicum; the mixture is rolled and steamed in plantain leaves.

In many areas maize is used as a carbohydrate source for the preparation of alcoholic and non-alcoholic drinks, for example, beer and 'mahewu' in South Africa. Mahewu is a fermented almost non-alcoholic drink made from maize meal or wheat flour or a combination of both. Maize starch is used as a thickening agent for sauces and soups, as a gel forming agent in confectionery and as a moisture retention agent in toppings and cake icings. Oil from maize is nutritionally excellent. It is highly digestible and a good source of essential fatty acids.

### *Animal feed*

The bulk of maize produced in industrialized countries is fed to livestock. For instance, in the United States, from 75–90 per cent of the maize crop is fed to livestock (Jugenheimer, 1976).

Wet milling of maize produces maize starch and maize oil, and residues that constitute the basis of several feed products. These include:

1. The solubles known as steep water, steep liquor or condensed fermented maize extractives.
2. The pericarp and other fibrous materials which may contain bits of unrecovered endosperm.
3. The gluten, which is the proteinaceous portion of the endosperm.
4. The germ residue remaining after oil refining.
5. The mother liquor from dextrose crystallization.

The feed products derived from the distilling process are mainly a mixture of fibre, gluten, germ, unfermented endosperm and yeast cells, which, after distillation, are separated from the solubles. Maize gluten feed has traditionally been added to ruminant rations. Condensed fermented maize extractives have been reported to be a source of growth factors for chicks. Maize gluten meal is used primarily in poultry feed because of its high xanthophyll content.

Maize has certain limitations for livestock feeding. It is primarily a source of energy, with poor quantity and quality of proteins and relatively low vitamin levels. Swine and other monogastric animals with a single digestive cavity require high quality protein maize, while cattle and other animals with three or four chambered stomachs can better utilize maize with lower quality protein.

## Composition

The composition of the maize kernel, as is the case with other cereals, varies depending on cultural practices, type of seed and soil, weather conditions and other factors. The main constituents of the kernel are shown in Table 6.1.

**Table 6.1 Composition of the maize kernel (dry weight)**

| Constituent | Composition (%) |
|---|---|
| Carbohydrates | 80.0 |
| Protein | 10.0 |
| Oil | 4.5 |
| Fibre | 3.5 |
| Minerals | 2.0 |

Maize is generally recognized to be high in carbohydrates, low in quantity and quality of protein and relatively low in oil. Maize is therefore usually referred to as a starchy product. Maize proteins have been shown to consist of albumin (7 per cent), globulin (10 per cent), prolamin or zein (39 per cent) and glutelin (35 per cent). Albumin, globulin and glutelin are the major proteins present in the germ, while zein and glutelin are concentrated in the endosperm. The maize proteins are deficient in lysine and tryptophan. Starch is a predominant carbohydrate component of maize. The ratio of amylose to amylopectin is usually 27 : 73. The essential amino-acid content, minerals and vitamin content of the maize kernel are shown in Table 6.2 and 6.3, respectively.

**Table 6.2 Essential amino-acid composition of maize (mg/gN)**

| Amino-acid | Maize grain (whole) | Maize gluten | Maize germ | Maize protein (zein) |
|---|---|---|---|---|
| Arginine | 300 | 194 | 506 | 113 |
| Histidine | 156 | 131 | 188 | 106 |
| Isoleucine | 400 | 319 | 263 | 465 |
| Leucine | 938 | 1000 | 444 | 1481 |
| Lysine | 144 | 94 | 363 | – |
| Tryptophan | 38 | 38 | 81 | 6 |
| Phenylalanine | 313 | 413 | 313 | 400 |
| Tyrosine | 375 | 394 | 313–375 | – |
| Cysteine | 94 | 94 | 75 | 63 |
| Methionine | 194 | 156 | 100 | 144 |
| Threonine | 231 | 250 | 275 | 188 |
| Valine | 331 | 356 | 188 | 188 |

**Table 6.3 Mineral and vitamin content of dry maize**

| Mineral or vitamin | Composition (mg/100 g) |
|---|---|
| Calcium | 6.0 |
| Phosphorus | 300.0 |
| Iron | 2.5 |
| Carotene (IU) or vitamin A | 0.015 |
| Ascorbic acid | 11.40 |
| Thiamine | 0.50 |
| Riboflavin | 0.08 |
| Niacin | 2.0 |

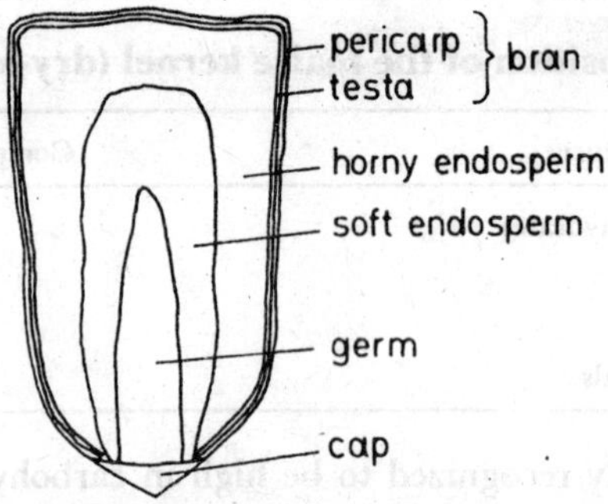


**Figure 6.3 Structure of the maize kernel.**

The main parts of a maize kernel are the hull, the germ and the endosperm. The hull or bran, about 5 per cent of the kernel, is usually used as feed. The germ or embryo, about 12 per cent of the kernel, contains most of the oil in maize and is a very nutritious portion of the kernel with a high protein, mineral and solubles content. The largest part, and the most important, is the endosperm, constituting about 82 per cent of the kernel mass. It consists largely of gluten and starch. The tip cap is the smallest portion of the kernel, amounting to 1 per cent. The structure of the maize kernel is shown in Figure 6.3.

## Processing of maize

At the time of harvest, maize has a moisture content ranging from 28–35 per cent. The combination of unfavourable climatic conditions (high ambient temperature and high humidity) and high moisture content is conducive to spoilage by micro-organisms, fungi and insect pests. Spoilage is prevented wholly or partly either by appropriate storage of maize or by processing it into various storable products. Traditional methods of processing maize into flour – using such equipment as mortars and pestles (Figure 6.4), grinding stones and manually operated mills – vary in detail depending on culture and geography. The main principles, however, are the same.

Maize processing essentially begins with soaking the grain and then grinding it between stones or pounding it in a mortar and pestle arrangement. During pounding or grinding the bran is removed. The grain is winnowed at intervals to remove the bran from the kernels. The dehulled grain is then pulverized into flour by further grinding or pounding. Processing of maize into desirable end products usually involves primary processing (cleaning, grading, soaking, dehulling, grinding and sieving) and secondary processing (blending, cooking, frying and baking).

### *Maize milling*

The process of milling maize can either be dry or wet. Dry milling may or may

**Figure 6.4 Pounding maize with wooden pestles and mortars.**

not include degerming as a preliminary step. Both processes involve the separation of the germ from the remainder of the grain in order to extract and recover the germ oil. After germ removal, dry milling employs roller mills to produce grits, meal and flour, whereas wet milling involves a steeping stage and the complete disintegration of the endosperm, to enable the recovery of the starch and protein as separate products.

## Wet milling

Wet milling is a process which commences with soaking the kernels, to prepare them for the subsequent separation of the various kernel components. Separation of components is achieved by grinding, screening and centrifugal action. Maize dried at 82°C or higher is less suitable for wet milling than kernels dried at lower temperatures (Watson and Hirata, 1962). The former results in a lower starch yield because of the presence of endosperm particles, which apparently are made resistant to milling by high-temperature drying. Maize is wet milled to obtain starch, oil, feed (gluten feed, gluten meal, germ cake) and the hydrolysis products of starch (solid and liquid glucose and syrup). The operations involved in maize wet milling are shown in Figure 6.5.

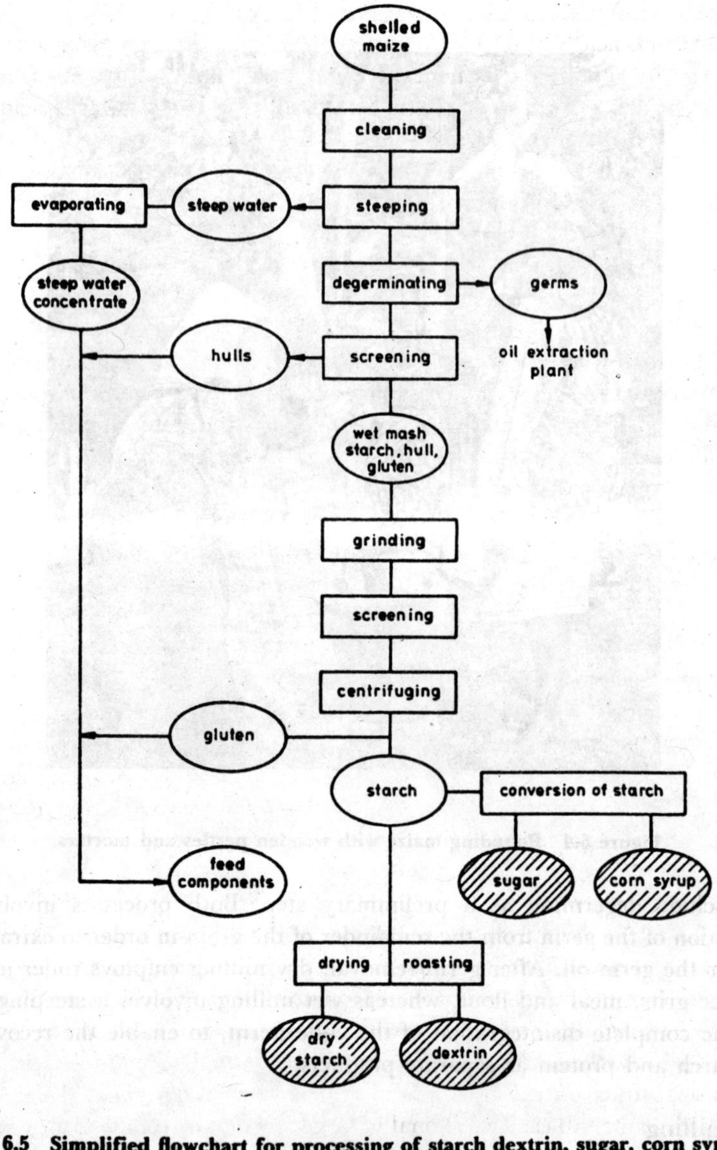


**Figure 6.5 Simplified flowchart for processing of starch dextrin, sugar, corn syrup and feed components.**

***Steeping***

In steeping or soaking, the maize is conditioned to produce optimum milling and separation qualities. This is done by soaking the maize grain in sulphurous acid water for 36–48 hours at a temperature of between 50 and 55°C.

The objectives of steeping are:

1. To soften the maize kernel for grinding.
2. To facilitate disintegration of the protein that holds the starch granules together in their kernels and to remove solubles, mainly from the germ,

making it more susceptible to recovery.

Sulphurous acid water (0.1–0.2 per cent $SO_2$ in water) prevents germination of the grain and limits fermentation and other undesirable microbiological changes but allows growth of lactic acid producing bacteria. This acid buffers the pH value and increases softening of the kernel. The by-product, the steep water, is concentrated in vacuum evaporators to 35–55 per cent solids, and finds use either in animal feeds or as a nutrient in certain fermentation media.

*Degermination*

Steeped maize with a moisture content of about 45 per cent and fully softened is passed to the degermination mill, where the oil-rich germ, or embryo, is separated from the starch, fibre, gluten and hulls. The degerminating mill is a machine consisting essentially of one stationary and one rotating metal plate with projecting teeth. It is designed for tearing the soft kernels apart and freeing the germs without crushing them.

The ground pulpy material containing germs, hulls, starch and gluten is passed through hydrocyclones or other separators, where the germ is separated from other components. The separated germs are washed free of starch, dried and conveyed to the oil extraction plant.

*Separation of hulls and fibre from starch and gluten*

The main constituents of the maize kernel remaining after the degermination process are the starch, gluten and hulls. These are fed into attrition or impact mills, where the wet mash is ground and starch is readily released. The milled slurry is sieved in screens to separate the fibre from the starch and gluten. The fibre is mixed together with the steep water and dried for feed products.

*Starch–gluten separation*

The starch and gluten slurry, commonly referred to as mill starch, contains 5–8 per cent protein. It is separated into individual components. Separation of gluten from the starch may be carried out on long, narrow, slightly inclined wooden troughs, also called starch tables. The starch–gluten slurry is allowed to flow slowly down the tables and sedimentation occurs with the denser starch depositing on the bottom and gluten being removed at the top of the lower end of the tables. Traditionally, large wooden or concrete tubs are used to hold the slurry until the starch settles at the bottom.

In modern production units these methods of separation are replaced by centrifugal separators in which, due to the difference in density, the comparatively heavier starch is separated from the lighter gluten by centrifugal force. From the centrifuge, the wet starch is dewatered and finally dried in tunnel driers. The gluten, after centrifugation, consists of a thin slurry, which is dewatered and destarched in another centrifuge, then filtered and dried to be used as corn gluten meal or corn gluten feed, or further processed to recover the protein, zein, which has many non-feed uses.

The products of wet milling are steep water, germs, gluten, hulls and starch (Figure 6.5).

*Corn syrup and corn sugar*
Corn syrup and corn sugar are produced by acid conversion of corn starch. Syrups are made by partial hydrolysis of starch, and corn sugar by complete hydrolysis. Any mineral acid is suitable for conversion, but hydrochloric acid is generally used to catalyse the reaction. The process involves cooking the starch slurry with the required amount of dilute acid until the desired degree of conversion is reached. The reaction is stopped by neutralization with soda ash (sodium carbonate). Suspended solids and other impurities are filtered out. The syrup is finally bleached and concentrated to the required density.

When corn sugar is being made, the conversion is allowed to proceed to completion. After conversion, the liquor is neutralized, filtered, clarified and concentrated, and finally the whole mass is allowed to crystallize into sugar. In past years, instead of acid, enzymes were used to convert corn starch into syrup or dextrose. In this way, the starch was first gelatinized by heating it to above the starch gelatinization temperature and simultaneously partially converting it with acid or alpha-amylase enzyme. At this point, the enzyme was inactivated and the solution was adjusted to pH 4.0–4.5. Further conversion was then accomplished by the addition of another enzyme, such as glucoamylase, and saccharification proceeded to the desired level. This type of enzyme conversion produces syrups which are sweeter and less viscous than acid-converted products. In addition, the side-effect reactions are reduced to a minimum and so the unpleasant tasting impurities are eliminated. Corn syrups and dextrose have many food applications and are also used in pharmaceutical, industrial and feed products.

## Dry milling

Maize is dry milled by two general systems – degerming or non-degerming. The non-degerming method produces a meal with little or no removal of germ. The resulting product has a rich, oily flavour because of its high fat content. The degerming method frees the kernel of its hull and germ and recovers these, as well as one endosperm fraction or several. In the conventional degerming system moisture is added to the corn under controlled conditions before degerming takes place. With the dry degerming process, on the other hand, corn having no more than 15 or 16 per cent moisture is degermed without prior addition of moisture to produce grits, meal and flour.

*Production of maize flour*
The production of low-fat maize flour is carried out using the *Buhler-Miag* method, in which dry kernels are separated into various parts (Figure 6.6). In this method, the kernels are crushed in an impact-type corn degermer. Due to differences in friability between the pericarp, the endosperm and the germ, the kernels are separated into various fragments. The germ maintains its original structure to a large extent. The corn fragments (hull, germ, endosperm) are separated by means of sifters; the separation is based on differences in particle size, density or aerodynamic characteristics. The hulls are then removed by air aspirators and the remaining fragments are sorted on gravity table separators.

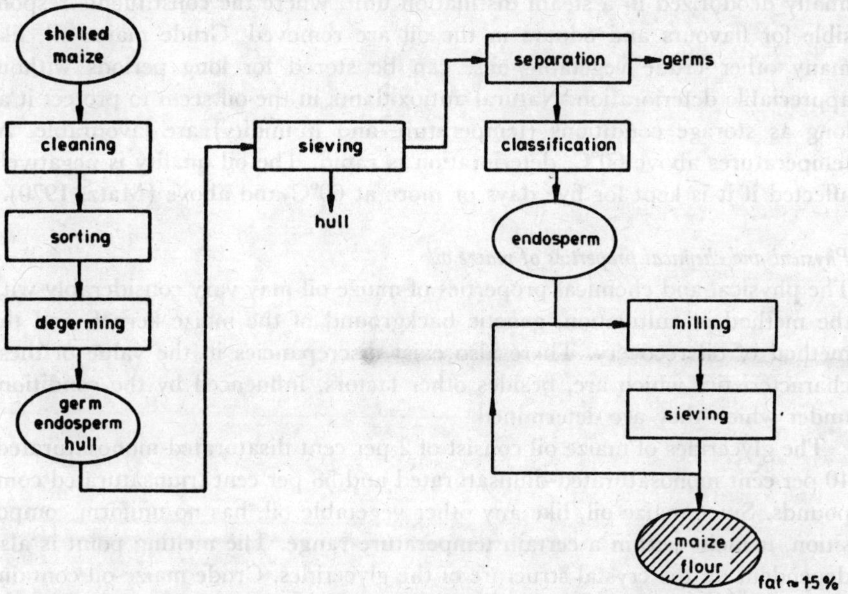


**Figure 6.6 Flowchart for the production of maize flour (dry degerming Bühler-Miag system).**

The germ-free fragments are milled into flour. Depending upon the hardness of the corn, about 65–80 per cent of the raw kernel mass can be utilized as maize flour with a fat content of less than 1.5 per cent, and approximately 10–14 per cent is used as germs with 16–24 per cent fat.

## Maize oil

The raw material for maize oil extraction is the germ or embryo after it has been separated from the whole kernel. The germ contains about 85 per cent of the oil in the kernel.

Prior to their delivery to oil extraction presses, the germs are washed, dried and heated. The oil may be expressed in hydraulic presses, or semi-continuous or continuous screw-type presses may be employed. When utilizing screw-type presses, the germs are first passed through flaking rolls, until they are crushed into a coarse meal. The meal is passed into steam-heated temperers and then into the expeller unit, where it is forced under high pressure through a slotted barrel made up of steel sections in the form of a rotating screen called an 'oil reel'. During this stage of processing, most of the oil is pressed out through the slots, while the fibrous portion, the foots, is discharged at the end of the barrel. The foots are usually returned to the expeller unit for repressing. Ultimately, the residue still retains about 5–8 per cent of its original oil content.

A combination of mechanical pressing and solvent extraction for removing oil from the germs may also be employed. Solvent extraction is capable of reducing the oil content of the residue to about 0.5 per cent.

After leaving the expeller units, the crude oil is filtered and clarified, and finally deodorized in a steam distillation unit, where the constituents responsible for flavours and odours in the oil are removed. Crude maize oil, like many other crude vegetable oils, can be stored for long periods without appreciable deterioration. Natural antioxidants in the oil seem to protect it as long as storage conditions (temperature and humidity) are favourable. At temperatures above 60°C, deterioration is rapid. The oil quality is negatively affected if it is kept for five days or more at 60°C and above (Matz, 1970).

*Physical and chemical properties of maize oil*

The physical and chemical properties of maize oil may vary considerably with the method of cultivation, genetic background of the maize kernel, and the method of oil recovery. There also exist discrepancies in the value of these characteristics which are, besides other factors, influenced by the conditions under which they are determined.

The glycerides of maize oil consist of 2 per cent disaturated-monosaturated, 40 per cent monosaturated-diunsaturated and 58 per cent triunsaturated compounds. Since maize oil, like any other vegetable oil, has no uniform composition, it melts within a certain temperature range. The melting point is also dependent on the crystal structure of the glycerides. Crude maize oil contains up to 3 per cent free fatty acid. After treatment with lye and bleaching agents and the deodorizing process, a light, storable oil with less than 0.1 per cent free fatty acids is obtained. The free fatty acid content of crude oil from wet milled maize is reported to usually be higher than that from dry milled maize (Matz, 1970).

The free fatty acid composition of maize germ oil is shown in Table 6.3.

**Table 6.3 The free fatty acid composition of maize germ oil**

| Fatty acid | Composition (%) |
|---|---|
| Myristic | 0.5 |
| Palmitic | 9.7 |
| Stearic | 3.6 |
| Hexadecenoic | 0.2 |
| Oleic | 30.4 |
| Linoleic | 55.6 |

Source: Ullmann's Enzyklopadie, 1976.

The largest part of the unsaponifiable matter of oils and fats consists essentially of sterines, which are either in the form of free fatty acids, esterified with fatty acids, or glycolipids (Ullmann's Enzyklopadie, 1976). The sterine content of crude maize oil is in the range of 0.58 and 1.0 per cent.

The viscosity of maize oil is similar to that of other vegetable oils with the same degree of unsaturation. At 25°C the viscosity is about 58 centiponds; this value is relatively high, as is the case for most oils, because of the intermolecular reciprocal action of the chains of fatty acids. The smoke, flash, and fire points of refined maize oil are similar to those of other refined vegetable oils of similar molecular weight. Typical values are smoke point 237.5°C, flash point

321°C and fire point 354.5°C. A number of other chemical and physical properties of maize oil, as compared with rice and wheat oils, are given in Table 6.4.

**Table 6.4 Some chemical and physical properties of maize, rice and wheat oils**

| Property | Maize oil | Rice oil | Wheat oil |
|---|---|---|---|
| Solidifying point in °C | -18—-10 | -5—-10 | 0 |
| Density at 15°C in g/cm$^3$ | 0.91—0.92 | 0.91—0.92 | 0.92—0.93 |
| Iodine value | 109—133 | 92—109 | 115—126 |
| Saponification value | 187—196 | 183—194 | 180—189 |
| Unsaponifiable matter in % | 1.3—2.0 | 3.5—5.0 | 3.5—6.0 |
| Refractive index at 40°C | 1.46 | 1.46 | 1.46—1.47 |
| Free fatty acid in % as oleic | 1.0 | 5—50 | — |

Source: Matz, 1970; Ullmann's Enzyklopadie, 1976.

## Cornflakes

Cornflakes are a hydrothermically treated maize product of world-wide popularity. They owe their success to their high nutrient value combined with low caloric content and good digestability. Flaking is a process consisting of cooking fragments of cereal kernel, grits, to a certain consistency, pressing the cooked mass between rollers to form flakes, and toasting the flakes at an appropiate temperature (Figure 6.7).

Cornflakes are produced exclusively from grits obtained from the major part of the horny endosperm of the maize kernel. The grits remaining after the removal of the germs and hulls are cooked for 2—2.5 hours at a temperature of 120°C, and mixed with ingredients such as corn syrup, sugar, salt and vitamins. The cooked grits are passed into driers through which preheated air is blown, to reduce the moisture content to about 15 per cent. The dried material is held in tempering tanks for 6—8 hours to permit the residual moisture to become equally distributed; this is important to ensure uniform toughness for flaking.

The tempered grits are then pressed into thin flakes, in flaking rolls revolving

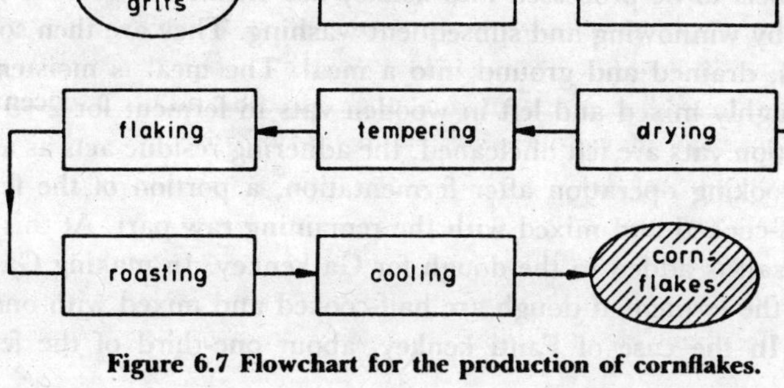


**Figure 6.7 Flowchart for the production of cornflakes.**

at a speed of about 180–200 rpm. The rolls are cooled by internal circulation of water, to avoid sticking of the flakes to the rolls. From the rolls, the flakes are passed to toasting ovens for 3–5 minutes at a temperature of 200–250°C. With less than 3 per cent moisture, the flakes finally emerge from the oven and are cooled to room temperature before packaging.

Grits with a moisture content of about 14 per cent yield 80–83 per cent cornflakes, with a moisture content of 3–5 per cent. During processing the raw material is subjected to the following physico-chemical changes:

1. The starch in the material is gelatinized and probably slightly hydrolized, or the starch granules are ruptured by temperatures which form a gel of soluble starch and dextrins.
2. The particles undergo a browning reaction due to the interaction of proteins and sugars; enzymatic reactions are halted, rendering the final product more stable.
3. Dextrinization and caramelization of sugars take place as a result of the high temperatures in the roasting oven. The flakes become crisp due to the reduction of its moisture content to a low level.

100 g of cornflakes contain approximately 7.5 g of protein, 0.7 g of fat and 85 g of carbohydrate. The calorie content of 100 g cornflakes is about 1590 joules. The bulk density of cornflakes is about 120–140 g/litre (Kunde, 1979).

## Fermented maize products

Fermented maize dough is an intermediate product which forms the basis for a variety of foods. Akasa, akple, banku, abolo, fonfon, ogi, koko, pozol, uji and kenkey are all basic maize products, which require maize dough with a particular degree of fermentation.

### *Kenkey*

In Ghana the most important maize product is kenkey due to its food value and its popularity amongst a large section of the population. Essentially, there are two types of kenkey, which differ in shape, constituents, and in methods of preparation. Accra or Ga kenkey has the form of a ball measuring about 9 cm in diameter or a cylinder 8–10 cm long wrapped in single layers of maize husks. The second type, Fanti kenkey is bigger and is wrapped in 5–6 layers of dried plantain leaves.

Maize kernels to be processed into kenkey are cleaned to remove dust and other debris by winnowing and subsequent washing. They are then soaked for 12–48 hours, drained and ground into a meal. The meal is moistened with water, thoroughly mixed and left in wooden vats to ferment for 2–3 days. If the fermentation vats are left uncleaned, the adhering residue acts as a starter.

In a pre-cooking operation after fermentation, a portion of the fermented dough is half-cooked and mixed with the remaining raw part. At this stage of the process, salt is added to the dough for Ga kenkey. In making Ga kenkey, two parts of the fermented dough are half-cooked and mixed with one part of raw dough. In the case of Fanti kenkey, about one-third of the fermented

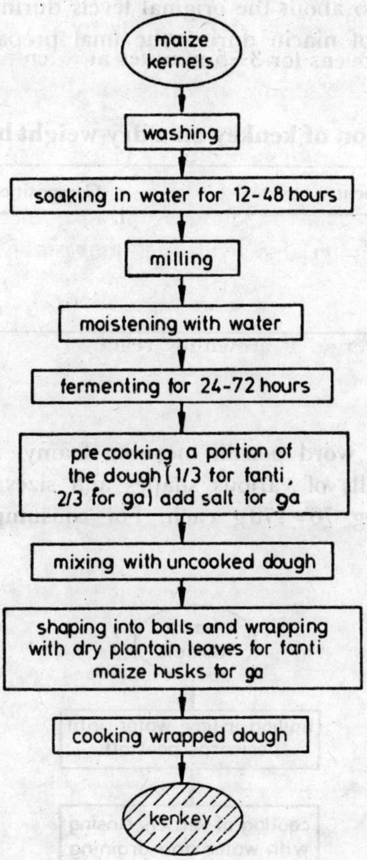


**Figure 6.8 Flowchart for the processing of kenkey.**

dough is pre-cooked and mixed with the remaining two-thirds. A flowchart for kenkey processing is given in Figure 6.8.

The mixture of half-cooked and raw dough is then moulded into balls, wrapped, and boiled until cooked. A representative composition of kenkey is shown in Table 6.4. Investigations into kenkey fermentation carried out by Christian (1970) indicated that at the onset of fermentation, fungi, principally *Aspergillus*, *Rhizopus*, and *Penicillium* species, and Gram-negative, catalase-positive coccoid forms of bacteria predominate. The fungi die out as fermentation proceeds but the presence of Gram-positive, catalase-negative cocci, grouped in pairs, is evident from about the 9th hour of fermentation. These acid-forming streptococci rapidly multiply, reaching a peak between 24–36 hours of fermentation. Yeasts, including *Saccharomyces* spp., start to develop on the surface of the fermenting dough after about 24 hours of fermentation.

Yeasts probably contribute largely to the aroma of kenkey by esterifying organic acids and alcohols. The production of the characteristic acid flavour in the dough – resembling a mixture of diacetyl and acetic acid – follows the growth of the Gram-positive cocci. The pH of the dough decreases from about 6.8 to 3.5–4.0. Levels of niacin and lysine increase during the period of dough fermentation, but fall to about the original levels during the partial cooking. There is further loss of niacin during the final preparation of the kenkey (Ofosu, 1971).

**Table 6.4 Composition of kenkey on a dry weight basis**

| Constituent | Composition (%) |
|---|---|
| Moisture | 47–54 |
| Protein | 8–12 |
| Fat | 2.7–4.1 |
| Starch | 63–88 |

## *Pozol*

Pozol, from the Aztec word pozolli meaning foamy, is a fermented maize dough shaped into balls of various shapes and sizes, 10–12 cm long and 5–8 cm wide, weighing 70–170 g each. For consumption, balls of pozol,

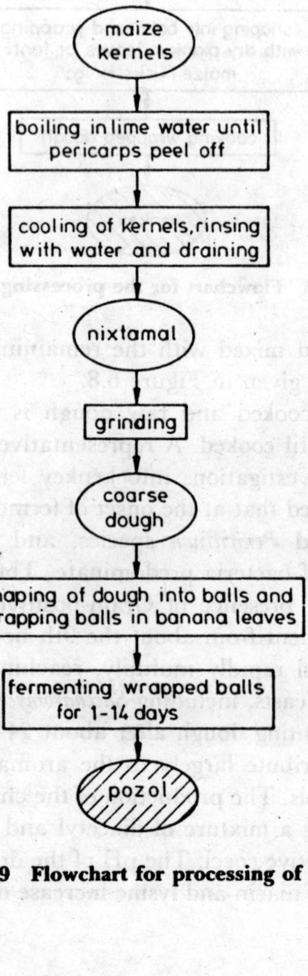


**Figure 6.9 Flowchart for processing of pozol.**

unfermented or fermented, are diluted with water in the proportion of 1 : 2 to 1 : 3 to obtain a sort of beverage which is drunk without cooking. Salt and toasted chilli pods, sugar, or honey may be added.

Pozol is largely consumed by various ethnic groups, irrespective of social class, in Mexico. Consumption is said to vary from 80 g to 1000 g of pozol dough per person per day.

To prepare pozol, maize kernels (1 – 1.5 kg) are boiled for about one hour in 2 litres of water to which has been added a handful of lime powder to give approximately 10 per cent $Ca(OH)_2$. When the kernels are swollen and their pericarps peel off, they are cooled, rinsed with water and drained to obtain what is called nixtamal (Figure 6.9) (Ulloa *et al.*, 1977). In the next step, the nixtamal is ground into a coarse dough, which is then shaped into balls by hand. The balls of pozol are finally wrapped in banana leaves and left to ferment for 1 – 14 days or more depending on the taste of the consumer. During the first stages of fermentation, bacteria are believed to outnumber the yeasts and moulds and are said to be responsible for most of the acid production during the first 24 hours. During this time, the pH decreases from 7.5 to 5.0 (Steinkraus *et al.*, 1983). The two most conspicuous changes that take place during pozol fermentation are the development of the acidic flavour and the characteristic aroma.

The content of protein, niacin, riboflavin, lysine and tryptophane in pozol has been shown to be higher than in maize.

### *Ogi and agidi*

Ogi is prepared by soaking maize in water for 1 – 3 days, depending upon the variety and maturity of the raw material. The softened maize is ground either

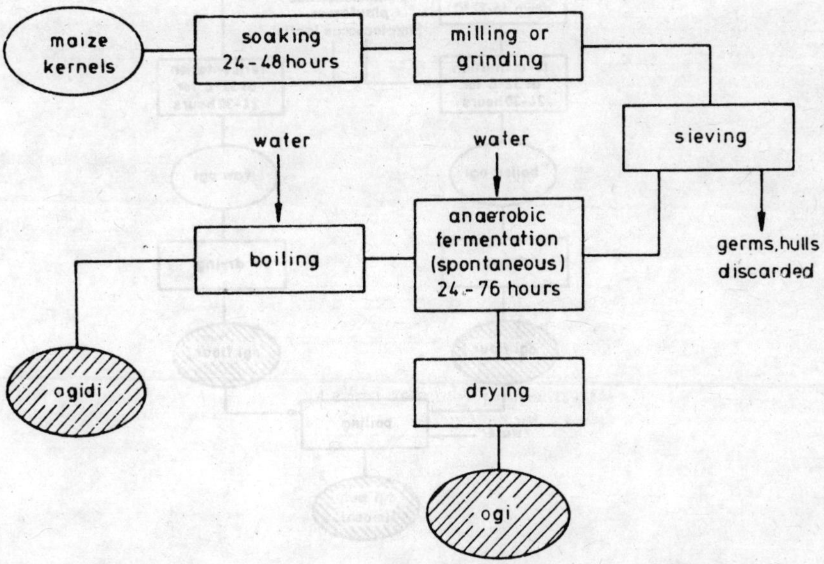


**Figure 6.10 Flowchart for traditional processing of agidi and ogi.**

between stones or in a mortar, and then sieved to remove the germs and hulls. Water is added to the product, which is almost pure starch, and this mixture is allowed to ferment anaerobically for 2–3 days. The fermented meal is dried to yield ogi which is boiled and eaten hot. Ogi is an important traditional weaning food and a breakfast cereal for adults. As a breakfast cereal, ogi is eaten with meat stew and fried plantain, fried bean cake (akara), steamed bean cake (moin-moin), or bread and fried egg. When the fermented mass is wrapped in banana leaves and cooked, it is known as agidi, which has a dry matter content of about 10 per cent (Muller, 1970).

Fermentation of ground maize to produce ogi, as with any other fermentation, is strongly influenced by temperature. The optimum temperature is found to be 32°C and the optimum taste is obtained at a pH value of 3.8, through lactic acid production and the formation of esters. The predominant organism in ogi fermentation is *Lactobacillus plantarum*, which is responsible for the production of lactic acid, the main acid present (Banigo and Muller, 1972b). A flowchart illustrating the necessary steps in the traditional production of ogi is shown in Figure 6.10.

The main drawback with this method of processing maize is the high loss of nutrients (up to 50 percent) in the soaking and sieving processes. Dried ogi

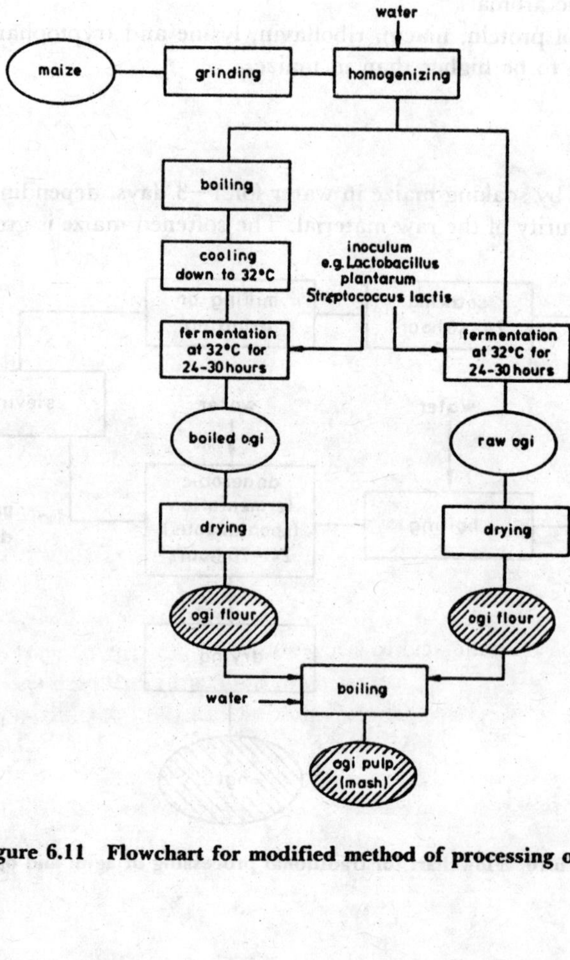


**Figure 6.11 Flowchart for modified method of processing ogi.**

can be stored for 18 weeks at a temperature of 30°C.

In a modified method of processing ogi the sieving process is eliminated (Figure 6.11) (Bergmann *et al.*, 1981). Acceleration of the fermentation and of the formation of acid is obtained through cooking the mixture of maize and water and inoculating with the micro-organisms *Lactobacillus plantarum* and *Streptococcus lactis*. Undesirable acetic and butyric acid fermentation can be avoided in this way through the swift formation of acid.

## Mahewu

Mahewu (magou) is traditionally made by mixing maize meal and water in a ratio of approximately 450 g maize to 3.8 litres of water (8–10 per cent solids), boiling for about 1.5 hours until the porridge is cooked, cooling, and adding a small amount of wheat flour or meal (about 5 per cent of the weight of maize meal). The wheat flour or meal serves as a source of inoculum and a source of growth factors for the subsequent spontaneous fermentation. The bacteria introduced into the wheat flour multiply rapidly in the maize porridge and convert part of the starch into lactic acid. The inoculated mahewu is incubated at 30–35°C for about 36 hours during which time the desirable sour flavour is developed (Schweigart *et al.*, 1960). Schweigart and Fellingham (1963) describe a method for producing mahewu under controlled isothermic conditions.

After wheat flour as a source of growth factors has been added to the maize porridge, the mixture is inoculated with either *Lactobacillus delbrückii* or *L. bulgaricus* and incubated at 45°C for rapid, uniform fermentation.

The predominant micro-organism in the spontaneous fermentation is believed to be *Streptococcus lactis*. Other acid-producing micro-organisms also develop, along with yeasts. The net result is that acids other than lactic acid are produced and traces of alcohol are also formed. The uncontrolled spontaneous fermentation, apart from being a long and irregular procedure, permits the development of undesirable bacteria that cause secondary fermentations and result in the production of acetic or butyric acids which are detrimental to the final flavour (Schweigart and Fellingham, 1963).

Research on the use of starter cultures, which produce a consistently high quality in the relatively short fermentation time of 8 to 12 hours, has shown that *Lactobacillus delbruckii* is the most suitable organism for producing lactic acid. At the recommended incubation temperature of 51°C, *L. delbrückii* grows rapidly. The growth of undesirable micro-organisms is inhibited at such a high temperature (Merwe *et al.*, 1964/65).

The most pronounced change in the substrate during mahewu fermentation is the production of lactic acid to a level of 0.4–0.5 per cent total acidity, with a pH of about 3.5. The protein content in mahewu (7–9 per cent) and its low food value can be improved by the addition of skimmed milk powder, whey protein, soya flour, yeast or fish flour.

Mahewu may be prepared in a more concentrated form by reducing its water content, or it can be dried to a powder (Merwe *et al.*, 1964/65). Spray drying and roller drying are reported to be the most appropriate drying methods. Using a Niro laboratory spray drier, Schweigart *et al.* (1960)

successfully dried mahewu of 8 per cent solids. The intake air temperature was 190–210°C and the exhaust air temperature was 90–100°C. The capacity was about 1.7 litre/hour. In experiments using a roller drier with a diameter of 304.8 mm and revolving at 1.2 rpm, mahewu of 8 per cent solids was also dried effectively. With steam pressure of 206.7 kN/m$^2$, the capacity of the drier was about 7 litre/hour.

During the spray drying process, where the temperature does not exceed 45°C, enzymatic changes such as fat-splitting can occur leading to rancidity of the product. Rancidity can be retarded by the inactivation of the enzymes through heat treatment of the liquid mahewu. Enzymatic action can also be inhibited by the use of a buffer, as buffered mahewus do not become rancid within one year of storage. Whey and phosphates serve as buffers. For small-scale production, roller drying is comparatively more cost-effective.

### *Uji*

A suspension of maize, millet, sorghum or cassava flour in water is known as akamu, or uji in East Africa. It is either fermented before or after cooking or is cooked in its fresh form to make a creamy soup. Uji is an important component

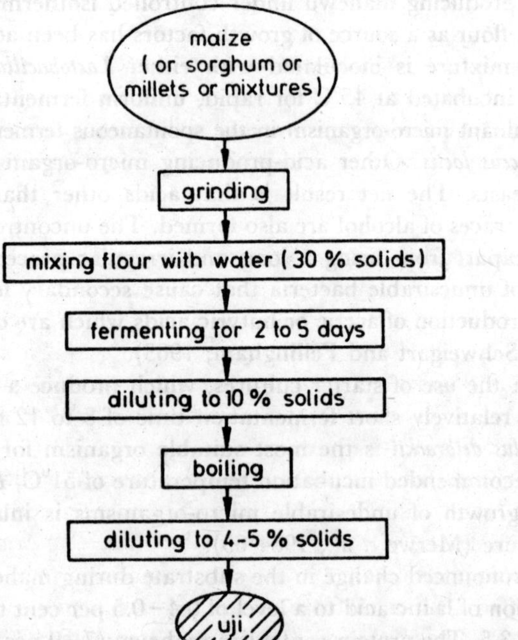


**Figure 6.12 Flowchart for the processing of uji.**

of both breakfast and lunchtime meals in East Africa. The basic cereal used for uji production is maize, although mixtures of maize, sorghum and millet are used too. Maize is ground and slurried with water. Without inoculation the slurry is left to ferment for 2–5 days at room temperature until the acidity reaches 0.3–0.5 per cent acid (as lactic acid). The fermented slurry is then diluted to about 8–10 per cent solids and brought to the boil (Figure 6.12). It is further diluted to 4–5 per cent solids, sweetened with sugar and consumed while still warm. The pH ranges from about 3.8 to 4.4.

The fermented millet, maize, or sorghum flour suspension may be sun-dried, and the granulated powder used as the need arises for subsequent uji preparation, by hydrating and then cooking.

In a typical spontaneous uji fermentation based on maize and sorghum flours, the coliforms dominate the fermentation for the first 16–24 hours. The lactobacilli then become dominant, producing sufficient acid to cause a decrease in the numbers of coliforms. Mbugua (1981) reported that *Lactobacillus plantarum* accounted for 72 per cent of the lactobacilli in fermenting uji.

## Other maize products

### Maize fritters

In Dahomey, maize fritters are made in the form of balls about the size of hen's eggs and deep fried in palm oil. Another product is made by kneading and rolling a mixture of maize flour, ground chillies, salt and water into small balls, which are then wrapped in banana leaves and steamed.

### Tortillas

Tortillas are a maize product, popular throughout Central America. Traditionally, tortillas are prepared by alkaline cooking of the maize kernels, the alkali of wood ashes or quicklime is used giving the product a characteristic flavour and unique physical texture. The alkali is washed out of the maize after the initial cooking by placing the maize in a pot or jar with perforated holes and washing with water.

The maize is ground on a metate to yield a dough. Following this, the dough or masa is made into bread by frying or baking on both sides on a hot griddle to give thin 'leathery' cakes known as tortillas. The protein content of a maize tortilla is about 9 per cent. When the tortilla is cooked and fried with olive oil and stuffed with a mixture of red peppers, onions, raisins, garlic, hard-boiled egg and grated cheese, it is called an enchilada (Inglett, 1970).

Another type of maize bread product is arepa which is popular in Columbia and Venezuela. A highly esteemed food in Central and South America is the tamale. A tamale is made from a piece of meat enclosed in a maize meal dough and seasoned with chilli peppers. It is generally wrapped in maize husks or in a leaf and then boiled or steamed.

## References


**Banigo, E.O.I.** and **Muller, H.G.** (1972b). Carboxylic acid patterns in ogi fermentation. *J. Sci. Food Agric.* **23**: 101–111.

**Bergmann, C.** *et al.* (1981). *Nutzungsmöglichkeiten moderner Erkenntnisse der Biotechnologie für Entwicklungsländer.* Forschungsberichte des BMZ, Band 18. Weltforum Verlag, Köln.

**Christian, W.F.K.** (1970). Lactic acid bacteria in fermenting maize dough. *Ghana Journ. Sci.* **10**: 22–28.

**FAO Production Yearbook** (1984). Vol. 38, Rome.

**Hoffman, L.A.** (1965). *Economic Geography.* Ronald Press Co, New York.

**Inglett, G.E.** (1970). *Corn: Culture, Processing, Products.* Westport, Connecticut, AVI Publishing Co. Inc.

**Jugenheimer, R.W.** (1976). *Corn improvement, seed production and uses.* Wiley-Interscience Publication, New York/London/Sydney. 216.

**Kent, N.L.** (1983). *Technology of cereals,* 3rd Edition, Pergamon Press, Oxford.

**Kunde, K.H.** (1979). Cornflakes, ein zeitgemasses und beliebtes Nahrungsmittel, in *Muhle und Mischfuttertechnick.* Heft 20, (Vol. 116).

**Matz, S.A.** (1970). *Cereal Technology.* Westport, Connecticut, AVI Publishing Comp. Inc.

**Mbugua, S.K.** (1981). *Microbiological and biochemical aspects of Uji fermentation and its enhancement through application of lactic acid bacteria.* Ph.D. thesis, Cornell University.

**Muller, H.G.** (1970). Traditional cereal processing in Nigeria and Ghana. *Ghana Journal Agric. Sci.* **3**: 187–195.

**Ofosu, A.** (1971). Changes in the levels of niacin and lysine during the traditional preparation of kenkey from maize grain. *Ghana J. Agric. Sci.* **4**: 153.

**Schweigart, F., van Bergen, W.E.L., Wiechers, S.G.** and **de Wit, J.P.** (1960). *The production of Mahewu.* South African Council for Sci. + Industrial Research, C.S.I.R. Research Report No. 167, Pretoria, S. Africa.

**Schweigart, F.** and **Fellingham, S.A.** (1963). *A study of fermentation in the production of mahewu, an indigenous sour maize beverage of Southern Africa.* Milchwissenschaft **18**: 241–246.

**Steinkraus, K., Cullen, R.E., Pederson, C.S., Nellis, L.F.** and **Gavitt, B.K.** (1983). *Handbook of indigenous fermented foods.* Marcel Dekker, Inc. New York and Basel.

**Ullmann's Enzyklopädie der technischen Chemie**. (1976). Bd. 11., 4. Auflage, Verlag Chemie Weinheim/BergstraBe.

**Ulloa, M., Herrera, T.** and **Taboada, J.** (1977). *Pozol, a fermented maize dough consumed in southeastern Mexico.* Symposium on Indigenous Fermented Foods, Bangkok, Thailand.

**Van der Merwe, A.le.R., Schweigart, F.** and **Cachia, V.A.** (1964/1965). (Sonderdruck). *Mahewu – its industrial production and its value as a nutrient.* Vitalstoffe Zivilisationskrankheiten, 7:

**Watson, S.A.** and **Hirata, Y.** (1962). Some wet milling properties of artificially dried corn. *Cereal Chem.* **39**: 35–43.

**Wellhausen, E.J., Alejandra, F.A., Corzo, A.H.** and **Mangelsdorf, P.C.** (1957). *Races of Maize in Central America.* National Academy of Sciences, 1957, 511.

7

# Oil palm

## Classification and production

The oil palm, *Elaeis guineensis* Jacq., is a monocotyledonous plant (Figure 7.1). Within the species *E. guineensis*, different varieties and types can be distinguished that show definite characteristics with respect to the quantity and quality of the extractable oil. The simple classification of the oil palm fruit is based on its internal structure, especially the thickness of its shell, and the fruit form may be described as belonging to one of the following groups:

1. *Dura*. Shell (endocarp) 2–8 mm thick, comprising 20–25 per cent weight of fruit; medium mesocarp content of 35–55 per cent mass, but up to 65 per cent in the deli palms.
2. *Tenera*. Shell thin, 0.5–4 mm thick, medium to high mesocarp content (60–95 per cent).
3. *Pisifera*. Shell-less. It is of little commercial value. Its importance is found in the breeding of commercial palms.

The oil palm is largely cultivated in the equatorial regions of Africa, South-east Asia and America (Purseglove, 1975). Of all oil-bearing plants, oil palm is the highest yielding. In the humid regions of West Africa, yields amount to about 4.5 tonnes of oil per hectare per annum. Global production data for palm oil indicate that in 1985 slightly in excess of 7.6 million tonnes were produced; major contributing countries are Malaysia, the Philippines and Nigeria. Palm kernel production figures for the same year reached 2.6 million tonnes (FAO, 1986).

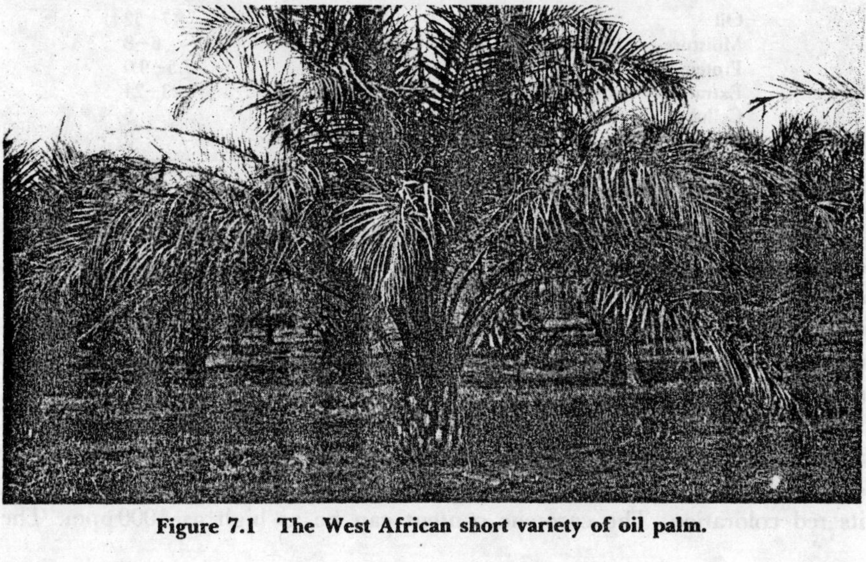

**Figure 7.1 The West African short variety of oil palm.**

## *Uses of palm oil*

The fruit of the oil palm is a drupe, the outer pulpy layer of which provides the palm oil. Within the pulp or mesocarp lies a hard-shelled nut containing the palm kernel (endocarp), which gives two useful products, palm kernel oil and the residue, palm kernel cake. Palm oil is an important part of the diet in West Africa. In addition to supplying cooking fat, it is a rich source of β carotene, a precursor of vitamin A. Palm oil, as well as palm kernel oil, is used in the manufacture of soap.

The ash obtained by burning the refuse which remains after stripping the fruit from the bunches is rich in potash, and is used as one of the ingredients in the making of soap in rural areas. In many regions, sap from the palm tree itself is tapped for making wine; apart from having a commercial value more than double that of the oil and kernels from similar palms, palm wine is an important source of B group vitamins.

In industrialized countries the main uses of palm oil and palm kernel oil are for shortening, margarine type products, cooking oil and salad oil. They are also used in the manufacture of grease, soap, candles, lubricants, cosmetics and chemical cleaning products (detergents). Palm kernel cake, the cake that remains after the extraction of oil from the kernels, is used as a livestock feed.

## Composition of the palm fruit, palm oil and palm kernel oil

The palm fruit is composed of a thick fibrous layer of pulp on the outside and a hard shelled kernel. It is rich in oil and extractable non-nitrogenous substances (Table 7.1).

**Table 7.1 Composition of palm fruits**

| Constituent | Composition (%) |
|---|---|
| Oil | 47–52 |
| Moisture | 6–8 |
| Protein | 7.5–9.0 |
| Extractable non-nitrogenous substances | 23–24 |
| Cellulose | 5 |
| Ash | 2 |

Source: Hartley, 1977.

Palm oil, which is derived from the outer pulpy layer of the palm fruit, is composed mainly of palmitic acid (a saturated fatty acid) and oleic acid (an unsaturated fatty acid). Both acids are roughly in equal proportions. Generally, on harvesting, fats in the mesocarp constitute 70–75 per cent of the dry matter. The percentage composition of the fatty acids of palm oil and their melting points are shown in Table 7.2.

In addition, there is about 1 per cent palmitoleic acid, and traces of lauric and linolenic acids. Apart from the fatty acids, the major constituents of palm oil are the fat soluble carotenoids, chiefly carotene, which give the palm oil its red coloration. The carotene content can be as high as 1000 ppm. The

**Table 7.2 The average fatty acid composition of palm oil**

| Fatty acid | Composition (%) | Melting point (°C) |
|---|---|---|
| **Saturated** | | |
| Stearic acid ($C_{18}$) | 4.3 | 69 |
| Palmitic acid ($C_{16}$) | 41.2 | 64 |
| Myristic acid ($C_{14}$) | 2.3 | 58 |
| **Unsaturated** | | |
| Linoleic acid | 9.6 | –5 |
| Oleic acid | 42.5 | 14 |

Source: Hartley, 1977; Ullmanns Enzyklopaedie, 1976.

carotenoids are responsible for the high amount of vitamin A in palm oil.

Two main products are obtained from palm kernels besides the hard shells. These are palm kernel oil and palm kernel cake. The characteristic feature of palm kernel oil is its high content of saturated fatty acids, primarily lauric acid. Due to this high proportion of saturated fatty acids, palm kernel oil does not readily turn rancid. The percentage composition of the fatty acids and their melting points are listed in Table 7.3.

**Table 7.3 The average fatty acid composition of palm kernel oil**

| Fatty acid | Composition (%) | Melting point (°C) |
|---|---|---|
| **Saturated** | | |
| Caprylic acid ($C_8$) | 3.0 | 16.5 |
| Capric acid ($C_{10}$) | 6.0 | 31 |
| Lauric acid ($C_{12}$) | 50.0 | 48 |
| Myristic acid ($C_{14}$) | 16.0 | 58 |
| Palmitic acid ($C_{16}$) | 6.5 | 64 |
| Stearic acid ($C_{18}$) | 1.0 | 69 |
| **Unsaturated** | | |
| Oleic acid | 16.5 | 14 |
| Linoleic acid | 1.0 | -5 |

Source: Hartley, 1977; Ullmanns Enzyklopaedie, 1976.

Palm kernel cake is obtained after extracting oil from the palm kernels, which have previously been freed from their hard outer shells. The composition of the cake varies with the type or variety of oil palm and the method and efficiency of palm kernel oil extraction. The average composition of palm kernel cake is found in Table 7.4.

**Table 7.4 Average composition of palm kernel cake**

| Constituent | Composition (%) |
|---|---|
| Carbohydrate | 48 |
| Protein | 19 |
| Fibre | 13 |
| Water | 11 |
| Oil | 5 |
| Ash | 4 |

Source: Hartley, 1977.

### *Characteristics of palm oil and palm kernel oil*

Palm oil and palm kernel oil, like other vegetable oils, are mixtures of a number of different gylcerides. For the identification of the characteristics or properties of different oils or fats, a number of constants have been established. These constants are used as a guide in the analysis and testing of oils and fats as well as in the assessment of their quality and purity. Those most commonly used to establish identity are melting and solidification temperatures, iodine value, saponification value, refractive index, viscosity, density, and the Reichert, Polenske and Kirschner values. Other data sometimes determined to assess quality include the free fatty acid content, peroxide value, moisture content and content of unsaponifiable matter.

Given the fact that oils and fats are natural products and as such are subject to a certain degree of variation in composition, most of their physical and chemical properties fall within a range. The smoke, flame and fire points of palm oil containing 0.006% free fatty acid are 223°C, 314°C and 341°C, respectively. Table 7.5 lists some of the properties of palm oil and palm kernel oil.

**Table 7.5 Properties of palm oil and palm kernel oil**

| Property | Palm oil | Palm kernel oil |
|---|---|---|
| Saponification value | 195–205 | 242–254 |
| Iodine value | 44–58 | 14–20 |
| Unsaponifiable matter | <0.5% | 0.2–0.8% |
| Solidifying temperature | 30°–37°C | 23°–30°C |
| Refractive index | 1.453–1.456 | 1.449–1.452 |
| Carotene content | 0.05–0.2 | – |
| Viscosity at 50°C (mPas) | 24.81 | 17.73 |
| Density at 15°C (g/cm$^3$) | 0.921–0.947 | 0.925–0.935 |

Source: Hartley, 1977; Ullmanns Enzyklopaedie, 1976.

### *Properties of oils and fats*

An assessment of the types and quantities of the various fatty acids present in oils or fats is closely related to the composition and so the identity of the fat or oil being investigated.

Melting and solidification temperatures of fats are difficult to measure in that, depending on their thermal history, triglycerides exhibit variations in crystal structure (polymorphism) and therefore in melting point. Also, since fats are a complex of triglycerides, definite melting and solidification points become more complicated to measure. For these reasons two common measurements are made to give a fair indication of melting and solidification temperatures of fats. These are the slip point, which measures the temperature at which the fat sample will move or slip in a capillary tube when heated gently in a water bath, and the titre, which is the solidifying point of the fatty acids prepared from the fat. In reality the titre is not a characteristic of a fat itself.

Determination of the impurities present in fats is in principle the measurement of solid matter which is not soluble in fat or fat solvents. The accuracy of measurement will depend on the amount of material which may be dispersed in the fat but is insoluble in fat solvents, and according to the degree of dilution, may not all be measured.

The refractive index is a physical characteristic of triglycerides determined by the angle through which a beam of light is refracted or bent when passing through a thin film of melted fat. The refractive index is usually used in testing purity or components of a mixture. It is temperature dependent and is normally measured at 104°C, a temperature at which most fats are liquid.

Unsaponifiable matter is a measurement of the water-insoluble components produced after heating the fat under investigation with potassium hydroxide. In this reaction the triglycerides are broken down to form glycerol and potassium salts of the fatty acids. Glycerol and potassium salts are water-soluble, and the water-insoluble hydrocarbons, sterols and fatty alcohols are extracted from the solution and their amount can be measured. High values of unsaponifiable matter may indicate contamination and adulteration of the fat. The measurement of unsaponifiable matter also allows a check to be made for the presence of fuel oil or lubricating oil in the fat.

Saponification value is used as a measure of the proportion of the fatty acids present in the fat by calculating the ratio of their mean molecular weights. The process of saponification entails the hydrolysis of triglycerides into glycerol and then the addition of potassium hydroxide. This process measures the amount of alkali which is required to combine with the fatty acids liberated by the hydrolysis of the fats, and from this the equivalent weight and molecular weight of the fatty acids can be determined. For the purpose of characterizing the fats in terms of a constant the actual molecular weight is not calculated, the saponification value being expressed as the number of mg of potassium hydroxide required to saponify 1 g of fat. Due to the fact that each fat has, within the limits of biological variation, a constant fatty acid composition, determination of the saponification value is a reasonably reliable means of characterizing the fat.

The iodine value is a measure of unsaturation, that is the proportion of unsaturated acids present in a fat or oil. Oils and fats do not contain iodine, but the test measures the amount of iodine which can be absorbed by the unsaturated acids. Unsaturated organic compounds have the ability to form addition compounds with halogens at the double bond. As the concentration and types of unsaturated acids present in a fat or oil is fairly constant, the iodine value will indicate the degree of unsaturation, expressed as the percentage of iodine absorbed by the fat. The iodine value in this way is a measure of fat stability and resistance to oxidation.

Some of the deterioration that occurs during storage of either the raw material from which the fat or oil is produced, or in the fat itself after extraction, leads to hydrolysis of triglycerides to yield free fatty acids. Acid value is a measure of the amount of free fatty acids present in an oil or fat.

The peroxide value is usually used as an indicator of deterioration of fats or oils. As oxidation takes place, the double bonds in the unsaturated fatty acids are attacked, forming peroxides. These decompose to produce secondary oxidation products which cause rancidity. The peroxide value is not a complete measure of oxidation since the compounds formed are unstable and oxidation proceeds further. However, it may be used to estimate oxidation.

Reichert, Polenske and Kirschner values are all determined at the same time. They measure the amount of steam-volatile fatty acids which can be recovered from the fats or oils under standard conditions. Low molecular weight acids are steam-volatile under test conditions, and are separated from the higher molecular weight acids. The acids concerned are lauric, capric and caprylic, which are water-insoluble; and caproic and butyric, which are water-soluble. The Reichert value measures the water-soluble acids, the Polenske value the water-insoluble acids and the Kirschner value the butyric acid, by separation from capric after precipitating the latter as a silver salt. The significance of these values lies in the testing of butter, coconut oil and palm kernel oil for adulteration, or in the examination of mixtures containing these oils.

## Processing of palm fruit

The most important part of the oil palm from the oil processor's point of view

Figure 7.2 Oil palm fruits waiting to be processed.

is the bunch of palm fruits. The bunch is a large ovate fruit cluster, which contains many drupes varying in colour from deep orange to a dark red-brown (Figure 7.2). When in full bearing, a palm may produce up to 12 bunches of fruits per year; the average weight of a bunch is 13–18 kg. The unit operations involved in the processing of palm fruits into palm oil and kernels are sterilization, threshing or stripping, milling and digestion, pressing, and clarification. The essential steps which are involved in the recovery of the kernels are depericarping, drying, cracking, separation and grading (Figure 7.3).

## Sterilization

Sterilization of palm fruits or bunches before oil extraction is achieved by boiling. The purposes of sterilization are to make stripping (separation of fruits from the bunches) easy, to disinfect the fruits by killing pathogens, and to inhibit the action of lipolytic enzymes thus avoiding formation of free fatty acids. Sterilization may be carried out in drums or in sterilization chambers.

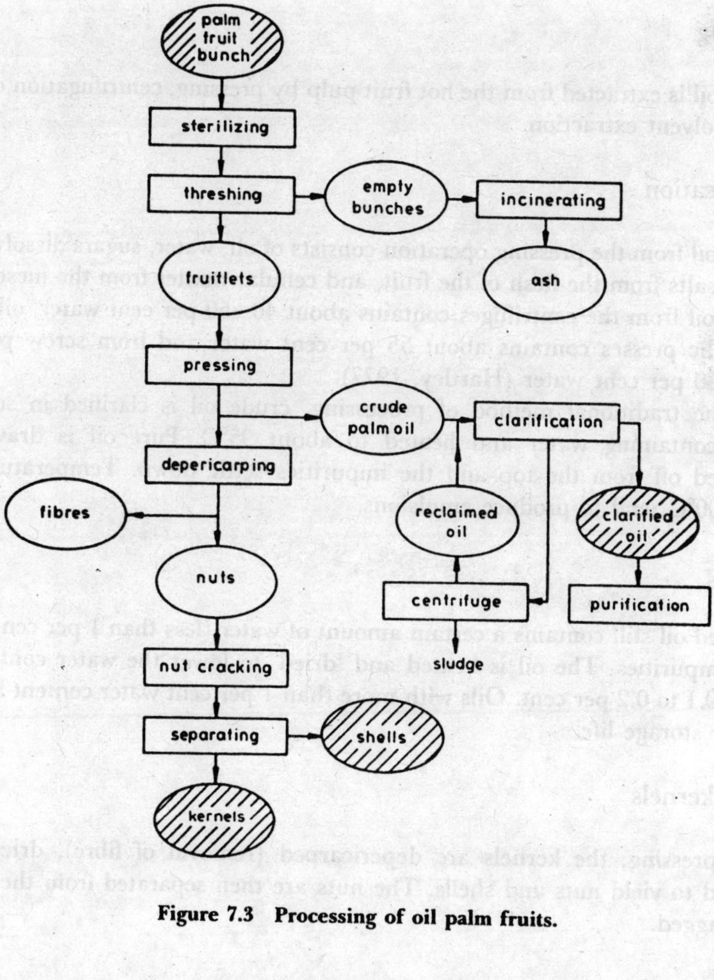


**Figure 7.3 Processing of oil palm fruits.**

### Threshing or stripping

Individual fruits are removed from the sterilized bunch. The fruits may be resterilized for about 40 minutes before digestion.

### Digestion

The purpose of digestion, as with pounding, is to crush the fruits physically to yield a hot homogeneous pulp. Digestion facilitates the liberation of oil from the cells of the mesocarp.

Modern digesters are steam-jacketed cylindrical vessels with a vertical rotating shaft in the centre, driven from the top. Attached to the shaft are pairs of stirring arms with which the fruits are mashed. The temperature in digesters must be maintained at around 95°C. It is important that the temperature does not rise to 100°C because the water in the mash must not be allowed to boil as this would cause turbulence in the digester and boiling water would spill over the top.

### Pressing

Crude oil is extracted from the hot fruit pulp by pressing, centrifugation or hot water solvent extraction.

### Clarification

Crude oil from the pressing operation consists of oil, water, sugars dissolved in water, salts from the flesh of the fruit, and cellular matter from the mesocarp. Crude oil from the centrifuges contains about 40–50 per cent water, oil from hydraulic presses contains about 55 per cent water and from screw presses about 60 per cent water (Hartley, 1977).

In the traditional method of processing, crude oil is clarified in settling tanks containing water and heated to about 95°C. Pure oil is drawn or skimmed off from the top and the impurities settle down. Temperatures of over 100°C tend to produce emulsions.

### Drying

Clarified oil still contains a certain amount of water (less than 1 per cent) and solid impurities. The oil is heated and 'dried' to lower the water content to about 0.1 to 0.2 per cent. Oils with more than 1 per cent water content have a shorter storage life.

### Palm kernels

After pressing, the kernels are depericarped (removal of fibre), dried and cracked to yield nuts and shells. The nuts are then separated from the shells and bagged.

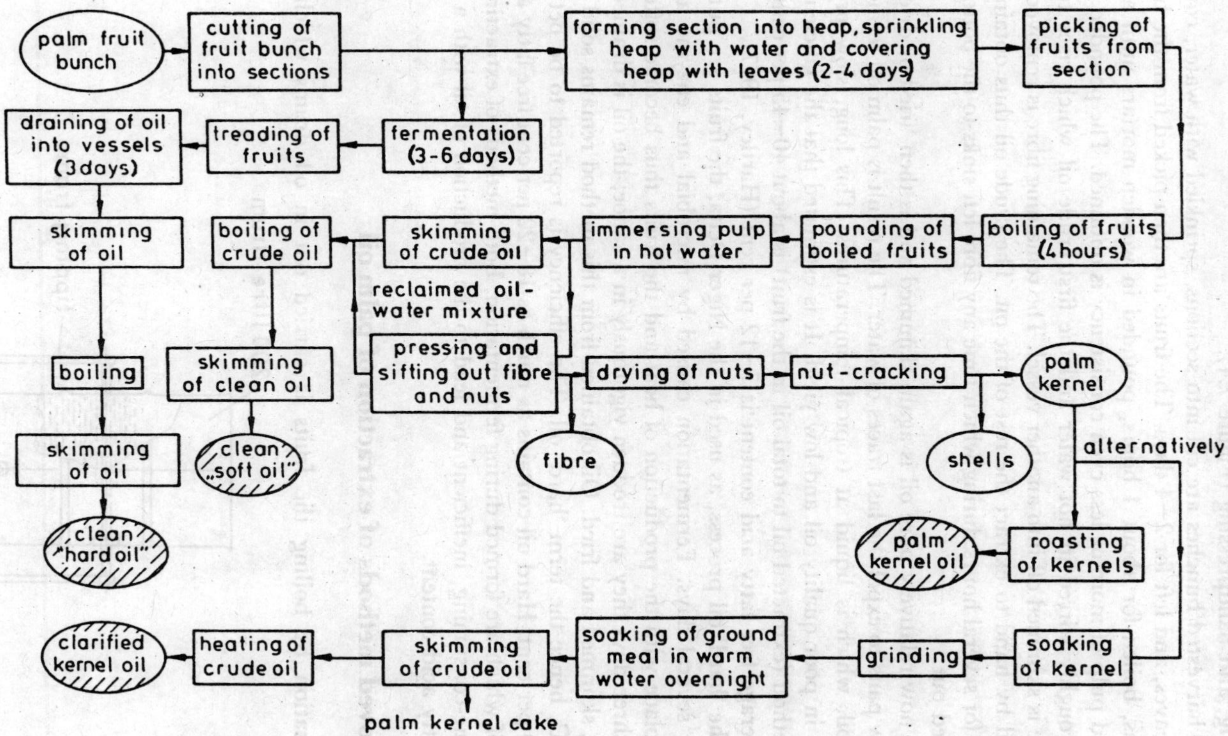


Figure 7.4 Flowchart for traditional processing of palm fruits.

## Traditional methods of extracting palm oil

Indigenous methods of extraction consist essentially of boiling, pounding or treading, and handpressing (Figure 7.4).

The harvested bunches are cut into sections, sprinkled with water, covered with leaves, and left for 2–4 days. The fruits are then picked from the bunch sections, boiled for about 4 hours, pounded in wooden mortars until a hot, crushed pulp of more or less even consistency is obtained. The pounded mass is thoroughly stirred in hot water and the first crude oil which rises to the surface is skimmed off into another vessel. The remaining fibre is removed and pressed by hand to extract the rest of the oil. The crude oil thus obtained is boiled for several hours during which time any fibre left sinks to the bottom or is sieved out.

The now relatively pure oil is again skimmed and is then 'fried or dried' in shallow pans to expel the last traces of water. The result is palm oil termed as 'soft' oil, which is liquid at tropical temperatures. This long, slow process results in poor quality oil and low yield. It is estimated that the efficiency of the method (extracted oil to total oil in the fruit) is about 40–45 per cent and the average free fatty acid content is 7–12 per cent (Hartley, 1977).

In the 'hard' oil process, as used in the Niger Delta, the fruits are put in a pit for several days. Fermentation caused by microbial and enzyme action takes place with the production of heat and the fruits thus become softened. After three days they are trodden vigorously in a canoe; the oil is drained off, boiled, skimmed and fried. Oil obtained from this method remains solid even at 32°C, hence the term 'hard' oil. The efficiency is reported to be between 20–30 per cent. Hard oil contains as much as 18–22 per cent free fatty acids, most of which are formed during fermentation. Both methods of extracting oil are time-consuming, inefficient and lead to the production of oil with a high free fatty acid content.

## Improved methods of extraction of palm oil

Sterilization by boiling the fruits is carried out in oil drums which are

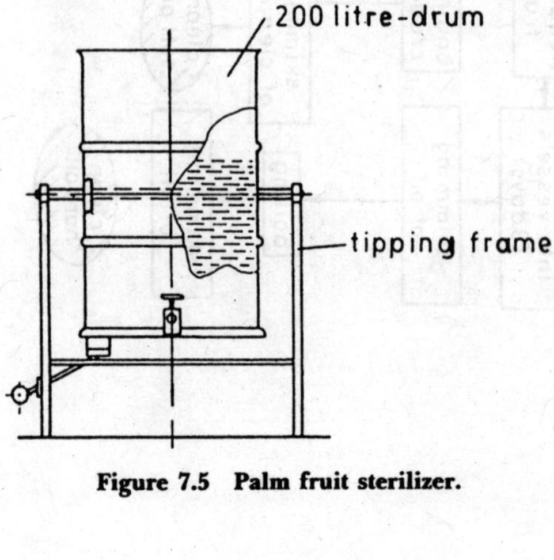


**Figure 7.5 Palm fruit sterilizer.**

welded together. As a source of heat, firewood is usually used. Boiling is continued for an hour or two after steam appears at the top of the drum. A fruit sterilizer of this type is designed to hold about 1 tonne of quartered bunches (Figure 7.5).

Stripping is done manually by beating the bunches to detach the fruits using wooden batons or pronged forks, or by the use of a hand-operated octagonal slatted drum (Figure 7.6).

Wood and Koo Pai Sing (1977) describe a process for loosening the fruit from the bunch prior to stripping by exposing the bunch to a gaseous mixture of air and an unsaturated aliphatic hydrocarbon, such as acetylene, ethylene, propylene or butylene, or a chemical derivative of these hydrocarbons. It is claimed that, by loosening the fruit in this way, much more limited treatment, such as heating to 80°C for 30 minutes, is sufficient for subsequent sterilization.

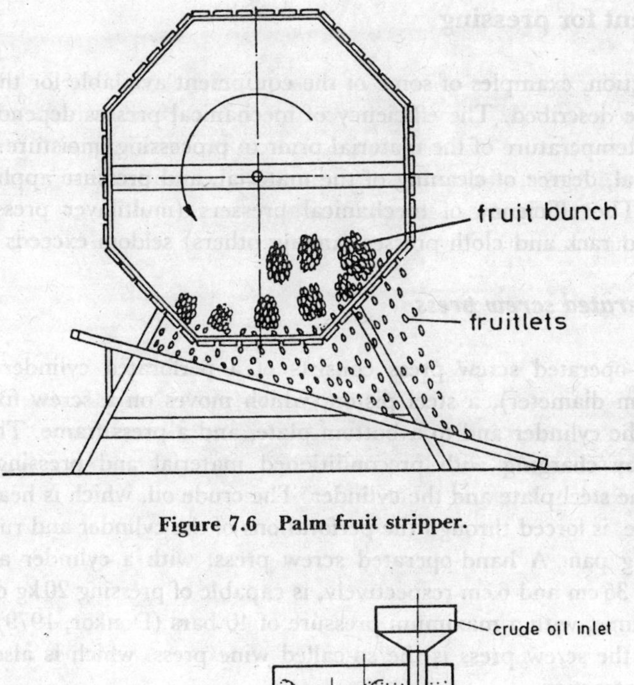


**Figure 7.6 Palm fruit stripper.**

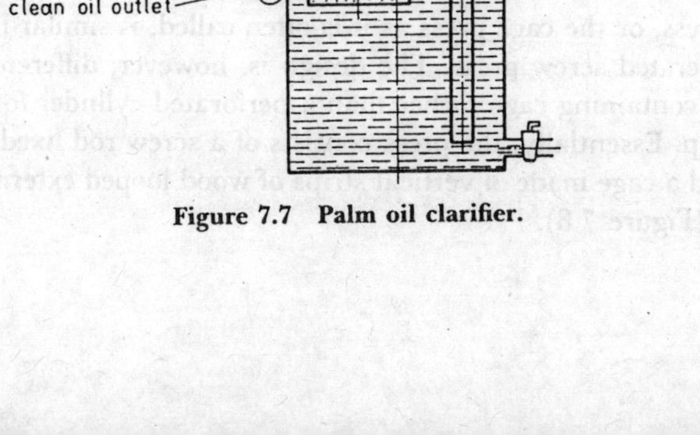


**Figure 7.7 Palm oil clarifier.**

Digestion is carried out traditionally by pounding stripped fruit in a large concrete or wooden mortar, 90 cm in diameter and 50 cm deep. The pounded material is then put in the so-called 'pounded fruit reheating drums', similar to the bunch sterilizer, which maintains the heat of the pulp till the time of pressing. Crude oil from the pressing operation is clarified in a simple clarifier consisting of a drum, a funnel and a tube (Figure 7.7).

A large drum measuring 90 × 100 cm is fitted with a 200 litre drum as an inner compartment. A funnel of about 25 cm diameter attached to a tube of 5 cm forms a feed pipe and extends downwards near the side of the large drum, to within 7.5 cm of the bottom. The inner 200 litre drum is supported in the centre of the larger drum by bars or an attachment to the feed pipe. The drum is initially one-quarter filled with water and brought to the boil. The oil-water mixture from the press is next poured through the funnel to mix with boiling water. The clean oil rises through the water and eventually overflows into the inner drum.

## Equipment for pressing

In this section, examples of some of the equipment available for the pressing process are described. The efficiency of mechanical presses depends on such factors as temperature of the material prior to processing, moisture content of the material, degree of cleaning of the material, and pressure applied during pressing. The efficiency of mechanical pressers (multilayer presses, screw presses and rack and cloth presses, among others) seldom exceeds 90%.

### *Hand-operated screw press*

The hand-operated screw press consists of a perforated cylinder (holes of about 5 mm diameter), a steel piston, which moves on a screw fixed at the centre of the cylinder and to a bottom plate, and a press frame. The press is operated by charging with preconditioned material and pressing it down between the steel plate and the cylinder. The crude oil, which is heavily laden with sludge, is forced through the perforations of the cylinder and runs off into a collecting pan. A hand-operated screw press, with a cylinder and piston measuring 35 cm and 6 cm respectively, is capable of pressing 20 kg of oil palm pulp at a time, with a maximum pressure of 40 bars (Donkor, 1979). Another version of the screw press is the so-called wine press, which is also used for palm oil extraction.

### *Wine press*

The wine press, or the cage press as it is often called, is similar in principle to the hand-operated screw press. The design is, however, different to allow for the use of a containing cage rather than a perforated cylinder for the digested oil palm pulp. Essentially, the press consists of a screw rod fixed in the centre of a plug and a cage made of vertical strips of wood looped externally with two iron bands (Figure 7.8).

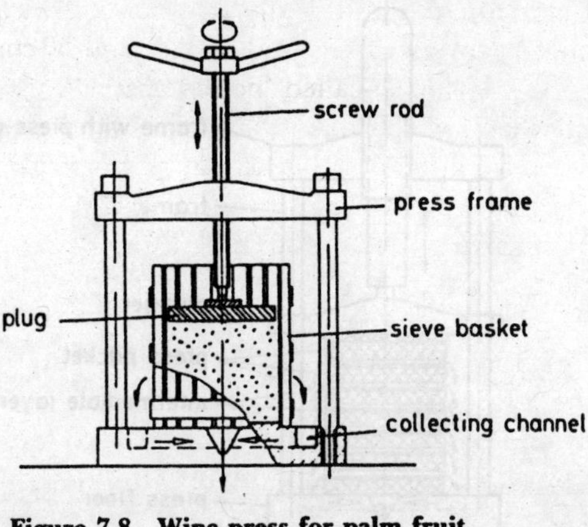


**Figure 7.8 Wine press for palm fruit.**

Pressure is applied by a force plug fixed to the screw rod. This is worked downwards by a crosshead turned manually. The oil is pressed out of the digested pulp between the strips, and into the vessel through the collecting channel. The cage can be opened in two halves and lifted off so that the pressed residue can be discharged and a fresh charge can be put in. An average charge of the press is about 70 kg fruit. Given the proper pretreatment of the oil palm fruits an extraction efficiency of about 60 per cent may be realized with the wine press (Hartley, 1977).

### *Hydraulic hand press*

The main components of this press are a hand-operated pump, a ram, which is the cylinder of the hydraulic mechanism and a perforated press cage. The press ram is moved downwards into the perforated cage filled with digested palm fruit when the hydraulic fluid pressure is increased by the hand operation of a two-piston pump. Pumping can be effected at the rate of 20–25 strokes per minute and must be continued till the required pressure has been reached, in which case the relief valve will open. The press can be operated by one man, and may be provided with two perforated press cages. During operation, one of these will be under the ram while the other is being filled or discharged. With two cages it is possible to complete 6–10 pressings per hour and thus to press 280–450 kg fruit.

The hand-operated hydraulic press is reported to be a suitable press for use in the improved rural method for processing palm oil. Under rural conditions an average extraction efficiency of 85 per cent is obtained (Hartley, 1977).

### *Multilayer press*

For the extraction of larger quantities of palm oil, the hydraulically operated multilayer press or *plate press* may be conveniently used (Figure 7.9). Using a square box the material to be pressed is put in filter cloths so that flat press packets are formed. About 10–25 such packets are piled between the press

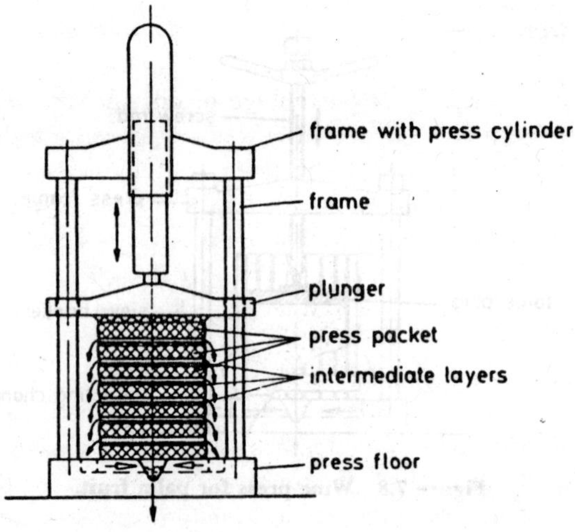


**Figure 7.9 Multilayer press (Ulrich, 1967).**

floor and the plunger. Between the individual packets, flat wooden boards are placed as intermediate layers to improve the drainage of the oil. During the pressing operation the plunger moves down and the entire assembly of packets is compressed.

The distance between the individual packets is about 10–15 cm and the pressure on the pulp is about 120 bars with the pressing plunger operating at pressures of 280–320 bars. The average time of the press cycle ranges from 30–36 minutes. The capacity, depending on the size and pretreatment of the raw material, is 2–10 tonnes per 24 hours. The power requirement is about 10 kW (Thieme, 1968).

## *Screw expeller*

For higher capacities, a power-driven screw expeller may be employed (Figure 7.10). The oil palm pulp is fed into a horizontal worm shaft via a feeding hole and pressed in the tapering space between the cylinder housing and the individual worm elements of the shaft. The oil comes out from the perforated housing and is collected in a trough set below the housing. This type of press is used over a capacity range from 30 kg/hour with 3.2 kW drive power up to 2500 kg/hour with 44 kW (Rossmann, 1979). The major disadvantage is the relatively rapid wear of the worm shaft and the cylinder.

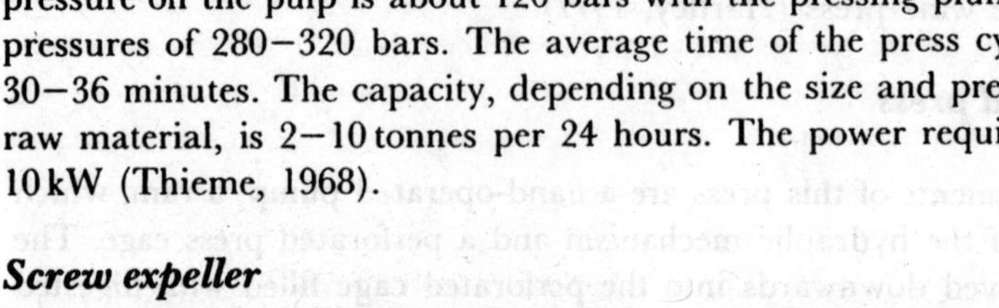


**Figure 7.10 Screw expeller (Ulrich, 1967).**

### *Solvent extraction*

For raw materials with a high percentage of oil, such as oil palm, solvent extraction is less appropriate. Solvent extraction presents technical problems in that optimum conditioning of the palm fruit mesocarp prior to solvent extraction is difficult to achieve at a reasonable cost. Other drawbacks of solvent extraction for small to medium-scale processors are the dependance on an outside supply of the solvent and the relatively high capital cost of solvent systems compared to mechanical extraction systems.

### *Processing in modern palm oil mills*

The technology of palm oil production has advanced in the last two decades or so, notably in Malaysia and South America. New improved technologies are being introduced in the processing chain to minimize losses, achieve optimum process efficiency and produce palm oil and palm kernel oil of good quality (Figure 7.11). The production processes in a modern mill consist of steam and power generation, sterilization, mechanical stripping, fruit digestion, oil extraction and oil clarification. Steam is used extensively in oil mills, and is generated on site using the oil palm by-products, fibre and shell, as fuel. Most palm oil mills generate all or at least part of their power by utilizing steam turbines.

Sterilization is carried out in a pressure vessel where the fruit bunch is subjected to steam of sufficient pressure for about 70 minutes or more. This facilitates softening of the pericarp and increases oil recovery during digestion

**Figure 7.11 A modern palm oil factory, Rivers State, Nigeria.**

and pressing. Separation of the individual fruits from the bunches is done mechanically. The mechanical stripper, as the equipment is called, is a rotating drum with perforations on the periphery just wide enough for the stripped fruits to pass through. The 'empty' bunch exits from the open end of the drum onto a conveyor that passes it to the incinerator. In the incinerator the 'empty' bunch is burnt to produce ash which may be used as fertilizer or as an ingredient in soap-making due to its high potash content.

The separated fruit is conveyed into the digester, a cylindrical vessel with a steam jacket. It is open at the top and is connected at the bottom to the press via a chute. The aim of fruit digestion is to break up the fruits physically by means of beater arms and to raise the temperature of the digested pulp to about 90°C. This facilitates easy extraction of oil in the presses. From the digester the hot, mashed pulp is fed into the press where oil extraction occurs.

Various types of press may be used. In the modern oil mill the double screw expeller is mostly used. It consists of a perforated cage, two helical screws, two cones, a hydraulic pump and a drive mechanism. The two screws are located inside the cage and revolve in opposite directions. The digested mash is pressed by the screws, while the cones operated by the pump provide the necessary pressure in the opposite direction. In this manner the oil-water mixture is squeezed out through the perforations in the cage.

The crude oil from the press or expeller is fed into a clarification tank where steam is added to raise the temperature and reduce the viscosity for better separation. Separation of oil from water and sludge occurs faster when the temperature of the crude oil is raised to about 95°C and as oil has a lower density it rises to the top of the tank. The top oil is recovered and then passed through a heater to raise the temperature for the purifier. In the purifier the remaining dirt is removed. The clean oil is then fed into a drier to reduce the moisture content. From the drier the oil is pumped to the main storage tank.

## Extraction of palm kernel oil

### *Traditional method*

The main residue from palm oil extraction consists of the palm kernel nuts which contain the kernels. Traditionally, the nuts are sun-dried and then cracked by hand one by one on a stone to free the kernels from the hard shells. After this they are roasted in shallow pans. In the roasting process a dark brown oil emerges from the kernels. Extraction of oil is completed when the kernels have turned quite black.

Another method of palm kernel oil extraction is to soak the kernels after they have been freed from their hard shells, and then crush or grind them finely, and soak the ground material in warm water overnight. The oil which rises to the surface is skimmed off the following morning. This crude oil is clarified by heating and the water vapour is driven off. Nutcrackers driven by small engines are used in small-scale mills. Engines of around 2.2 kW are required to drive the cracker at between 2000 and 2500 rpm and about 0.75 tonnes of nuts may be cracked per hour (Hartley, 1977).

### *Mechanized extraction of palm kernel oil*

The mechanized process of palm kernel oil extraction consists of the following stages:

1. Separation of the fibre from the nuts.
2. Nut cracking.
3. Kernel and shell separation.
4. Extraction of the oil.

Separation of fibre from the nuts is commonly known as *depericarping* and the machines used are called *depericarpers*. In the most commonly used type of fibre separator, the fibre-nut mixture is put into a drum rotating at about 15 rpm. Inside the drum are baffles which carry the mixture upwards, and allow it to fall. The current of air passing through the drum is sufficient to blow away the partially dried fibre.

Another type of mechanical fibre separator consists of a screened drum which is rotated and allows the separated fibre to fall through the screen.

The so-called *cage* type has a rotary cage bounded by rollers revolving in opposite directions in pairs, which remove the fibre to the outside of the cage but retain the nuts inside.

*Roller* or *centrifugal crackers* are used in modern mills. The nuts are fed into the cracker, thrown out of slots on the face of a shaft rotating at high speed and hurled against a cracking ring. The crackers with horizontal shafts are reported to be suitable for large dura or deli nuts whilst tenera nuts are best cracked using crackers with vertical shafts. The speed of nut crackers varies from 800–2500 rpm according to the diameter of the rotor.

Kernels can be separated from shells in three different ways. In dry separation the shells are blown away by an upward current of air. The principle of wet separation is based on the different density of the kernel (1.07 kg/m$^3$) and of the shell (1.30–1.35 kg/m$^3$). A mixture of water and clay to give a density of 1.17 kg/m$^3$ is suitable for wet separation. The shells being heavier sink to the bottom and the floating kernels are skimmed off, washed and dried. Clay bath separators commonly in use have capacities of 2–4 tonnes of nuts per hour.

The advent of the *hydrocyclone* has reduced the use of clay baths. In the hydrocyclone a fast-moving water stream within a cylinder-like vessel effects the separation of the lighter kernels from the heavier shells. After this separation, the kernels are dried (to about 5–7 per cent water content) on trays passing over hot air, and then put into a worm-type press for the extraction of oil. Sometimes roller mills are first used to grind the material, after which the ground material is put into hydraulic presses for oil extraction.

## The quality of palm oil and kernels

The objectives of oil palm processing are not only to extract the maximum quantity of oil but also to produce palm oil and kernels of high quality and good storability. The factors that largely determine quality and storability are the content of free fatty acids, contamination of the oil with water and other impurities and bleachability. The presence of water and other foreign contaminants encourages high rates of free fatty acid formation, through the activities of lipolytic enzymes (lipases).

## Free fatty acids

The fatty acids in palm oil may be formed by autocatalytic action, by the action of the lipolytic enzyme lipase from the palm fruit, or by microbial lipases. The formation of free fatty acids is mainly caused by the action of lipase before processing. Palm fruits contain active lipase which can only effect the breakdown of fats into fatty acids and glycerol when the cells of the fruit have been damaged. The fat is protected from the lipase in the fruit by the membranes of the vacuoles which can be ruptured either mechanically or by low temperature. The enzyme is completely inactivated at high temperatures, that is temperatures above 55°C. The rate of free fatty acid formation in palm fruits following crushing is shown in Figure 7.12. It can be seen that, with crushed fruit the speed of fatty acid formation is fast, and the fat splitting process, as indicated by the free fatty acid content, reaches its peak in 30–60 minutes.

Loncin and Jacobsberg (1963) show that, after destruction of enzymes, oil in storage can deteriorate through autocatalytic hydrolysis. The fatty acids, already present in small quantities, act as catalysts in the reaction between the triglycerides and water. Figure 7.13 shows the spontaneous hydrolysis of water-saturated palm oil at 70°C, yielding di- and monoglycerides and glycerol.

For the production of oil low in free fatty acids, rupturing of the fruits ought to be avoided, and to provide an oil which will maintain a low free fatty acid content during storage the final processing ought to be geared primarily to reducing the water content to less than 0.1 per cent, since below this value little autocatalytic hydrolysis is likely to occur.

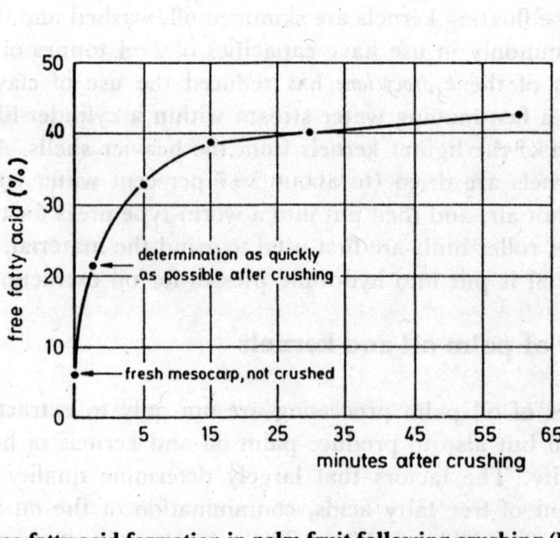


**Figure 7.12 Free fatty acid formation in palm fruit following crushing (Desassis, 1955).**

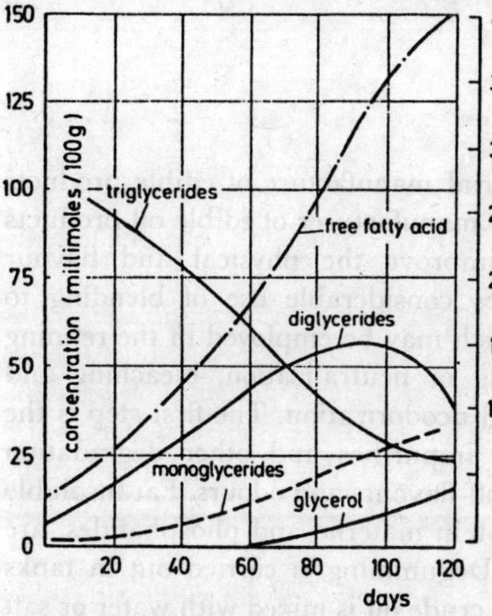


**Figure 7.13 Spontaneous hydrolysis of water-unsaturated palm oil at 70°C (Loncin and Jacobsberg, 1965).**

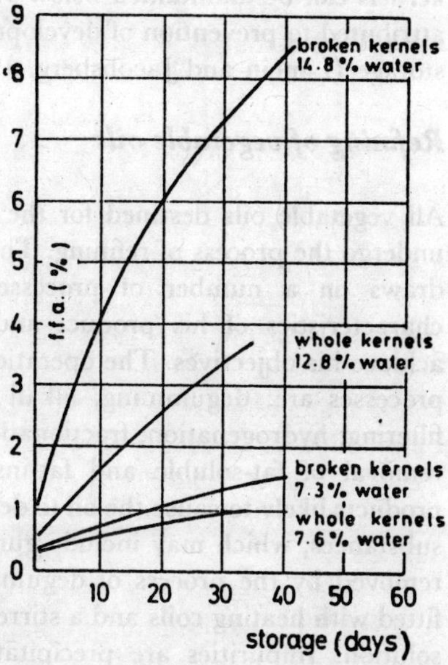


**Figure 7.14 Free fatty acid formation in broken and whole palm kernels at various moisture contents (Stock Oil Review, 1963).**

## Bleachability of palm oil

High bleachability is important and desirable for oils that are to be hydrogenated for margarine base oils, shortening and cooking oils. Studies carried out at the Tropical Products Institute London (Cornelius, 1977), suggest that poor bleachability of West African oil, which was largely produced by small-scale processing methods, might be due to oxidation of lipoxidases in bruised fruit stored for some time before processing, oxidation catalyzed by iron during processing and bulking, and mixing of different oils. It was also suggested that high carotene oils are more likely to deteriorate than oils of low carotene content.

Also, if fermentation is allowed to take place, as is the case in one of the traditional processing methods, bleachability will be affected, since oxidation of the unsaturated fatty acids gives rise to compounds responsible for colour fixation in fats. Good quality palm oil contains about 3 per cent free fatty acids and is easily bleachable.

For the production of a low free fatty acid oil from the kernels, low moisture content and low breakage are required, as illustrated in Figure 7.14. Apart from moisture content and breakage of kernels, a rise in free fatty acid content during storage has been found to be due to the action of heat-resistant lipases, undestroyed by drying or high temperatures. By successive injections of steam and cold air, thus sterilizing and drying the surfaces of the kernels before introducing them into the drier, the free fatty acid content of the oil in the kernels can be maintained below 1.5 per cent for 6 months of storage. This is attributed to prevention of development of micro-organisms before and during storage (Loncin and Jacobsberg, 1965).

## *Refining of vegetable oils*

All vegetable oils destined for the industrial manufacture of edible products undergo the process of refining. Thus, the manufacturer of edible oil products draws on a number of processes to improve the physical and flavour characteristics of his product and makes considerable use of blending to achieve his objectives. The operations which may be employed in the refining processes are: degumming; alkali refining or neutralization; bleaching and filtering; hydrogenation; fractionation; and deodorization. The first step is the removal of fat-soluble and fat-insoluble impurities and other degradation products likely to cause the oil to develop off-flavours and odours. Fat-insoluble substances, which may include gums, protein material and phosphatides, are removed by the process of degumming. Degumming is carried out in tanks fitted with heating coils and a stirrer. The crude oil is mixed with water or salt solution. Impurities are precipitated into an aqueous layer which is then drawn off.

In the next step, the oil is neutralized or alkali-refined to remove free fatty acids. Free fatty acids are undesirable and would oxidize to cause rancidity. In this process the free fatty acids are combined with an alkali, usually caustic soda, to form a soap solution, which is then separated from the oil. The oil is then washed and dried under vacuum. A slight bleaching effect is achieved by neutralization, in that the alkali solution absorbs some of the material responsible for colour (Ihekoronye and Ngoddy, 1985). The main bleaching effect is achieved by a specially treated natural bleaching earth, such as Fuller's earth. The earth absorbs colour pigments and residual oxidized material which would later affect flavour stability. For the removal of the Fuller's earth, the oil is passed through filter presses.

The bleached oil flows directly to either the hydrogenator or the deodorizer or to both in sequence, for conversion to edible products. For food purposes the bleached oil is hydrogenated to form margarine base oils, shortening, cooking oils and all-purpose salad and cooking oils.

Hydrogenation is the process of adding hydrogen in the presence of a catalyst, preferably nickel, to make the oil less liquid and more stable. Hydrogenated oils may have melting points between 40°C and 50°C. In comparison, palm oil melts at temperatures between 25°C and 40°C. For the purpose of producing two fats, one melting at a higher temperature than the other, the process of fractionation may be employed. Fats are a mixture of solid and liquid glycerides and fractionating is a means of changing the proportion of solid to liquid. To fractionate a fat it is first melted, then cooled under controlled conditions. The high-melting glycerides crystallize first and can be filtered off. Fractionation is gaining considerable importance for manufacturers of margarines and cooking fats because the solid part resulting from fractionation can fully or partly replace the incorporation of hydrogenated fats. Particularly for developing countries fractionation seems to be more attractive since hydrogen and catalysts are expensive and hydrogenation plants require expensive safety precautions.

From bleaching and filtering or from fractionation, the oil is passed on for deodorization, the last refining process. Deodorization is a process of steam distillation under vacuum, in which the odoriferous substances, since they are steam volatile, concentrate in the distillate. The oil is finally separated from the distillate. Deodorization also provides a further bleaching effect, which is of special importance for palm oil.

After refining, oils and fats may be blended to yield different finished products. For example, cooking fats are blends of oils, chilled and textured. The oil blend can be made out of 100 per cent palm oil base, of which some part is hardened palm oil.

## Palm wine production

Palm wine is produced from the sap induced by tapping the male inflorescence. A piece of the front spathe is removed and the main stem of the spadix is cut horizontally to form a so-called 'tapping panel' (Tuley, 1965). The cut is covered with a piece of 'felt' made from the fibrous leaf sheath fabric. A new slice is cut once or twice a day until the sap begins to flow. A bamboo funnel is inserted in the felt cover and is set in position. The wine is allowed to flow into a calabash or bottle. It is collected morning and evening and a new slice is cut from the tapping panel at each collection. The flow of sap can continue for a month.

Bassir (1962) investigated the composition and fermentation of palm sap from the male inflorescence stalk. From samples taken from one palm he analysed the chemical constituents of fresh palm sap (Table 7.6).

**Table 7.6 Composition of palm sap**

| Constituent | Palm sap (g/100 ml) |
|---|---|
| Sucrose | 4.3 |
| Glucose | 3.4 |
| Ammonia ($NH_3$) | 0.038 |
| Lactic acid | present |
| Amino-acids | present |

The pH value of fresh palm wine is 7.4; it falls to 6.8 in the initial stage of fermentation and then to 4.0 in the final stage, when fermentation virtually ceases. The wine is usually drunk at pH 5.5–6.5 after about 12 hours fermentation. According to Bassir (1968) the organic acid content of the wine includes acetic, lactic and tartaric acids. Thirteen amino-acids have also been identified as well as a considerable content of both vitamin C and vitamin D.

Experiments have suggested that fermentation of palm wine proceeds in two stages; in the first, bacterial action is believed to be responsible for the production of organic acids; in the second, sucrose is reported to be inverted by natural yeasts and ethyl alcohol and more organic acids are produced (Soda Ayernor and Mathews, 1971).

Fresh palm wine is sweet, and appears milky due to the high concentration of yeasts. It has a slightly sulphurous odour because of sulphur-containing amino-acids in the proteins.

The production of so-called '*down wine*' from the terminal cabbage of a felled palm is also widely practised. The major difference between the wine from the tapped immature male inflorescence and that from a felled palm (down wine) is the alcohol content. Fermented inflorescence palm wine contains only ethanol whereas down wine contains some methanol and propanol, an undesirable characteristic because methanol is poisonous. Palm wine has long been distilled into spirit or gin in Ghana, locally known as 'akpeteshie'.

## References


**Bassir, O.** (1962). Observations on the Fermentation of Palm Wine. *W. African J. Biol. Chem.* **6**: 2, 20.

**Bassir, O.** (1968). Some Nigerian Wines. *W.A.J. Biol. Chem*, **10**: 12, 42.

**Cornelius, J.A.** (1977). Palm oil and palm kernel oil. *Progress in the Chemistry of Fats and Other Lipids*, **15**: 5–27.

**Desassis, A.** (1955). Le dètermination de la teneur en huile de la pulpe de fruit d'*Elaeis guineenis. Ole agineux*, **10**: 739 and 823.

**Donkor, P.** (1979). A hand-operated screw press for extracting palm oil, in *Appropriate Technology* Vol. 5 No. 4.

**Eckey, E.W. (1954).** *Vegetable fats and oils*. Reinhold Publishers Inc., New York.

**FAO** (1986). *FAO Mon. Bull. Stat.* Vol. 9. Rome.

**Hartley, C.W.S.** (1977). *The Oil Palm.* 2nd Edition, Longmans, London.

**Ihekoronya, A.I.** and **Ngoddy, P.O.** (1985). *Integrated Food Science and Technology for the tropics.* Macmillan Publishers Ltd. London and Basingstoke. Chapter 5.

**Loncin, M.** and **Jacobsberg, B.** (1963). Studies in Congo Palm oil. *J. Am. Oil Chem. Soc.*, **40**: 18.

**Loncin, M.** and **Jacobsberg, B.** (1965). *Research on Palm Oil in Belgium and the Congo.* Paper presented at the TPI Oil Palm Conference, London, 85. Ministry Overseas Development, UK.

**Oil Palm News** (1980). TPI No. 24, London. Dr J.A. Cornelius, (ed.), TPI.

**Purseglove, J.W.** (1975). *Tropical Crops: Monocotyledons.* Longman, London.

**Rossman, G.** (1979). Oil Press, in *German Appropriate Technology Exchange* (GATE), Eschborn.

**Soda Ayernor G.K.** and **Mathews, J.S.** (1971). The Sap of the Palm as Raw Material for Alcoholic Fermentation in Ghana. *Trop. Science*, 13, 71.

**Stock Oil Review.** (1963). *Kernel Recovery*, **3**: 4–5.

**Thieme, J.G.** (1968). *Coconut oil processing*, FAO, Rome.

**Tuley, P.** (1965). How to Tap an Oil Palm. *The Niger.* Fld. **30**: 28.

**Ullmańs Enzyklopädie per technischen Chemie**. (1976). 4. Auflage, Band 11.

**Ulrich, H.** (1967). *Mechanische verfahrenstechnik*, Springer Verlag, Berlin.

**Wood, B.J.** and **Koo Pai Sing**. (1977). British Patent Specification. **1**: 463: 65, 2.

# 8

# Sorghum and millet

## Production and varieties

Sorghum and millet form a group of cereals in the family Gramineae, belonging to different genera; within the group are a number of species and varieties differing in structure and chemical composition. Various terms are used to refer to sorghum and millet according to the local name or the place of cultivation. *Sorghum* is a large variable genus with many cultivars (Figure 8.1) and millet is a general term given to the many grasses yielding small-seeded edible grains (Figure 8.2), widely grown in the drier areas of Africa and Asia. Table 8.1 lists the main species of cultivated sorghum and millet.

Sorghum and millet constitute a major source of energy and protein for millions of people in Asia and Africa; they are in fact the staple foods of many of the world's poorest and least privileged people. In the Sahel zone of Africa and in India, where rainfall distribution and frequency are variable and unpredictable as exemplified by the severe droughts in the early 1970s, it is believed that 90 per cent of the rural people rely upon sorghum and millet as their principal source of calories (Hulse *et al.*, 1980).

**Figure 8.1 Tall, white-seeded variety of sorghum, *Sorghum bicolor*.**

**Figure 8.2 Pearl millet, *Pennisetum typhoides*.**

Sorghum and millet belong to the oldest group of cultivated plants in the world; today they are found in the semi-arid tropics of Asia and Africa. Of the millets, pearl millet is the most widely grown. The most important characteristic of sorghum and millet is their unique ability to tolerate and survive under adverse conditions of continuous or intermittent drought; they will also stand temporary waterlogging (Purseglove, 1972). Their drought-resistance is attributed to the low rate of transpiration of their very dense, far-reaching root systems and their ability to cease metabolism and remain dormant during periods of drought. Another physiological characteristic of the plants is the silica deposits in the endodermis of their roots, which prevent collapse during drought stress.

The area under sorghum and millet cultivation, and the production data are only estimates, since sorghum and millet production is concentrated on small subsistence farms where grain is largely consumed by the producer. According to FAO statistics, the average annual global production of sorghum in 1984 was about 71.7 million tonnes. The area harvested for sorghum was about 49 million hectares. The annual millet production during the same year was 31 million tonnes from an area of 42.35 million hectares (FAO Production Yearbook, 1985). Sorghum and millet rank fifth and sixth as major cereal crops in terms of total world production.

Sorghum and millet can be used in many ways, but above all, they are excellent dual-purpose crops; the grain may be milled, cooked, baked, or fermented in various ways as nutritious food, and the remaining parts can be used for feed, building materials and fuel.

**Table 8.1 Varieties of sorghum and millet**

| Common name | Botanical name |
|---|---|
| Sorghum | *Sorghum bicolor* |
| Pearl millet | *Pennisetum typhoides* |
| Proso or common millet | *Panicum miliacium* |
| Japanese millet | *Echinochloa frumentacea* |
| Foxtail millet | *Setaria italica* |
| Finger millet or ragi | *Eleusine coracana* |
| Teff | *Eragrostis tef* |
| Kodo millet | *Paspalum scrobiculatum* |
| Job's tears | *Coix lacryma-jobi* |
| Acha | *Digitaria exilis* |

## Utilization

By contrast to their use as an animal feed in the developed countries, sorghum and millet are used almost entirely for human consumption in developing countries. Sorghum and millet form the staple food in many parts of Africa, Asia, Central America and the Arab countries of the Middle East, and also serve as major sources of beverages in many countries. About 300 million people rely on sorghum for their subsistence and about 400 million on millet (Kent, 1983).

## *Food uses of sorghum*

As a major foodstuff in many areas of Africa, from Ethiopia and the Sudan southwards through East and Central Africa to South Africa, sorghum grain provides an ingredient for many unique indigenous foods and beverages. Food preparation methods are generally simple and differ from region to region. A porridge or a stiff paste, the basic diet in most of East Africa, is prepared by adding pounded flour to hot water. In Ethiopia and the Sudan, a flat cake is made, or the grain may be parched, popped, or boiled whole.

The preparation of sorghum flour is much influenced by grain type. The soft brown grains are generally ground whole, using a stone on a flat rock, and the flour is most often mixed with cassava or sweet potato flour, forming about one-quarter of the mixture, making a very palatable food. The soft brown grain may also be malted by soaking in water, mixing with ash and germinating. It is then pounded to remove the radicle, and dried. The soaking removes some of the pigment and some of the bitter substance from the grain, and germination results in the production of sugar so that a sweet porridge can be made.

In West Africa sorghum is used to prepare many foods, of which the most popular are *tuwo*, *akamu*, *kunu* and *koko*. Tuwo is made by stirring flour in hot water and allowing the thick paste to cool and gel. When cold, pieces of tuwo are cut and eaten with soup. Akamu is a flour paste made with cold water to which boiling water is added, and then stirred to form a thick gruel. Kunu is a thin gruel formed by stirring meal in hot water, and is often flavoured with potash, tamarind or groundnut flour. To prepare koko, meal is soaked in cold water, filtered through a cloth, and the expressed mass is boiled and made into balls.

Sorghum beer is also popular. It may be drunk as *burkutu*, an alcoholic gruel, or as *pito*, when the sediment has been removed. In most parts of Asia, particularly in India, sorghum grain is ground, made into a dough and baked as flat unleavened bread.

## *Food uses of millet*

Millet, like sorghum, can be used in many ways. The whole grain may be milled, cooked, baked or fermented in many forms as a nutritious food. Millet may also be consumed in the form of porridge made from dry parched grain, or may be cooked with sugar, groundnuts or other pulses to make desserts.

In Nigeria, the dry grain of pearl millet, for example, may be ground to make either a meal or flour, from which a porridge (tuwo) is prepared, or grits, from which *burabusko* (a food similar to couscous) is prepared. The dry grains may also be roasted and then ground to make roasted meal or flour, from which *guguru adun* and other cookies or snacks are prepared. In Ethiopia and in the Sudan the flour or meal from pearl millet is used for making flat cakes. In Uganda, millet grain is mostly used for preparing porridge and beer. The grain is malted and sprouted. Next, the radicle is removed to eliminate some of the pigment and the bitter tannins. The radicle-free grain is then dried. As a result of malting, sugars are produced, so that a sweet porridge can be made. In India, millet has been used to prepare analogues of pasta, rice, chapaties and other local foodstuffs traditionally made from wheat, rice and legumes (Desikachar, 1975). In French-speaking Africa millet is used extensively for the preparation of couscous, pap and fritters. A summary of the major traditional foods made with sorghum and millet is presented in Table 8.2.

**Table 8.2 Major traditional foods made from sorghum and millet**

| Type of food | Indigenous name | Countries, continents | Method of preparation |
|---|---|---|---|
| Thin porridge | Ogi, oko, akamu, kafa | Ghana, Nigeria | Sorghum grain or millet is steeped in water for 2–4 days at room temperature, and then fermented. Fermentation is followed by grinding and wet-sieving to remove the bran. The sediment, ogi, is boiled to produce a sour porridge; it is mainly used as a weaning food |
| Thin porridge | Uji, akamu, ambali, edi, eko, koko, kamu, nasha, obungi, bwa, kala, obushera | India, Africa | Sorghum or millet flour is mixed with boiling water and continuously stirred until a thin paste is formed. Part of the flour may be fermented one day prior to cooking as is sometimes done in India, Sudan and Uganda. It is served with milk, sugar, honey, or roasted groundnuts, depending on local eating habits. The porridge should be smooth and pourable |
| Thick porridge | Ugali, tuwo, atap, eko tutu, tuo, zaafi, sangati, kafa dalaki | India, Africa | Sorghum flour is mixed with boiling water and then stirred continuously until the sorghum flour is completely gelatinized. Additional flour is added to the gelatinized mass and stirred until a thick paste is formed. Colour change in the paste indicates sufficient cooking. The stiff porridge may be moulded or shaped and eaten either hot or cold with a sauce. Sometimes an acid or base is mixed into the cooking water for a particular desired taste of the end product |
| Leavened bread | Injera | Ethiopia | Injera is made from teff alone or from mixtures or sorghum and teff flour. The desired flour is mixed with water; a little flour dough from a previous batch, as yeast, is added and the mixture is kneaded into a dough which is left to ferment for 2 days. After fermentation, about 10 per |

## Table 8.2 continued

| Type of food | Indigenous name | Countries, continents | Method of preparation |
|---|---|---|---|
| | | | cent of the fermented dough is mixed with water and the mixture is cooked. The cooked portion is added to the remaining dough, and left to stand for 2 hours, during which vigorous fermentation takes place. The batter is poured in a thin layer onto a greased hot griddle and baked |
| Leavened bread | Kisra, dosa dosai | Sudan, India | Sorghum flour is mixed with water and starter (9:2:1 ratio respectively), to yield a paste which is fermented overnight. A starter is made by mixing yeast and warm water, or as with the preparation of injera, part of the previous batter may be used. The fermented batter is spread very thinly on a hot griddle and left to cook for 30–60 seconds. Properly prepared kisra is paper thin with a sour taste and a fermented aroma. Kisra batter is thinner than that of Injera |
| Unleavened bread | Chapati, roti, rotti | India | Sorghum flour (95–100 per cent extraction) is mixed with water, kneaded into a dough, and the dough is formed into a thin circular piece about 200–250 mm in diameter. This dough is put on a griddle at a temperature of about 350°C for 25–30 seconds on one side. Then the outside is sprinkled with water, the dough piece is turned over and left to cook for another 30 seconds, or until it puffs. With respect to texture, taste, colour, and keeping quality, white sorghum is found to make the best chapatis |
| Unleavened bread | Tortilla | Central and South America | Sorghum grain is cooked in a 0.8–1 per cent lime solution until soft. The cooked grain is then washed with water to about 8.5 pH and ground wet into a dough from which flat, thin cakes – tortillas – are rolled. Tortillas are cooked on each side for about 30 seconds at about 350°C on a griddle. Sorghum tortillas are more elastic than maize tortillas |
| Couscous | Couscous | North and West Africa | Finely ground sorghum or millet flour is mixed with water until the flour particles stick together. The particles are then |

**Table 8.2 continued**

| Type of food | Indigenous name | Countries, continents | Method of preparation |
|---|---|---|---|
| | | | pushed through a coarse screen and put in a perforated container which is placed on top of boiling water. Steam penetrates through the agglomerated particles. The couscous is stirred several times during cooking. The cooked couscous is eaten with a sauce |
| Cooked dehusked grain | Sorghum 'rice', acha, baksi, dafa, duka owa | India, Africa | Sorghum is dehusked to remove the pericarp, cooked, and served like rice |

## Injera

Injera or enjera is the Amharic name for a fermented sour leavened bread consumed as a staple in Ethiopia. The flat loaves are usually 55–60 cm in

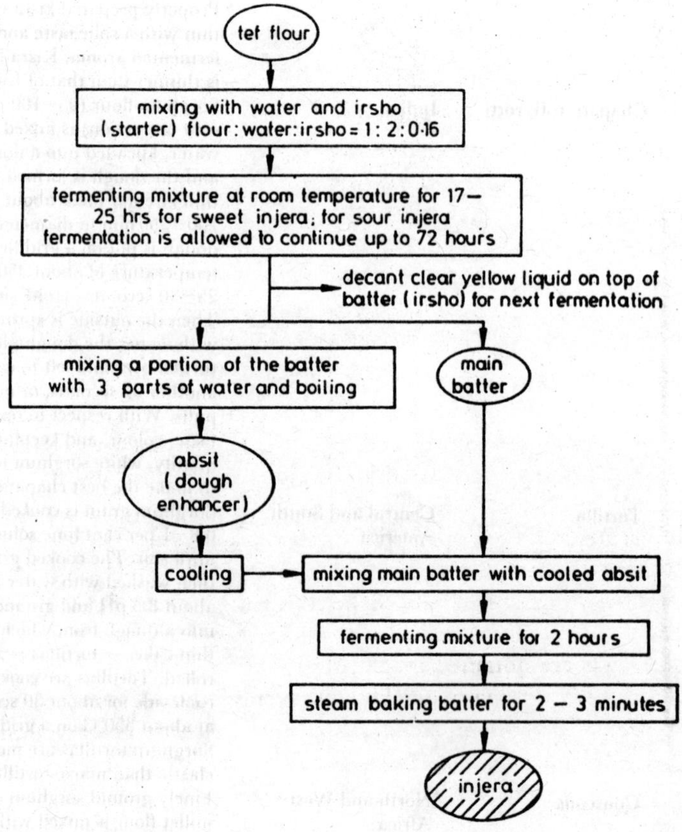


**Figure 8.3 Flowchart for traditional processing of injera (Vogel *et al.*, 1977).**

diameter and weigh between 250 and 700 g depending on the thickness of the batter and the type of the flour used.

Injera can be made from sorghum, maize, millet or barley, but the much preferred injera is made from teff millet. Though teff is a relatively expensive crop to cultivate because of high labour input and low yield, teff milling involves much less work than milling other grains, and teff injera has better keeping qualities. While sorghum, barley and maize need dehusking, pounding, winnowing and drying before milling, teff, needs only to be washed, dried and milled. Teff injera remains soft and pliable even after 3 days of storage, whereas wheat injera is sticky immediately after baking and maize injera becomes dry and brittle after several hours. Sorghum injera becomes dry after a day or two. Teff is a tiny grain with a large surface area per g; 1000 seeds weigh 0.3–0.4 g as compared to 50 g for 1000 seeds of wheat. A flowchart for the traditional preparation of injera is shown in Figure 8.3 (Vogel *et al.*, 1977).

Teff flour is mixed with water and irsho (starter). The mixture, which is a thin, watery batter, is allowed to ferment for 17–72 hours, depending upon the type of injera desired. During fermentation a clear, yellow liquid accumulates on the surface of the batter. This liquid, also called irsho, is collected and serves as an inoculum for the next fermentation.

A portion of the batter is mixed with three parts of water and boiled to yield absit. The absit is then mixed with the main part of the batter and the mixture is allowed to ferment for another 1.5–2 hours during which time the batter becomes thoroughly leavened and acidic. The batter is then steam-baked in a thin layer on a covered clay griddle for 2–3 minutes over a fire. Chemical changes occurring in teff and in flours of other cereals during injera fermentation are shown in Table 8.3.

## Kisra

Kisra is a bread made from fermented sorghum flour. It is the most popular bread in Sudan. Kisra may be made in the form of thin baked sheets, sometimes called kisra rhaheefa, or as a porridge called kisra aseeda.

To prepare kisra (Figure 8.4), sorghum grain is dry milled and sieved, and the flour is mixed with water in a ratio of 2 : 3. Sorghum flour may be fortified with other cereal flours if desired (Abel Gadir and Mohamed, 1977). A starter which has been taken from a previous fermented dough is added to the mixture and left to ferment for 12–18 hours at a temperature of 35–38°C. The fermented dough is diluted with water to obtain a consistency suitable for baking. Kisra sheets are obtained by placing the diluted dough on one side of the hot plate and spreading very quickly and evenly onto the whole surface of the plate. Baking takes approximately 1–2 minutes. Aseeda is made from less diluted and incompletely fermented dough and is baked in a cooking pot.

The initial pH of the dough decreases from between 4.9 and 5.2 to between 3.5 and 3.7 after about 18 hours of fermentation. The dough volume increases as a result of $CO_2$ production, and a desirable fine texture develops in the later stages of fermentation. The predominant micro-organisms believed to be in

**Table 8.3 Chemical changes in the substrate during injera fermentation***

| Constituent | White teff | | Barley | | Corn | | Millet (pearl) | |
|---|---|---|---|---|---|---|---|---|
| | Flour | Injera | Flour | Injera | Flour | Injera | Grain | Injera |
| Energy (joules) | 1580 | 1692 | 1574 | 1550 | 1701 | 1663 | 1550 | 1534 |
| Nitrogen (g) | 2.1 | 2.4 | 2.2 | 2.5 | 1.7 | 1.6 | 1.5 | 1.3 |
| Protein (g) | 12.4 | 14.1 | 11.9 | 12.9 | 9.2 | 9.6 | 8.2 | 6.6 |
| Fat (g) | 2.7 | 1.5 | 2.6 | 1.5 | 5.2 | 3.2 | 1.6 | 0.8 |
| Carbohydrate (g) | 82.1 | 81.7 | 81.8 | 82.9 | 84.3 | 85.6 | 87.7 | 89.2 |
| Fibre (g) | 3.3 | 2.2 | 2.5 | 5.2 | 2.3 | 3.2 | 6.4 | 6.6 |
| Ash (g) | 2.8 | 3.6 | 2.5 | 3.0 | 1.6 | 1.6 | 3.8 | 3.4 |
| Calcium (mg) | 174 | 383 | 80 | 119 | 10 | 39 | 439 | 243 |
| Phosphorus (mg) | 408 | 439 | 482 | 285 | 297 | 295 | 250 | 332 |
| Iron (mg) | 21.1 | 37.5 | 17.5 | 6.7 | 11.4 | 15.6 | 96.8 | 58 |
| Thiamine (mg) | 0.71 | 0.68 | 0.17 | 0.35 | 0.33 | 0.27 | 0.23 | 0.26 |
| Riboflavin (mg) | 0.16 | 0.14 | 0.17 | 0.15 | 0.12 | 0.11 | 0.10 | 0.11 |
| Niacin (mg) | 2.0 | 1.9 | 7.3 | 0.2 | 2.1 | 2.1 | 0.9 | 5.5 |

*All values are per 100 g of product on a dry weight basis; injera normally contains 56–62 % moisture.
Source: Ethiopian Nutrition Institute (ENI); Agreen and Gibson (1968).

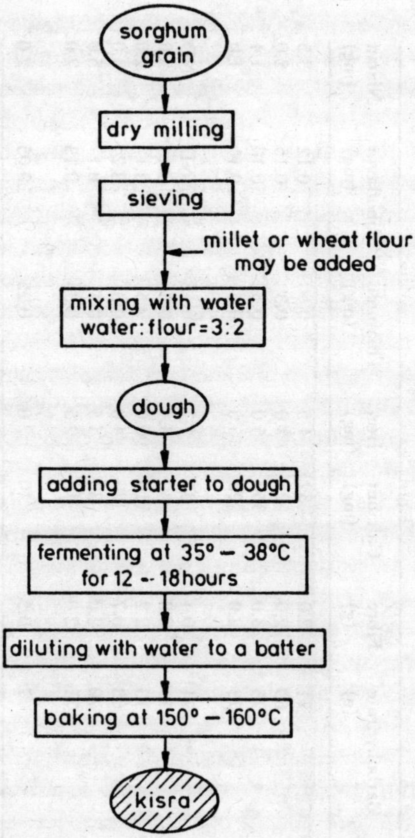


**Figure 8.4 Flowchart for traditional processing of kisra.**

kisra are *Lactobacillus* spp., *Acetobacter* spp. and *Saccharomyces* spp. The yeasts (mainly *Saccharomyces cerevisiae*) also produce $CO_2$, which leavens the dough. Lactic acid bacteria increase in numbers for the first 15 hours of fermentation.

## Chemical composition

The main components of whole grain sorghum and millets are carbohydrates, protein and lipids. Lesser quantities of fibre, vitamins and minerals are also present. The mean nutrient composition of sorghum and millets and various other cereals is presented in Table 8.4. The nutrient composition is influenced both by environment and genetics, although the most common sources of variation are factors such as soil fertility, soil moisture and cultural practices.

Sorghum and pearl millet have naked caryopses (hulls), but other millets usually have covered caryopses, which explains their higher fibre and ash content. Generally, the protein content of sorghum and millet is similar to that of rice and wheat although the quality of the former is somewhat inferior. In sorghum and millet, lysine is the primary limiting amino-acid. An overall comparison of the amino-acid scores suggests, however, that the protein composition of pearl millet, common millet, finger millet and kodo millet is

**Table 8.4 Representative nutritient values of sorghum and millet in comparison with other cereals***

| Cereals | Moisture (%) | Energy (joules) | Protein (g) | Fat (g) | Carbohydrate (g) | Fibre (g) | Calcium (mg) | Iron (mg) | Thiamine (mg) | Riboflavin (mg) | Niacin (mg) |
|---|---|---|---|---|---|---|---|---|---|---|---|
| Sorghum | 12 | 1468 | 10.4 | 3.4 | 71 | 2.0 | 32 | 4.5 | 0.50 | 0.12 | 3.5 |
| Pearl millet | 12 | 1520 | 11.0 | 5.0 | 69 | 2.0 | 25 | 3.0 | 0.3 | 0.15 | 2.0 |
| Finger millet | 12 | 1407 | 6.0 | 1.5 | 75 | 3.0 | 350 | 5.0 | 0.3 | 0.10 | 1.4 |
| Common millet and foxtail millet | 11 | 1486 | 10.0 | 2.5 | 73 | 2.0 | 20 | 5.0 | 0.6 | 0.1 | 1.0 |
| Kodo millet | 12 | 1478 | 11.5 | 1.3 | 74 | 0.4 | 35 | 1.7 | 0.15 | ND | ND |
| Acha | 10 | 1382 | 7.7 | 1.8 | 71 | 6.8 | 30 | 3.4 | 0.3 | 0.1 | 3.0 |
| Teff | 11 | 1444 | 8.5 | 2.2 | 73 | 2.2 | 110 | 90.0 | 0.5 | 0.1 | 2.0 |
| Job's tears | 12 | 1528 | 14.0 | 4.0 | 68 | 0.7 | 20 | 4.0 | 0.3 | 0.2 | 3.0 |
| Wheat | 13 | 1440 | 11.5 | 2.0 | 70 | 2.0 | 30 | 3.5 | 0.4 | 0.1 | 5.0 |
| Maize (Whole and parboiled) | 12 | 1520 | 10.0 | 4.5 | 71 | 2.0 | 12 | 2.5 | 0.35 | 0.13 | 2.0 |
| Rice (parboiled) | 12 | 1482 | 8.0 | 1.5 | 77 | 0.5 | 10 | 2.0 | 0.25 | 0.05 | 2.0 |

*per 100 g of edible portion
ND = not determined
Source: Platt, quoted in Hulse *et al.*, 1980.

nutritionally superior to normal sorghum. The prolamine of finger millet, eleusine, has a high biological value, with a good content of cystine, tyrosine, trytophane and methionine, which are important in the prevention of kwashiorkor. However, they are low in lysine.

In most sorghum, the amylose content is reported to range from 20–30 per cent, whereas in pearl millet, amylose represents between 18 and 25 per cent and in common millet between 23 and 27 per cent of the starch present (Hulse *et al.* 1980). Purseglove (1972) quoted Deshaprabhu (1966) as stating that the starch of pearl millet consists of 68 per cent amylopectin and 32 per cent amylose.

## Structure of sorghum and millet

The sorghum kernel is a caryopsis (Figure 8.5). Its major components are the pericarp (6–8 per cent of the dry weight of the kernel), the endosperm (80–84 per cent of the dry weight of the kernel) and the embryo or germ (7.8–12.1 per cent of the dry weight of the kernel) (Wall and Blessin, 1969). The endosperm consists of a single layer of aleurone cells and the starchy endosperm. The relative proportions of the endosperm may vary with genetic background, environment and degree of maturity.

The embryo is rich in protein, lipids, minerals and B group vitamins. Removal of the outer pericarp proportionally increases the protein and reduces the cellulose, lipid and mineral content of the grain. The grain structure of millets has been less thoroughly studied. However, some general structural similarities exist between the grains of sorghum and pearl millet. Pearl millet kernels are about one-third as large as grains of sorghum. The kernel of pearl millet appears to contain a higher proportion of germ than the kernel of sorghum. The pericarp of sorghum and pearl millet is reportedly lower in average protein content than the whole grain, whereas the pericarp of finger millet generally contains about twice as much protein as the whole grain. An important feature of the minor types of millet is the high cellulose content of their outer seed layer (hull, seed coat).

Kernel size, shape, and density, of sorghum and millet vary, depending on variety and the environmental conditions in which they mature. The

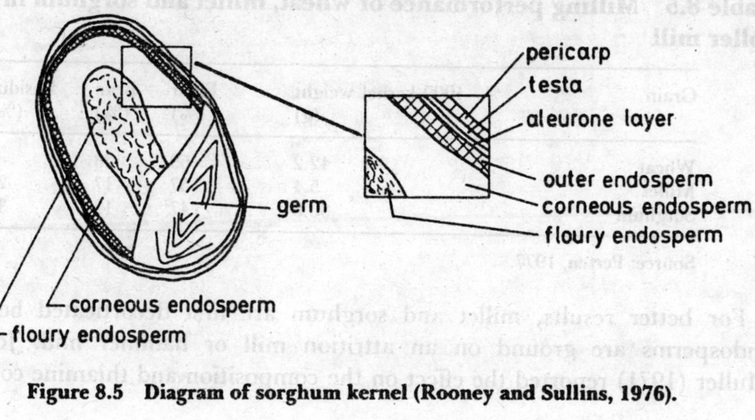


**Figure 8.5 Diagram of sorghum kernel (Rooney and Sullins, 1976).**

sorghum kernel weight varies from 5–80 mg or more, the bulk density varies from 1.2–1.45 g/cm$^3$, and the shape varies from flat seashell to spherical. The thousand kernel weight of sorghum is 25–35 g whereas that for pearl millet is 5.5 g.

The pronounced peripheral endosperm of sorghum is believed to effect the milling properties of the grain and is also a major reason why sorghum grain is generally considered to be less digestible than maize and other cereals.

## Milling of sorghum and millet

Milling aims at separating the kernel into three factions: bran, germ and endosperm. Removal of the pericarp lowers the fibre and ash content and decreases the pigmentation of the meal or flour. Elimination of the germ (embryo) and aleurone layer reduces the fat content and ensures longer stability of the milled endosperm. Nutritionally, the main advantage of milling and sifting sorghum and millets is the elimination of indigestible glucide materials which can hinder the digestion of other constituents of the diet. On the other hand, there is evidence to suggest that retaining some of the bran fraction is nutritionally desirable.

From an experiment on milling of wheat, sorghum and millet, Perten (1977) reported that wheat milling technology is not the most appropriate for the milling of millet and sorghum. Whereas the wheat bran, through its suppleness, resists crushing and allows itself to be separated easily by sieving as coarse particles, the bran coat and part of the germ of millet and sorghum easily pulverize to a fine powder, which is difficult to separate from fine flour.

The differences in milling properties of millet and sorghum compared with those of wheat are shown by a milling test in a Bühler laboratory roller mill. All three grains, Senegalese millet, sorghum and French soft wheat, were milled under the same conditions. The results of the milling test are shown in Table 8.5. The low yield of millet and sorghum flours is apparent. The high ash and fat contents of the flour fractions of millet and sorghum in contrast to wheat, after regrinding the residue, confirm the differences in milling properties of the grain.

**Table 8.5 Milling performance of wheat, millet and sorghum in a roller mill**

| Grain | 1000 kernel weight (g) | Flour (%) | Bran (%) | Residue (%) |
|---|---|---|---|---|
| Wheat | 42.2 | 69 | 28 | 3 |
| Millet | 5.4 | 57 | 17 | 26 |
| Sorghum | 30.0 | 47 | 15 | 38 |

Source: Perten, 1977.

For better results, millet and sorghum are first decorticated before the endosperms are ground on an attrition mill or hammer mill. John and Muller (1971) reported the effect on the composition and thiamine content, of milling sorghum to varying extraction rates by the traditional pestle and mortar method, followed by grinding, and roller milling in a Bühler laboratory mill. The results (Table 8.6) indicate that the nutrient content decreased with decreasing extraction rate and that the nutrient loss was greater in the Bühler mill than the traditional pestle and mortar method. Roller milling removed more germ and bran than traditional processing.

**Table 8.6 Composition and thiamine content of roller milled and traditionally milled sorghum flour of various extractions**

| Method of milling | Extraction (%) | Protein (%) | Fat (%) | Fibre (%) | Ash (%) | Thiamine (mg/100 g) |
|---|---|---|---|---|---|---|
| Roller milled | 100 | 8.2 | 2.6 | 1.3 | 1.3 | 0.36 |
| | 90 | 6.0 | 1.3 | 0.8 | 0.7 | 0.20 |
| | 70 | 3.5 | 0.3 | 0.3 | 0.5 | 0.09 |
| | 50 | 2.0 | 0.3 | 0.3 | 0.5 | 0.05 |
| | 30 | 1.0 | 0.2 | 0.1 | 0.2 | 0.03 |
| Pestle and mortar milled | 100 | 8.2 | 2.6 | 1.3 | 1.3 | 0.36 |
| | 90 | 7.3 | 2.2 | 1.1 | 1.1 | 0.32 |
| | 70 | 5.5 | 1.7 | 0.9 | 0.9 | 0.26 |
| | 50 | 3.4 | 1.2 | 0.6 | 0.5 | 0.18 |
| | 30 | 1.6 | 0.6 | 0.3 | 0.3 | 0.09 |

Source: John and Muller, 1971.

Favier *et al.* (1972) report the effect of milling methods – traditional pestle and mortar and motor driven disc mill – on the chemical composition of a local market sample of Cameroon sorghum, and of kourou (Table 8.7). Kourou is made by first steeping decorticated sorghum grain in water for 20 hours to 4 days and then wet grinding it on a stone mill. The grinding is followed by washing and straining to yield a white flour. It is either used immediately to prepare a gruel or porridge, or sun dried for later use.

## *Types of mills*

Various techniques have been developed in past years for the processing of different grains. In this section a number of crushers are illustrated, which differ in the way they affect the grain during processing (Figure 8.6). In most cases the moisture content of the grain to be processed ranges between 11 and 15 per cent. Variations outside this range will require adjustments in equipment used.

### Mills which employ compressive force with friction

*Stone mills*

Stone mills consist of two millstones which are arranged either horizontally, vertically or inclined. The product to be ground flows axially between the stones, is distributed through the rotation of one of the millstones and is discharged radially after crushing. The millstones are fluted. Loncin (1969) suggests a running speed of about 100 rpm for horizontal stone mills. These

**Table 8.7 Nutrient content (dry weight basis) of traditionally milled and disc milled sorghum and of kourou**

| Constituent | Traditional milling (pestle and mortar) | | | | | | Disc milling | | Kourou | | |
|---|---|---|---|---|---|---|---|---|---|---|---|
| | Whole sorghum | Bran | Decorticated sorghum | Semolina | Fresh flour | Sun dried flour | Flour | Semolina | Fresh kourou | Dried kourou | Kourou semolina |
| Moisture (%) | 13.2 | 43.8 | 29.0 | 23.0 | 30.0 | 13.2 | 19.4 | 19.6 | 48.8 | 13.0 | 46.5 |
| Energy (joules) | 1603 | 1059 | 1721 | 1675 | 1729 | 1729 | 1725 | 1725 | 1733 | 1725 | 1712 |
| Protein (%) | 9.6 | 10.4 | 9.5 | 9.5 | 9.6 | 9.4 | 8.5 | 12.8 | 7.9 | 7.4 | 11.3 |
| Lipids (%) | 3.0 | 5.7 | 1.4 | 0.1 | 1.8 | 1.7 | 1.3 | 2.1 | 1.1 | 0.6 | 0.1 |
| Carbohydrates (%) | 85.6 | 79.3 | 88.0 | 90.1 | 87.4 | 88.0 | 89.3 | 83.7 | 90.7 | 91.8 | 88.4 |
| Formic acid insoluble matter (%) | 2.5 | 8.9 | 1.0 | 1.5 | 0.9 | 0.9 | 0.7 | 1.8 | 0.4 | 0.5 | 1.2 |
| Ash (%) | 1.8 | 4.6 | 1.0 | 0.3 | 1.2 | 0.9 | 0.9 | 1.4 | 0.3 | 0.2 | 0.1 |
| Calcium (mg/100 g) | 12.8 | 15.9 | 13.5 | 6.6 | 11.8 | 10.4 | 11.6 | 16.9 | 5.6 | 3.0 | 3.4 |
| Phosphorous (mg/100 g) | 335 | 644 | 192 | 42 | 227 | 208 | 159 | 262 | 56 | 62 | 35 |
| Calcium/Phosphorus | 0.04 | 0.02 | 0.07 | 0.15 | 0.05 | 0.05 | 0.07 | 0.06 | 0.1 | 0.05 | 0.1 |
| Sodium (mg/100 g) | 5.7 | 8.6 | 4.2 | 4.6 | 5.9 | 4.7 | 4.8 | 4.2 | 4.6 | 4.8 | 5.5 |
| Potassium (mg/100 g) | 361 | 682 | 165 | 78 | 199 | 184 | 190 | 262 | 16 | 18 | 21 |
| Thiamine (mg/100 g) | 0.54 | 1.05 | 0.30 | 0.04 | 0.37 | 0.32 | 0.21 | 0.04 | 0.07 | 0.07 | 0.01 |
| Niacin (mg/100 g) | 4.6 | 9.9 | 1.9 | 0.5 | 2.2 | 2.2 | 1.2 | 2.9 | 1.6 | 1.5 | 0.8 |
| Riboflavin (mg/100 g) | 0.10 | 0.20 | 0.05 | 0.03 | 0.05 | 0.04 | 0.04 | 0.07 | 0.05 | 0.07 | 0.01 |

Source: Favier *et al.* (1972).

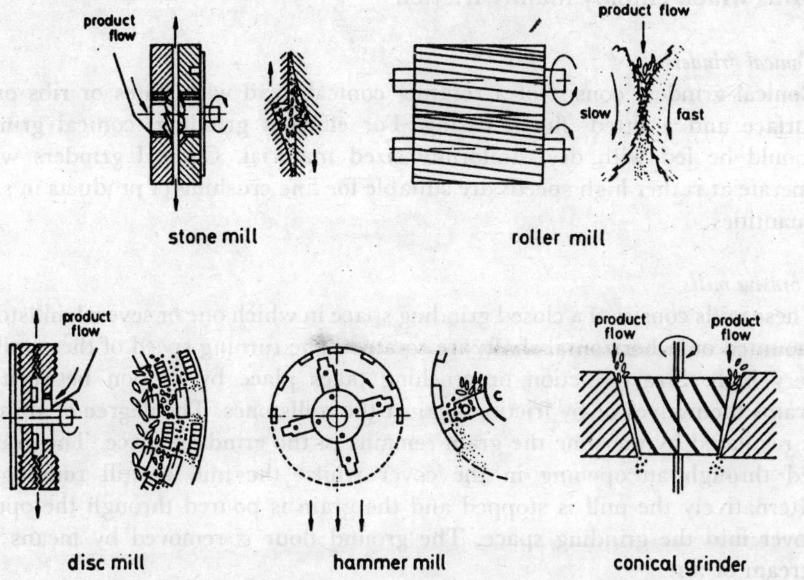


**Figure 8.6 Working principles of various mills.**

are suitable for fine crushing of products in a single operation. Stone mills for coarse meal production have a diameter range of 200–1500 mm with a running speed of 250–1100 rpm. The fineness of the meal is determined or regulated by the clearance between the millstones. The increase in temperature of the meal amounts to 12–20°C (AEL, 1976).

*Disc mills*
Disc mills are similar in contruction to stone mills; they employ two saucer shaped discs mounted on a horizontal shaft. There are disc mills with either one or two rotating millstones. In the first case, one of the discs is fixed to the casing, and in the latter both discs rotate in opposite directions. The material is introduced and discharged after crushing as is in the case of the stone mill. The disc, having a disc diameter of 200–350 mm, rotates at 500 rpm. The increase in meal temperature amounts to 12–20°C (AEL, 1976).

*Roller mills*
The roller mill operates on the principal of two chilled cast iron rolls rotating at different speeds, for example a 3:1 ratio, in opposite directions. The crushing action occurs when the product enters the mill and is fed between the rolls. The increase in meal temperature amounts to 5–10°C. The rolls have a diameter of 250–300 mm, the quicker roll having a running speed of about 480 rpm. Roll pairs with a smaller diameter have a shorter milling zone than those with a larger diameter. The former produce proportionally more grit, and the latter more flour.

## Mills which employ mainly friction

*Conical grinders*

Conical grinders consist of a rotating conical head with flutes or ribs on its surface and a fitted, fluted casing. For efficient grinding, conical grinders should be fed with dry, uniformly sized material. Conical grinders which operate at rather high speeds are suitable for fine crushing of products in small quantities.

*Abrasive mills*

These mills consist of a closed grinding space in which one or several millstones, mounted on a horizontal shaft, are rotated. The turning speed of these mills is very high. Size reduction or crushing takes place by friction between the grains themselves or by friction against the millstones. The degree of abrasion is regulated by the time the grain remains in the grinding space. The grain is fed through an opening in the cover whilst the mill is still running, or alternatively the mill is stopped and the grain is poured through the opened cover into the grinding space. The ground flour is removed by means of a stream of air.

## Mills which employ impact and collision

*The hammer mill*

The hammer mill utilizes fixed or free swinging hammers, mounted on a shaft or rotor, and turning at a running speed of 1400–6000 rpm. These mills usually incorporate a perforated metal retaining screen which the ground material must pass through. The basic grinding action of a hammer mill is by impact between the rapidly moving hammers and the grains.

The main advantages of the hammer mill are its large throughput and the minimum wear of the hammers. Since a strong current of air is produced through the turning of the rotor, the ground material is cooled considerably. The increase in flour temperature during the grinding process is about 8–12°C.

## Hand-operated crushers

Table 8.8 shows a series of commercially available hand-operated and/or engine-powered crushers, which are reported to be suitable for use in the rural areas of developing countries (ITDG, 1973). Among them are stone mills, disc mills and hammer mills. The power requirements range from 0.18–11 kW and the hourly throughput varies from 60–1100 kg.

# ***Traditional milling***

The two types of traditional mills generally used are the stone mill and the pestle and mortar mill.

## Table 8.8 Commercially available hand-operated crushers

| Name and/or producer (country of origin) | Diameter of disc (mm) | Hourly through-put (kg) | Remarks |
|---|---|---|---|
| Dunia (Kenya) | 150 | – | Flour is not as fine as in a hammer mill or stone mill |
| Atlas (Great Britain) | | | |
| No. 1 | 140 | 6.8 | – |
| No. 2 | 165 | 9.1 | – |
| No. 3 | 191 | 15.9 | – |
| Nova mill (France) | – | 35 | – |
| Junior mill (France) | – | 7 | For household use |
| Corn crusher (France) | 95 | 22–30 | Half-steel plate, milling fineness adjustable |
| CECOCO, hand mill (Japan) | – | 5–50 | For coarse and fine milling. Also for 0.18–0.37 kW electric engine, 60–150 kg/hour |
| Diamant (Denmark) | – | 7 | Also with electric engine 0.37–0.74 kW, 20 kg/hour |
| Simon Freres (France) | – | 65–100 | With 1.1 kW electric power, 85–200 kg/hour |

Source: UNICEF/Kenya, ITDG, 1967/68 and 73; FAO, 1979.

*Hand-operated stone mills*

The stone mill consists of a stone slab, often fluted, and usually placed on levelling stones so that it slopes away from the operator who sits, kneels or

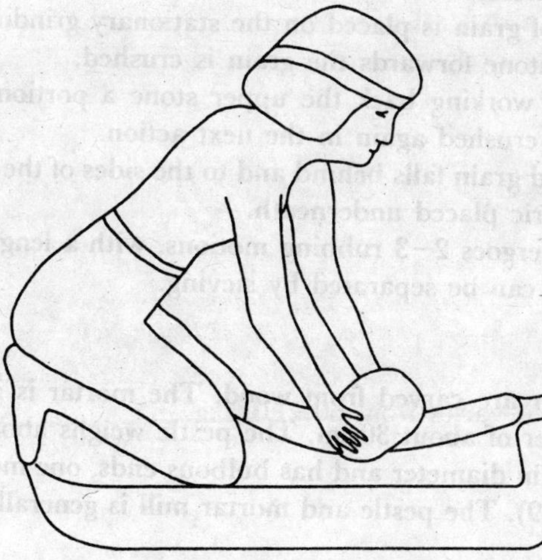

**Figure 8.7 Working on a stone mill.**

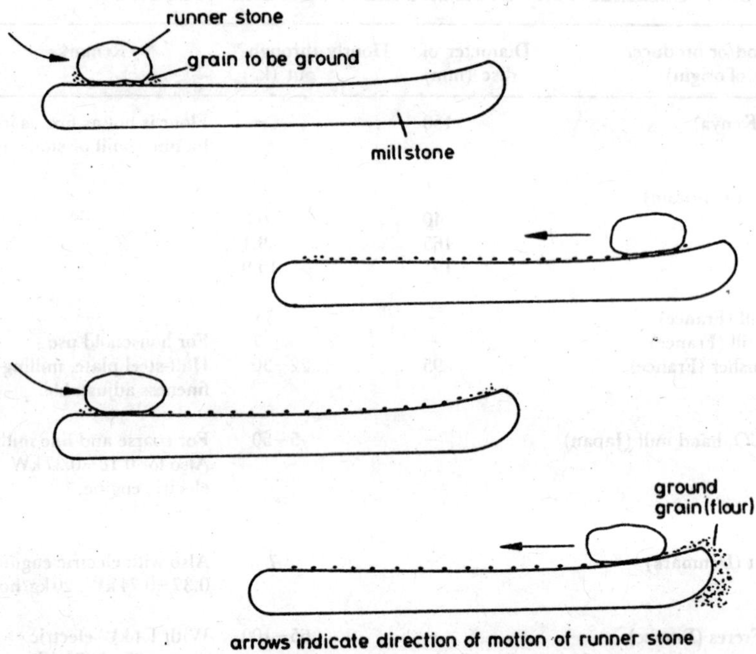


**Figure 8.8 Direction of motion of runner stone during milling (Gast, 1965).**

squats behind it, and rolls the grain with a round stone roller (Figure 8.7). A container or mat is placed in front of the base stone to receive the ground material.

The milling process showing the direction of movement of the runner stone is illustrated in Figure 8.8. It consists of:

1. I – a handful of grain is placed on the stationary grinding stone and by working the upper stone forwards the grain is crushed.
2. II + III – by working back the upper stone a portion of the grain is dragged back to be crushed again in the next action.
3. IV – the ground grain falls behind and to the sides of the stationary stone on to a piece of fabric placed underneath.

The material undergoes 2–3 rubbing motions, with a length of 15–25 cm. The flour and bran can be separated by sieving.

*Pestle and mortar mills*

A pestle and mortar are carved from wood. The mortar is about 60–80 cm high with a diameter of about 30 cm. The pestle weighs about 3 kg, is about 1.2 m long, 6–8 cm in diameter and has bulbous ends, one more pointed than the other (Figure 8.9). The pestle and mortar mill is generally used for softer grains.

Carr (1956, 1961) describes the traditional way of milling sorghum and millet in Zimbabwe. Red sorghum was placed in the mortar with about 20 per

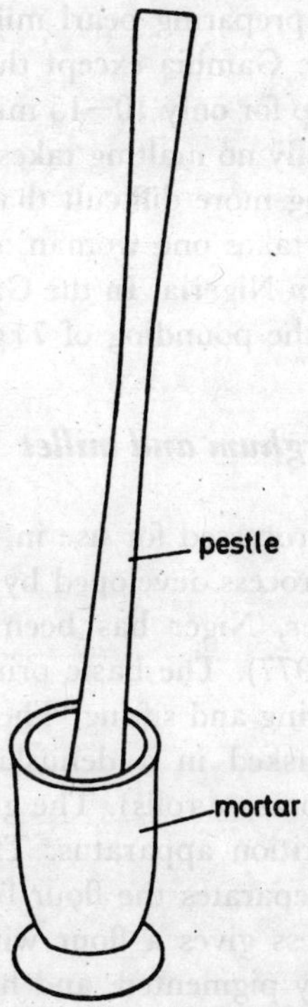


**Figure 8.9 Pestle and mortar mill.**

cent of its weight of water, stamped with the pestle for about 20 minutes, winnowed and restamped. The pounded grain was then sun dried for about 10 minutes, slightly roasted in a pan, then ground on stones and sifted through a fine sieve, the coarser fraction being repeatedly winnowed and reground. The final meal contained 15 per cent moisture, and the offals 36 per cent moisture. Finger millet was ground without water, and winnowed, sifted, and reground, to yield 80 per cent meal, 18 per cent offals and 2 per cent waste.

Pearl millet was ground as described for sorghum, with 20 per cent of its weight of water added. After stamping, the grain was winnowed, lightly roasted, and ground on stones without sieving. The extraction rate for pearl millet was about 75 per cent. Relatively high losses of nutrients in the case of sorghum were recorded. Losses from grinding pearl millet were comparatively low.

The traditional method of preparing pearl millet flour in the Gambia is by pounding the grain in a wooden mortar with a small amount of water for about 10 minutes, after which the bran is removed by winnowing. The separated grain is washed and left to stand for several hours (8–24 hours). During this time a considerable degree of malting takes place before the final pounding into flour. Evaporation occurs during pounding and winnowing and as a result both flour and coarse endosperm are drier than the original malted grain. Typical moisture contents are: grain 30 per cent, flour 27 per cent and coarse endosperm 24 per cent.

The traditional way of preparing pearl millet flour in Northern Nigeria is similar to that used in the Gambia except that, after washing to remove the bran, the grain is left damp for only 10–15 minutes before being pounded into a fine flour. Since practically no malting takes place, individual grains remain hard, making the pounding more difficult than in the Gambian method. It is reported that it normally takes one woman about 4 hours to process 7 kg of grain into flour in Northern Nigeria. In the Gambia, on the other hand, about one hour is sufficient for the pounding of 7 kg malted flour.

### *Industrial milling of sorghum and millet*

Sorghum and millet are processed for use in food and industrial products by dry milling processes. A process developed by SOTRAMIL (Société de transformation du mil), Zinder, Niger has been used to mill *Pennisetum* millet (Goussault and Adrian, 1977). The basic principles of this process consist of cleaning, dehusking, grinding and sifting. The grain is first cleaned to remove impurities and then dehusked in a dehusking machine with a horizontal millstone (decorticating abrasive rolls). The grain is then cleaned and ground in a so-called *ultrafine* attrition apparatus. The ground millet is sifted on a revolving bolting, which separates the flour from the semolina.

The SOTRAMIL process gives a flour with a 65–75 per cent extraction rate. The flour is relatively pigmented, and has a mean granulation (particle size) of 40 μm.

The mechanical milling of millet by SEPIAL (Société d'études et d'exploitation de procédés pour l'industrie dimentaire, Clichy) makes use of 'peeling-decortication'. In the SEPIAL procedure, the grain is superficially wetted and 'peeled' in an apparatus with a vertical axis and paddles, the action of which is sufficient to detach the moistened pericarp, leaving the remaining layers intact. The pericarp is then separated from the grains by brushes or aspiration. This step of the process leads to an intermediate product, the husked grain. In the second stage, the husked grain is decorticated by vigorous rubbing in a brushing machine. This procedure gives a protein-rich fraction and decorticated grain. The latter, representing about 84 per cent of the whole grain, may be sold as such, or reduced to semolina and flour in a hammer mill or roller mill. The products of the SEPIAL process, as a percentage of the whole grain, are:

(a) Whole grain: 100 %
(b) Peeled grain: 92.5 %
(c) Decorticated grain: 84.0 %
(d) Protein-rich fraction: 8.5 %

### *Comparative studies on mechanical and traditional milling of sorghum and millet*

Reichert and Youngs (1976) analysed samples of red-brown sorghum and pearl millet from Maiduguri, Nigeria, and determined the proportion of protein, oil and crude fibre after the following processes:

1. Traditional dehulling using a wooden pestle in a metal beaker, 20 per cent by weight of water added before pounding in the mortar, air drying and winnowing to remove the bran and other fines. The process was repeated twice to remove approximately 10, 25 and 45 per cent of the kernel.
2. Mechanical dehulling in a *Strong-Scott barley pealer*.
3. Mechanical dehulling with a *Palyi compact milling system* (attrition type).
4. Mechanical dehulling with a *Hill grain thresher* (abrasion type).

The dehulling efficiency of the last three were compared; dehulling with a Strong-Scott barley pearler and with a Hill grain thresher was superior to dehulling with a Palyi compact milling system. The effect of the four types of milling on the protein, oil and ash contents of sorghum and millet in relation to the amount of kernel removed is illustrated in Figure 8.10.

The three mechanical dehullers all reduced the oil and ash contents to a

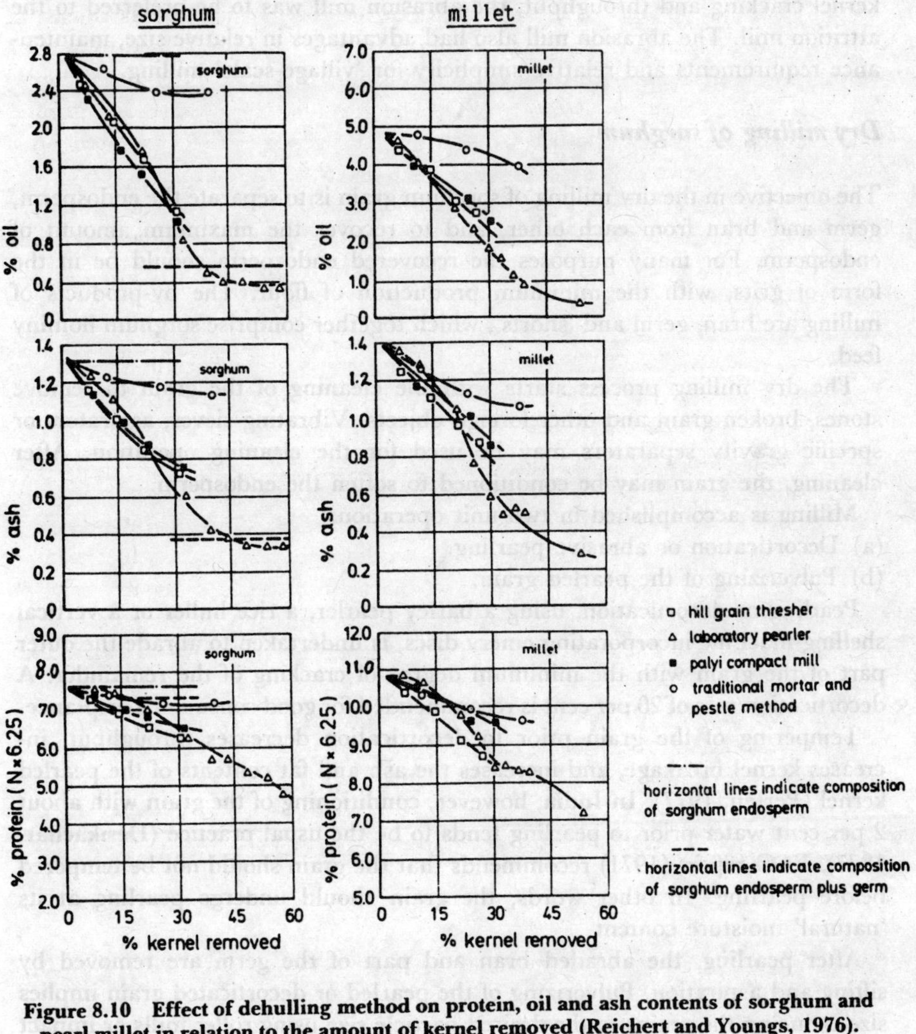


**Figure 8.10 Effect of dehulling method on protein, oil and ash contents of sorghum and pearl millet in relation to the amount of kernel removed (Reichert and Youngs, 1976).**

marked degree and the protein to a lesser degree, but between the dehullers there were only small differences in losses of protein, ash and oil when the grain was dehulled to the same extraction level.

Sorghum and millet dehulled by traditional methods contained higher oil and ash contents than the mechanically dehulled grain – the germ is firmly embedded in the sorghum endosperm and is not readily removed by pounding. Traditionally and mechanically dehulled grains did not differ as markedly in protein content as in oil and ash contents. At 75 per cent extraction, the crude fibre values were 0.8, 0.9, 1.2 and 1.2 per cent for sorghum and 0.6, 0.7, 1.0 and 1.1 per cent for millet in dehulling with the barley pealer, abrasion-type mill, attrition-type mill, and by the traditional method, respectively. The whole sorghum and millet had crude fibre contents of 2.4 and 2.0 per cent, respectively.

On the basis of relative efficiencies in terms of final colour, material removal, kernel cracking and throughput, the abrasion mill was to be preferred to the attrition mill. The abrasion mill also had advantages in relative size, maintenance requirements and relative simplicity for 'village-scale' milling.

## *Dry milling of sorghum*

The objective in the dry milling of sorghum grain is to separate the endosperm, germ and bran from each other, and to recover the maximum amount of endosperm. For many purposes the recovered endosperm should be in the form of grits, with the minimum production of flour. The by-products of milling are bran, germ and 'shorts', which together comprise sorghum hominy feed.

The dry milling process starts with the cleaning of the grain to remove stones, broken grain and other foreign objects. Vibrating sieves, aspirators or specific gravity separators may be used for the cleaning operation. After cleaning, the grain may be conditioned to soften the endosperm.

Milling is accomplished in two unit operations:

(a) Decortication or abrasive pearling.
(b) Pulverizing of the pearled grain.

Pearling or decortication, using a barley pearler, a rice huller or a vertical shelling machine incorporating emery discs, is undertaken to abrade the outer part of the grain with the minimum degree of cracking of the remainder. A decortication rate of 20 per cent is recommended for good consumer acceptance.

Tempering of the grain prior to decortication decreases throughput, increases kernel breakage, and increases the ash and fat contents of the pearled kernel (Perten, 1977). In India, however, conditioning of the grain with about 2 per cent water prior to pearling tends to be the usual practice (Desikachar, 1977). FAO-report (1971) recommends that the grain should not be tempered before pearling. In other words, the grain should undergo pearling at its 'natural' moisture content.

After pearling, the abraded bran and part of the germ are removed by sifting and aspiration. Pulverizing of the pearled or decorticated grain implies size-reducing the grains to the desired particle size using roller mills or impact mills. The fine flours needed for bakery products can be better obtained in attrition or hammer mills fitted with fine screens. Sorghum grains are degermed using either Beall degerminators or roller mills and sifters, or a circular wire brush rotating within a perforated cylinder. The detached germs are separated from the sorghum grits of similar particle size with a specific gravity separator. Product yields of 60–90 per cent are obtained in the dry milling of sorghum.

In an experimental milling in a conventional roller mill, using fluted and smooth rollers, the products of an 80 per cent extraction milling of sorghum were as follows: break flour 12 per cent, fine reduction flour 29 per cent, medium-sized grits 34 per cent, coarse grits 5 per cent, and bran and germ 19 per cent (Shoup *et al.*, 1970b).

## Processing of sorghum for starch

Sorghum is wet milled essentially to produce dry starch or starch hydrolysed to glucose. Watson (1970) describes the basic process. Clean sorghum is steeped in water for about 60 hours, then crushed to separate the embryos which float to the surface of the starch suspension. Edible oil can be extracted from the embryo and the low fat residue is used for animal feed. The remaining grist is ground and separated through bolting screens, the fibre and coarse fractions going to animal feed. The starch and water-insoluble protein (gluten) are centrifuged, the gluten fraction, containing about 60–65 per cent protein, is also used in animal feeds. The starch fraction is recentrifuged, washed and either dried to yield dry starch or hydrolysed to glucose.

## Composites of sorghum or millet flour used in bread making

Composite flours are mixtures in which cereal flours, predominantly wheat flour, are combined with other starch and protein sources including those derived from other cereals, roots, food legumes and oil-seeds. The successful use of composites of sorghum and millets with other flours for bread making would undoubtedly save many of the less developed countries from importing wheat.

Perten (1977) reported from the Sudan the satisfactory production of French-type bread from composites of 85 per cent wheat flour (87 per cent extraction) and 15 per cent sorghum flour. Bhatia *et al.* (1968) reported from Mysore, India, bread made from composites of 80 per cent wheat flour and 20 per cent sorghum flour, and stated that the calculated protein content of the composite was essentially the same as from 100 per cent wheat flour and the lipid, mineral and fibre contents were higher. Sumner and Nielsen (1976) prepared a traditional Nigerian-type bread using a 80 : 20 wheat–sorghum composite flour blend. The proving time for the dough was increased from 2 to 2.5 hours and the loaves produced were compared with a 100 per cent wheat flour bread. Although loaf volumes were similar, the specific volume of the composite flour bread was slightly less. Major differences were a darker brown external and internal colour, increased coarseness and firmness, and a stronger cereal flavour in the 80 : 20 wheat–sorghum bread.

Crabtree and Dendy (1977) reported making satisfactory bread with an optimum volume at 10 per cent millet, without a negative effect on flavour. Canadian hard wheat and unpearled proso millet (*Panicum miliaceum*) were conditioned individually to 16 per cent moisture content and then co-milled in a single operation in a Buhler laboratory mill. In the bread produced the crumb colour was dark. The crumb structure was, however, close and even.

In composite flours, the wheat flour is essentially used as the irreplaceable breadmaking factor. Using pentosan as a universal baking factor, Casier *et al.* (1976) reported the production of good quality bread from pure millet and sorghum flour as well as mixtures of maize and cassava starch. Well-structured bread from pure red sorghum flour of 100 per cent extraction was obtained, using 3 per cent rye pentosan. Other ingredients used in this investigation were yeast, water and bacon fat. The sorghum grain, with a moisture content of 18 per cent, was milled in a hammer mill without sieves. The bread produced was a faint pink colour, and was reported to have shown no signs of staling for at least one week. Bread of similar quality was obtained using flour from the same red sorghum with 53 per cent extraction and 4 per cent crude rye pentosan. Using 250 g of pure millet flour and 4 per cent crude rye pentosan, yellow coloured bread with good staling resistance was obtained. The weight of the bread produced was 445 g with a volume of 1050 ml. A mixture (50 : 50) of pure millet and sorghum flour, together 4 per cent crude rye pentosan, produced brown coloured bread. Crumb structure was better than for pure sorghum bread, but a somewhat lower figure for crumb structure was obtained than for pure millet bread.

## Malting and brewing of sorghum and millet

In some African countries and in India, sorghum and, to a lesser extent, finger millet, are processed into beer. Beer making essentially involves malting, brewing, fermentation and finishing. The process of malting consists of the controlled germinating and usually the subsequent drying of a grain. The conversion of the grain to malt is essentially physiological in nature and is a result of the action of enzymes.

Since sorghum has no more than traces of alpha- and beta-amylases, germination is essential for the production of these enzymes which are so important in brewing. The ratio of alpha- to beta-amylase in sorghum may vary from 2 : 1 to 3 : 1 (Novellie, 1977). The amylases of sorghum malt may be soluble, or partially or totally insoluble depending on the absence or presence of polyphenols in the grain. The presence of high concentrations of tannins (polyphenols) in sorghum gives an agricultural advantage ('birdproofness') but the nutritional disadvantages of poor digestibility and growth retardation. High polyphenol concentrations can also inhibit the enzymic reactions and microbial activity needed in brewing of sorghum beer. Novellie (1977) suggests that it is possible to improve the digestibility of high tannin sorghum by treating the grain with a predetermined amount of a non-acidic carbonyl containing compound, preferably formal-aldehyde, thus neutralizing the undesirable polyphenol in the grain.

Malting develops the amylolytic enzymes which change starch to fermentable sugars, the source of the alcohol in beer. It is dependant upon the alpha- and beta-amylases and the maltose splitting enzyme, maltase, that develops in the sorghum during fermentation.

The initial process in brewing is mashing. The dried malt is ground so that the starches can dissolve more easily. During mashing the length of time the mash remains at temperatures favouring beta-amylase activity (60–65°C) determines the amount of maltose formed and thus the alcohol content of the final beer. Starch is hydrolysed by both alpha- and beta-amylases. In addition to starch hydrolysis, an important function of mashing is to degrade the original high molecular weight soluble proteins into simpler compounds, as otherwise the high molecular complex proteins will cause turbidity and haziness in the beer. There are two types of mashing; decoction and infusion.

In decoction mashing, a part of the mash is withdrawn, boiled and then returned to the mash vat to raise the temperature of the whole mash. In infusion mashing conversion and extraction of wort is carried out in a single stage and at a single temperature in the mash vat. During mashing the boiled adjunct, usually maize (Figure 8.11). or sorghum, is added and thoroughly mixed to yield a mash mixture called wort. The wort is next boiled (Figure 8.12). Boiling kills most of the micro-organisms in the wort, coagulates some of the hydrolysed proteins, inactivates any remaining enzymes and sterilizes the liquid. The hot wort must be cooled down to about 6–16°C before fermentation, as high temperatures will inactivitate the yeast.

After the wort has been cooled it is left to ferment; the major change occurring during fermentation is the conversion of fermentable sugar (monosaccharides and disaccharides) to alcohol and carbon dioxide. Fermentation is considered to be complete when the desired degree of sugar conversion, called attenuation, has taken place. The pH value falls due to formation of carbon

**Figure 8.11 During the brewing process, maize is added and mixed into the mash to yield wort.**

**Figure 8.12 In traditional brewing the wort is boiled to kill micro-organisms.**

dioxide and organic acids, mostly lactic acids.

According to Novellie (1977) sorghum beer making depends upon two fermentations involving bacteria that produce lactic acid and yeasts that produce alcohol.

In the modern brewing process, a mixture of sorghum malt and water is first soured by thermophilic lactobacilli, then boiled with the grain adjunct – sorghum or maize brewer's grits – cooled to 60°C and saccharified with more sorghum malt. The mixture is strained to remove the coarsest particles and fermented with a top fermenting yeast, *Saccharomyces cerevisiae*. The resultant beer is sour (lactic acid), with a pH value ranging from 3.3–3.5. It is opaque because of suspended solids and yeast, and is still fermenting and lightly alcoholic, 3.0–3.5 per cent alcohol by weight (Novellie, 1977). Sorghum beer differs in many ways from that of barley (European beer) (Table 8.9).

**Table 8.9 Comparison of sorghum beer and barley beer**

| Beer | Raw materials used | | | Type of fermentation |
|---|---|---|---|---|
| | Malt | Adjunct | Hops | |
| Sorghum | Sorghum malt | Maize or sorghum | None | Lactic and alcoholic |
| European | Barley malt | Cereals or syrups | Some | Alcoholic |

Source: Novellie, 1977.

The nutritional value of sorghum beer is reported to be higher than that of most other beers because of the yeast and other suspended material. The value depends very much on the ingredients used; beer made from highly refined maize as an adjunct and unrefined sorghum malt has a lower vitamin value than that made from the same malt with ground whole unrefined sorghum grain as an adjunct.

Novellie (1977) reports that both of the millets, *Eleusine coracana* and *Pennisetum typhoides*, can be malted and used wholly or partly in place of sorghum malt in brewing. Both millets are capable of giving good diastatic powers comparable with those obtained from sorghum. However, the small size of these grains raises so many technical problems in modern malting plants that the millet in malted form has virtually been replaced by sorghum malt in brewing. Table 8.10 presents a summary of the use of sorghum in beer and alcohol production in various parts of the world.

**Table 8.10 Sorghum in beer and alcohol production**

| Country or region | Product | Method of processing |
|---|---|---|
| West Africa | Beer or pito – a relatively clear red liquid with a sweet pleasant taste and a low solids content. Continued fermentation produces a sour taste. About 1–5% alcohol | Sorghum is steeped and germinated for 4–6 days and dried. Mash is prepared by mixing the ground sorghum malt with water, the mixture is brought to boiling, the material is filtered. The wort is boiled, cooled, and inoculated with yeast from a previous batch of beer. Fermentation occurs overnight and actively fermenting beer is drunk the next day |
| Ethiopia | Tala (Ethiopian beer) | Made from red sorghum. It is much more dilute (1.39% solids) than the beer made in South Africa (6–8% total solids) |
| Mexico | Typical European beer. Pilsner type of beer | Sorghum grits are used as a brewing adjunct. Must be high in starch and low in fat and fibre content. |
| South Africa | Bantu or Kaffir Beer. (Alcohol content 1–8% by volume.) High solids content with a sour taste. Bright pink colour, the consistency of a milk shake. Alcohol content depends on length of fermentation time | Ground sorghum malt is mixed with water, allowed to sour, and then boiled with an adjunct (corn grits), cooled to 60°C, and saccharified with more sorghum malt. The mixture is filtered to remove the coarsest particles and fermented with a top-fermenting yeast |
| China | Distilled spirits (called sorghum wine in China) | Sorghum is the substrate for alcoholic fermentation. The 'sorghum beer' is distilled and the distillate is used after some ageing |
| World-wide | Alcohol | Sorghum is used as the substrate for alcoholic fermentation, followed by distillation. About 12 litres of alcohol are produced from 25 kg (1 bushel) of sorghum |

### *Traditional brewing of bantu or kaffir beer*

Bantu beer is different from beer made from barley malt in that the usual process of starch conversion and alcohol fermentation is preceded by a souring process which influences the body, stability, and alcohol content of the final product. The malt for Bantu beer is made principally in the conventional way, but while barley for malt is germinated at 18°C, sorghum is germinated after the soaking process at about 30°C and the development of roots and plumules in the germinating grain is far greater than occurs in barley malt. As a result, conversion of the endosperm is more advanced while diastatic activity is comparatively lower (Novellie, 1977).

Germination is stopped by drying in the sun, after which the malt is ground and mashed in water kept at a temperature of about 50°C, and inoculated with *Lactobacillus delbrückii* for the production of lactic acid. When the souring process, which usually lasts 8–12 hours, is advanced to a stage of about 0.4 g of lactic acid per 100 ml of liquid, a source of starch is added. Traditionally ground sorghum is used. The mixture produced is cooked for 30–45 minutes to sterilize the mass and to gelatinize the starch. After cooling to 30°C a fresh batch of ground sorghum is added to the gelatinous mixture for liquefying and conversion to sugar by the amylase in the malt. When the conversion is advanced sufficiently, the mixture is strained and brewers' yeast is added for alcohol fermentation, which usually takes 3–5 days. Typical Bantu beer is an

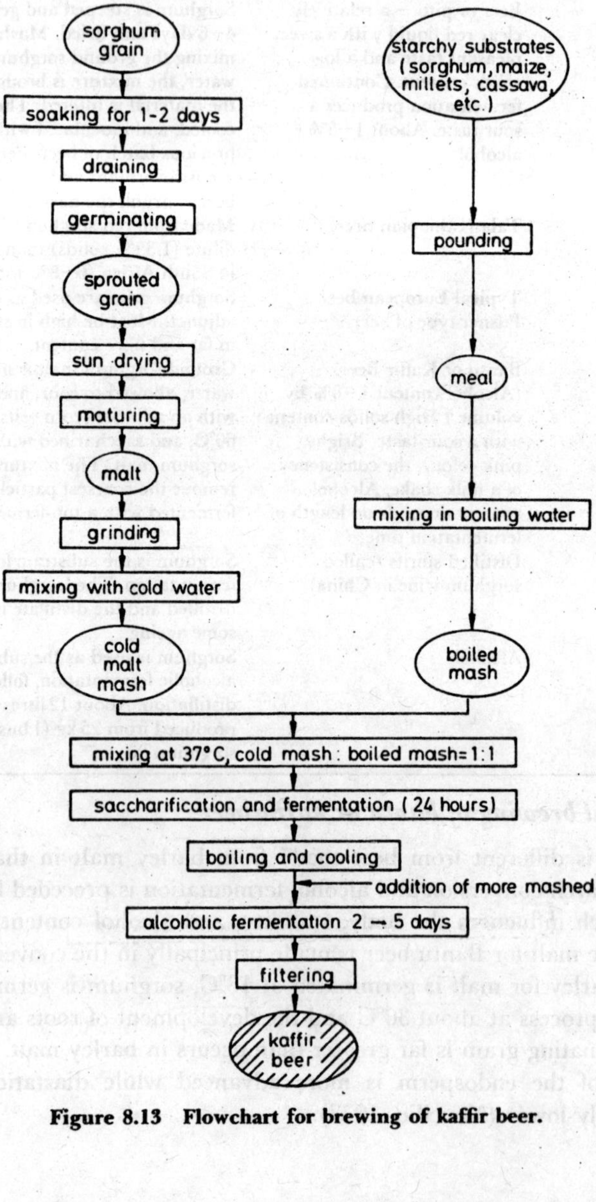


**Figure 8.13 Flowchart for brewing of kaffir beer.**

opaque fluid with a yeasty sour taste, low alcohol content (about 3 per cent) but high nutritional value (Figure 8.13).

The beer may contain 4–6 per cent of total solids, about half of which will be starch. The average protein content is about 0.6 per cent while the content of thiamine, riboflavin, and nicotinic acid is high, averaging 32, 36, and 189 µg per 100 ml of beer, respectively. To make each litre of beer requires 180–360 g of grain.

### *The making of pito*

Pito making involves the process of soaking, germination (malting), grinding and fermentation. The complete process of pito making requires 10–11 days.

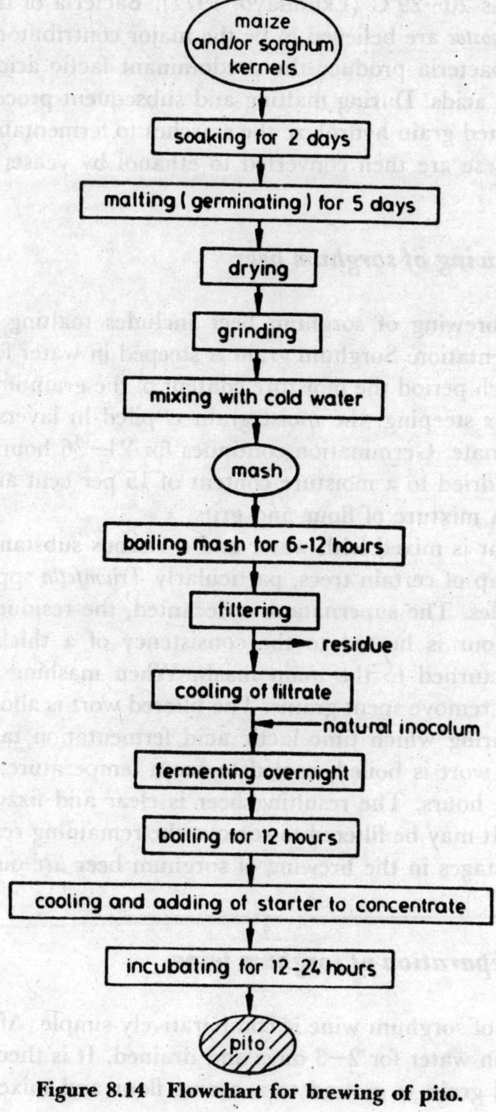


**Figure 8.14 Flowchart for brewing of pito.**

Maize and/or sorghum grain is soaked in water for up to 48 hours. Following soaking, the grain is drained, and held in a moist chamber for about 5 days for germination to take place. The malted grain is then sun dried, ground and mixed with water to the desired consistency. The mash is then boiled, sieved and left to cool. In the next stage, the cooled filtrate is left to ferment overnight. After this initial fermentation the filtrate is boiled for several hours to concentrate it, and is then cooled. Sediment from a previous brew is added as a starter to the cooled concentrate, and after overnight incubation the pito is ready for consumption (Figure 8.14).

The pH of the filtrate from the mash prepared from ground malt ranges from 6.2–6.6. It decreases to 5.4 during the first fermentation, and falls further to about 4.4 during subsequent fermentation. The optimum temperature for fermentation is 20–29°C (Ekundayo, 1977). Bacteria of the genera *Lactobacillus* and *Leuconostoc* are believed to be the major contributors to the acidity. The lactic acid bacteria produce the predominant lactic acids and traces of acetic and formic acids. During malting and subsequent processing, amylases from the germinated grain hydrolyze the starches to fermentable sugars (maltose, glucose). These are then converted to ethanol by yeasts from the genus *Saccharomyces*.

## *Traditional brewing of sorghum beer*

The traditional brewing of sorghum beer includes malting, mashing, wort boiling and fermentation. Sorghum grain is steeped in water for 24–72 hours, at the end of which period the moisture content of the grain increases to about 40 per cent. After steeping, the moist grain is piled in layers of about 4 cm, and left to germinate. Germination continues for 24–36 hours. The resulting malt is then sun dried to a moisture content of 15 per cent and subsequently ground to yield a mixture of flour and grits.

The coarse flour is mixed with water and a viscous substance from *Hibiscus esculenta*, or the sap of certain trees, particularly *Triumfetta* spp., to precipitate suspended particles. The supernatant is decanted, the residue containing undissolved malt flour is heated to the consistency of a thick paste and the supernatant is returned to the main mash. When mashing is complete the wort is filtered to remove spent grains. The filtered wort is allowed to stand for 20–24 hours, during which time lactic acid fermentation takes place. After fermentation the wort is boiled, cooled to room temperature, and allowed to ferment for some hours. The resulting beer is clear and fizzy and has a low alcohol content. It may be filtered to remove the remaining residue before it is consumed. The stages in the brewing of sorghum beer are outlined in Figure 8.15.

## *Traditional preparation of sorghum wine*

The preparation of sorghum wine is comparatively simple. After washing, the grain is steeped in water for 2–3 days and drained. It is then sun dried for a day. The treated grain is ground to a coarse flour and mixed with water to

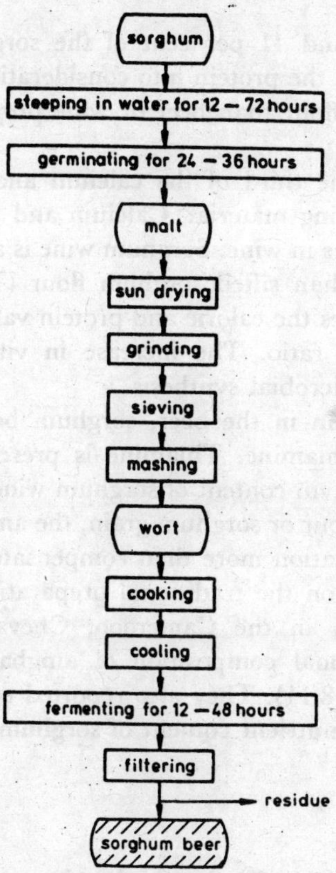


**Figure 8.15 Traditional brewing of sorghum beer.**

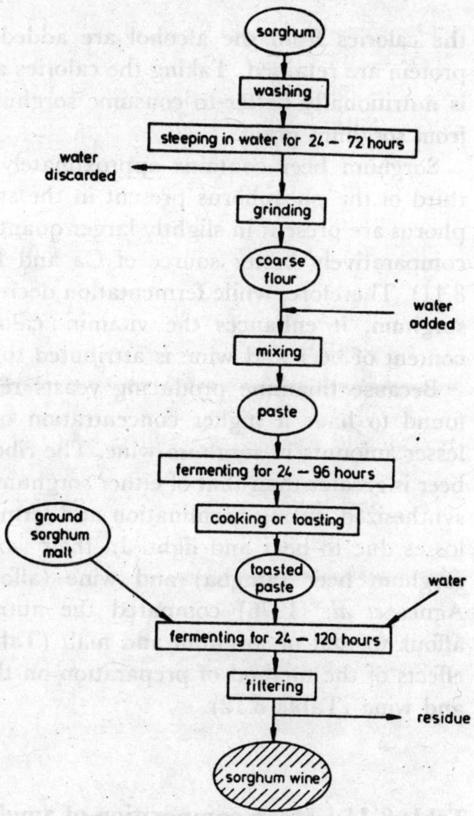


**Figure 8.16 Traditional preparation of sorghum wine.**

form a paste, which is left to stand for up to 4 days, or until a sour lactic acid taste develops. The paste is toasted to a firm consistency and then cooled.

In the next step, the toasted paste is mixed with ground sorghum malt and water. The malt amylases and micro-organisms rapidly bring about liquefaction and the mixture is allowed to ferment for a couple of days, 5 days at the most. For an increase of diastatic activity a small amount of malt may be added and the mixture is filtered to give sorghum wine (Figure 8.16).

## Nutrient content of sorghum beer and wine

In the preparation of sorghum beer and wine, considerable amounts of solids are lost, particularly in the grinding and decanting operations. Sorghum beer and wine contain respectively only 20 and 30 per cent of the original sorghum solids.

Sifted sorghum flour contains 75 per cent of the calories and 69 per cent of the protein of the original grain; equivalent figures for sorghum beer are 20 per cent of the calories (29 per cent if the calories from the alcohol are included), and 19 per cent of the protein. The preparation of wine also results in nutrient losses, as only 33 per cent of the sorghum calories (55 per cent if the calories from the alcohol are added) and 41 per cent of the sorghum protein are retained. Taking the calories and the protein into consideration, it is nutritionally better to consume sorghum flour than beer or wine prepared from sorghum grain.

Sorghum beer contains approximately one third of the calcium and one third of the phosphorus present in the starting material. Calcium and phosphorus are present in slightly larger quantities in wine. Sorghum wine is also a comparatively better source of Ca and P than sifted sorghum flour (Table 8.11). Therefore, while fermentation decreases the calorie and protein value of sorghum, it enhances the vitamin : calorie ratio. The increase in vitamin content of beer and wine is attributed to microbial synthesis.

Because thiamine producing yeasts remain in the beer, sorghum beer is found to have a higher concentration of thiamine. Thiamine is present in lesser amounts in sorghum wine. The riboflavin content of sorghum wine and beer is greater than that of either sorghum flour or sorghum grain; the amount synthesized during germination and fermentation more than compensates for losses due to heat and light. In their work on the traditional preparation of sorghum beer (amgba) and wine (affouk) in the Cameroon, Chevassus-Agnes *et al.* (1976) compared the nutritional composition of amgba and affouk to that of sorghum and malt (Table 8.11). They also reported on the effects of the method of preparation on the nutrient content of sorghum beer and wine (Table 8.12).

**Table 8.11 Mean composition of amgba and affouk calculated on a dry weight bases**

| Constituent | Sorghum | Malt | Amgba (beer) | Affouk (wine) | Residue |
|---|---|---|---|---|---|
| Moisture | 9.6 | 24.7 | 90.7 | 86.3 | 69.2 |
| Calories (joules) | 1595 | 1591 | 1649** | 1708** | 1009 |
| Protein (%) | 9.4 | 9.8 | 8.7 | 12.6 | 17.7 |
| Lysine (% total protein) | 3.3 | 3.7 | 7.2 | 4.6 | 2.4 |
| Lipid (%) | 2.8 | 2.2 | 0.3 | 3.3 | 3.2 |
| Total carbohydrate (%) | 85.6 | 86.2 | 86.1 | 80.4 | 77.5 |
| Cellulose (%) | 2.3 | 3.7 | 0.3 | 0.9 | 11.1 |
| Ash (%) | 2.14 | 1.75 | 4.06 | 3.7 | 1.6 |
| Minerals (mg/100 g) | | | | | |
| Calcium | 11.0 | 9.3 | 20.7 | 30.1 | 8.6 |
| Phosphorus | 319 | 327 | 630 | 684 | 260 |
| Phytic phosphorus | 166 | 85 | 112 | 320 | 85 |
| Vitamins (mg/100 g) | | | | | |
| Thiamine | 0.41 | 0.43 | 3.44 | 0.89 | 0.10 |
| (range) | (0.17–0.54) | (0.17–0.56) | (1.69–5.24) | (0.31–1.10) | (0.07–0.12) |
| Riboflavin | 0.10 | 0.23 | 0.76 | 0.44 | 0.18 |
| Niacin | 4.3 | 5.3 | 8.0 | 5.9 | 6.9 |
| Alcohol (°GL) | – | – | 2.7 | 5 | – |

* Carbohydrate is calculated and reported by difference, that is, by substracting the analysed combined percentages of moisture, protein, lipids, fibre and mineral matter from 100.
** Values include calories provided by the alcohol.
Source: Chevassus-Agnes *et al.*, 1976.

**Table 8.12 Comparison of nutritional composition of sifted flour, amgba and affouk as a percentage of the whole grain**

| Constituent | Sorghum | Sifted flour* | Amgba (beer) | Affouk (wine) |
|---|---|---|---|---|
| Solids | 100 | 70 | 20 | 31 |
| Calories (alcohol not included) (joules) | 419 | 314 | 84 | 138 |
| Total calories (alcohol included) (joules) | 419 | 314 | 121 | 230 |
| Protein | 100 | 69 | 19 | 41 |
| Lysine | 100 | ND | 41 | 56 |
| Lipid | 100 | 35 | 0 | 28 |
| Total carbohydrate | 100 | 72 | 19 | 29 |
| Cellulose | 100 | 28 | 0 | 12 |
| Ash | 100 | 39 | 36 | 42 |
| Calcium | 100 | 59 | 35 | 81 |
| Total phosphorus | 100 | 41 | 38 | 65 |
| Phytic phosphorus | 100 | ND | 13 | 47 |
| Thiamine | 100 | 39 | 160 | 65 |
| Riboflavin | 100 | 34 | 145 | 141 |
| Niacin | 100 | 30 | 35 | 41 |

*Obtained by traditional milling methods (Favier *et al.*, 1976).
ND-No values determined
Source: Chevassus-Agnes *et al.*, 1976.

Table 8.13 shows the composition of sorghum grain, sorghum malt, wort, and beer traditionally brewed by Moba woman in Togo. The difference in values of nutrients in the original grain and in the resulting beer is striking. The beer was higher in riboflavin and vitamin $B_{12}$ but lower in all the other nutrients determined.

**Table 8.13 Composition of sorghum grain, sorghum malt, wort and beer, traditionally brewed in Togo**

| Constituent | Sorghum grain (per 100 g) | Sorghum malt (per 100 g) | Wort (per 100 ml) | Beer (per 100 ml) |
|---|---|---|---|---|
| Moisture (g) | 15.3 | 15.0 | ND | ND |
| Calories (joule) | 1340 | 1340 | 168 | 146 |
| Protein (g) | 8.0 | 8.1 | 0.3 | 0.3 |
| Fat (g) | 2.3 | 1.6 | ND | ND |
| Carbohydrate (g) | 72.1 | 73.4 | 10.3 | 3.3 |
| Ash (g) | 2.3 | 1.9 | 0.2 | 0.2 |
| Calcium (mg) | 33 | 28 | 1.6 | 1.4 |
| Thiamine (mg) | 0.56 | 0.44 | 0.03 | 0.03 |
| Niacin (mg) | 4.9 | 7.1 | 0.6 | 0.5 |
| Riboflavin (mg) | 0.08 | 0.25 | 0.02 | 0.04 |
| Vitamin $B_{12}$ (μg) | traces | traces | 0.01 | 0.03 |
| Panthothenic acid (mg) | 0.53 | 0.96 | 0.07 | 0.09 |
| Alcohol (g) | – | – | – | 3.03 |

ND = no values determined.
Source: Perisse *et al.*, 1959.

## Sweet sorghum juice

Sorghum syrup or 'molasses' is made from the juice of certain varieties of sweet sorghum, also called 'sorgo'. Compared with grain sorghums it has juicy stalks with a high sugar content. A good variety of sweet sorghum yields at least 22.5 kg of juice from 45 kg of stalks when passed through a roller mill.

The traditional method of processing sorghum syrup includes stripping the leaves from the stalks and topping the seed heads. The stalks are then cut and carried to hand-operated or animal-powered mills, where the juice is expressed. The juice is then boiled in open pans over wood fires until it reaches the desired consistency for syrup. Production units are small, from 0.5–2 ha. This method of production is still practised in many rural areas. For large-scale production units, more efficient processing equipment is used, but the basic unit operations are the same. The application of improved equipment has undoubtedly resulted in syrups with better flavour, colour and stability.

Juice is stored in shallow tanks immediately after milling to allow heavy particles of trash, bagasse and starch to settle to the bottom, thus enhancing clarification. This stage of the processing chain is important because the large quantities of starch extracted during milling of the stalk are undesirable as they cause jelling of the syrup.

Clarification is usually done by heat alone, but clay and filter acids can be used to assist in the operation. Simple open-type evaporators are used for clarification, evaporation and concentration of the syrup. After concentration, the syrup is cooked at high temperatures under normal air pressure long enough for the characteristic sorghum flavour to develop. The minimum temperature for this flavour development appears to be about 106°C. To prevent formation of sucrose crystals in stored syrup, an enzyme, invertase, can be added either to the juice or to the thin syrup. This enzyme degrades the sucrose to simple sugars, dextrose and levulose. Similarly, problems with starch can be reduced by the addition of amylase or malt diastase to the syrup after the starch has been gelatinized by heat.

## *References*


**Abdel Gadir, A.M.** and **Mohamed, M.** (1977). *Kisra: Sudanese lactic/acetic acid fermented sorghum bread*. Symposium on Indigenous Fermented Foods, Bangkok, Thailand.

**AEL** (1976). *Schrot – und Mischanlagen im landwirtschaftlichen Betrieb*, Arbeitsgemeinschaft für Elektrizitätsanwendung in der Landwirtschaft e.v., Heft 7.

**Agren, G.** and **Gibson, R.** (1968). *Food composition table for use in Ethiopia*. I. Children's Nutrition Unit (CNU) Report No. 16. Addis Ababa, Ethiopia.

**Bhatia, B.S., Chakrabarty, T.K., Mathur, V.K., Siddish, C.H.** and **Raghavan, P.K.V.** (1968). Use of maize and milo flours in the preparation of bread. *Indian Fd. Pckr.* 22–35.

**Carr, W.R.** (1956). The preparation and analysis of some African foodstuffs. *Centr. Afric. J. Med.* **2**: 334–339.

**Carr, W.R.** (1961). Observations on the nutritive value of traditionally ground cereals in Southern Rhodesia. *British Journal of Nutrition*, **15**: 339–343.

**Casier, J.P.J., de Paepe G., Willem, H., Goffings, G.** and **Noppen, H.** (1976). *International Association for cereal chemistry*, Vienna 11–12th May 1976. (Dendy, D.A.V., ed.), Tropical Products Institute, London.

**Chevassus-Agnes, S., Favier, J.C.** and **Josef, A.** (1976). Traditional technology and nutritive value of sorghum beer from Cameroon (in French), *Cah. Nutr. Diet.* **11**: 89–104.

**Crabtree, J.** and **Dendy, D.A.V.** (1977). Co-milling of wheat and millet grain, in *Proceedings, symposium sorghum and millets for human food*, Tropical Products Institute, London.

**Deshaprabhu, S.B.** (1968) *The Wealth of India: raw materials*. Vol. 7. Publication and Information Directorate, CSIR, New Dehli, India.

**Desikachar, H.S.R.** (1975). Processing of Maize, Sorghum and Millets for food uses, *Journal of Scientific and Industrial Research*, **34**: No. 4, 231–237.

**Desikachar, H.S.R.** (1977). Processing of sorghum and millets for versatile food uses in India, in *Symposium on sorghum and millets for human food*, International Association for Cereal Chemistry 1976, S. 41–46.

**Ekundayo, T.A.** (1977). *Nigerian pits*. Symposium on Indigenous Fermented Foods, Thailand.

**FAO** (1971). Food and Agriculture Industries Service Division of FAO. The current situation in relation to the milling of sorghum and millets and their utilization for flour. Article IV. *Agric. Econ. Bull. Africa* (E/CN 14/AGREB), **13**:73.

**FAO** (1985). *FAO Production Yearbook*, 1985. Rome.

**Favier, J.C., Chevassus-Agnes, S., Joseph, A.** and **Gallon, G.** (1972). Traditional Technology of Sorghum in Cameroon – influence of pestling on nutritional value (in French). *Annls. Nutr. Aliment.* **26**, 221–250.

**Freeman, J.E.** and **Bocan, B.J.** (1973). Pearl millet. A potential crop for wet milling. *Cereal Science Today*, **18**: 69.

**Gast, M.** (1965). Etude Ethnologique, in (Adrian, J. and Gast, M., eds.), *Mils et sorhgo en Ahaggar*, 13–37.

**Goussault, B.** and **J. Adrain** (1977). *The milling of Pennisetum millet and the value of the protein in the products*. International Association for cereal chemistry, Vienna, 11–12th May 1976. Tropical Products Institute, London.

**Hulse, J.H., Laing, E.M.** and **Pearson, O.E.** (1980). *Sorghum and the millets: Their composition and nutritive value*. Academic Press, Toronto.

**ITDG** (1973). *Tools for agriculture*, Guide to hand-operated and animal drawn equipment, Intermediate Technology Development Group, London.

**John, S.W.** and **H.G. Muller** (1971). Sorghums – have we overlooked their real value? *Milling*, **153**: 9, 21–24.

**Kent, N.L.** (1983). *Technology of Cereals*, 3rd Edition, Chapter 17. Pergamon Press. Oxford.

**Loncin, M.** (1969). *Die Grundlagen der Verfahrenstechnik in der Lebensmittelindustrie*, Verlag, Sauerländer, Aarau und Frankfurt/M.

**Novellie, L.** (1977). *Beverages from sorghum and millets*. Proceedings of a symposium on sorghum and millet for human food. 9th Congress, I.C.C., Tropical Products Institute, London.

**Perisse, J., Adrian, J., Rerat, A.** and **Le Berre, S.** (1959). Nutrient balance in conversion of sorghum into beer. Preparation, composition and consumption of a Togo beer. *Annls. Nutr. Aliment*, **13**: S. 1–15 (cited from Hulse *et al.*, 456).

**Perten, H.** (1977). *Specific characteristics of millet and sorghum milling*. International association for cereal chemistry, Vienna, 11–12th May 1976. Tropical Products Institute, London.

**Platt, B.S.** (1955). Some traditional alcoholic beverages and their importance in indigenous African communities. *Proc. Nutr. Soc.* **14**: 115–124.

**Purseglove, J.W.** (1972). *Tropical crops, Monocotyledons*. Longman, London.

**Reichert, R.D.** and **Youngs, C.G.** (1976). Dehulling cereal grains and grain legumes for developing countries. I. Quantitative Comparison between attrition – and abbrasive-type mills, *Cereal Chemistry*, **53** (6): 829–839.

**Rooney, L.W.** and **Clark, L.E.** (1968). The chemistry and processing of sorghum grains. *Cereal Science Today*, **13**: (7).

**Rooney, L.W.** and **Sullins, R.D.** (1976). The structure of sorghum and its relation to processing and nutritional value, in *Sorghum and Millets for Human Food*, 91–110.

**Shoup, E.K., Deyoe, C.W., Farrell, E.P., Hammond, D.L.** and **Miller, G.D.** (1970b). Sorghum grain milling. *Food Technology*, **24**: 1028–1032.

**Sunner, A.K.** and **Nielsen, M.A.** (1976). *Food legume utilization*. Progress Report No. 4 July 1, 1975 to June 30, 1976. IDRC Project File No. 3-P-73-0032. Ottawa, Canada.

**Vogel, S., Gobezie, A.** and **Gifawesen Ć.** (1977). *Enjera – a fermented pancake-like bread Ethiopia*. Symposium on Indigenous Fermented Foods, Bangkok, Thailand.

**Wall, J.S.** and **Blessin, C.W.** (1969). Composition and structure of sorghum grains. *Cereal Science Today*, **14** (18): 264–270. 276.

**Watson, S.A.** (1970). Wet milling process and products, in *Sorghum production and utilization* (J.S. Wall and W.M. Ross, eds.), 602–626. AVI Publishing Co. Inc., Westport, Connecticut.

# 9

# Soya bean

## Characteristics

The soya bean, *Glycine max* (L.) Merrill, is a grain legume (Figure 9.1). It is cultivated in many areas of the world, from the tropics to temperate regions. World production of soya beans amounts to 100 839 000 tonnes. At present the United States of America is the largest producer, accounting for about 60 per cent of world production, followed by China and Brazil (FAO, 1986).

The seed yield is between 1400 and 2000 kg/hectare. Soya, unlike most oil seed species, is a dual purpose plant; the seeds (beans) of the high yielding variety contain about 18 per cent oil and 38 per cent protein, and the 'extraction residue' represents more than 40 per cent of the utilization value of the plant. Without this, cultivation of soya bean with such a low oil content would not be lucrative.

**Figure 9.1 Mature soya beans.**

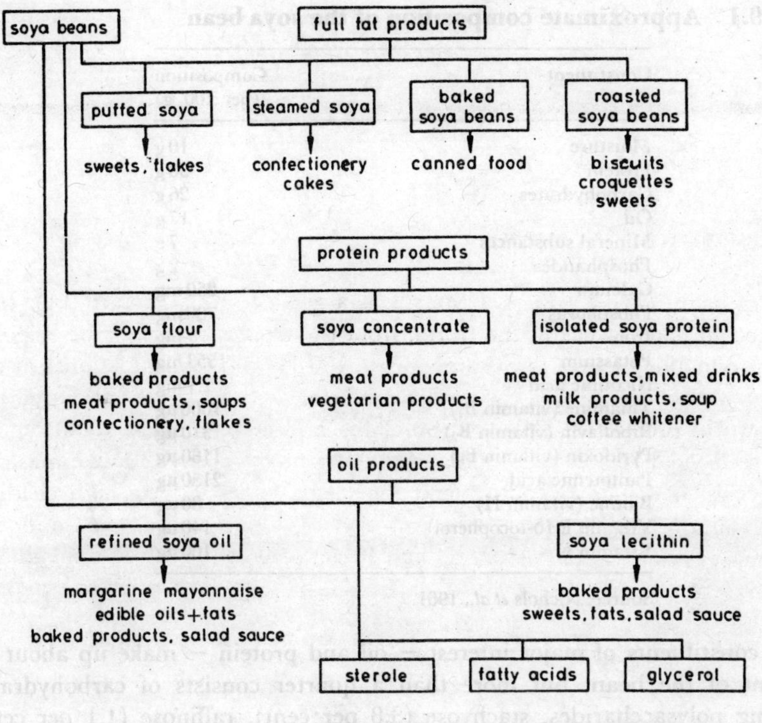


**Figure 9.2 The food uses of soya beans (American Soybean Association, 1981).**

For centuries the soya bean has played a major role in the diet of the population of eastern Asia, particularly China, where it was considered to be one of the five sacred 'grains', the other four being rice, wheat, barley and millet. It was not until the beginning of the 19th century that people in Western countries recognized the importance of the soya bean as a potential source of high quality protein and oil.

## Uses

The uses of soya beans are numerous (Figure 9.2). Unprocessed, they have an undesirable flavour and bitterness; moreover, they contain the toxic proteins hemagglutinin and antitrypsin. These substances must be destroyed or inactivitated to make the beans palatable and digestible both for human and animal consumption. Soya beans can be processed into soya milk, soya sauce, tofu (soya bean curd), yoghurt, soya bean sprouts, tempeh (soya steak), soya flour and other products. Defatted soya beans can be used for the manufacture of protein isolates and concentrates.

## Chemical composition

The approximate composition of soya beans is given in Table 9.1.

**Table 9.1 Approximate composition of the soya bean**

| Constituent | Composition (per 100 g) |
|---|---|
| Moisture | 10 g |
| Protein | 38 g |
| Carbohydrates | 26 g |
| Oil | 17 g |
| Mineral substances | 7 g |
| Phosphatides | 2 g |
| Calcium | 250 mg |
| Phosphorus | 700 mg |
| Iron | 7 mg |
| Potassium | 1553 mg |
| Nicotinic acid | 2.1 mg |
| Thiamine (vitamin $B_1$) | 1000 μg |
| Riboflavin (vitamin $B_2$) | 310 μg |
| Pyridoxin (vitamin $B_6$) | 1180 μg |
| Pantotemic acid | 2150 μg |
| Biotine (vitamin H) | 80 μg |
| Vitamin E (δ-tocopherol) | 140 μg |
| Vitamin K | 100 μg |

Source: Nichols *et al.*, 1961.

The constituents of major interest – oil and protein – make up about 55 per cent of the bean, but more than a quarter consists of carbohydrates including polysaccharides, stachyose (3.8 per cent), raffinose (1.1 per cent) and sucrose (5.0 per cent). Soya oil belongs to the group of oils most valuable for human nutrition, with the highest content of polyunsaturated fatty acids, linoleic acid and linolenic acid (Table 9.2). The high content of linoleic acid (essential fatty acid) makes soya oil superior in nutritive value to other vegetable oils, since this acid cannot be synthesized by the body and must be provided by the diet. The other advantage of linoleic acid is its ability to reduce blood cholesterol levels, thus decreasing the risk of developing heart and blood vessel diseases. On the other hand, the presence of linolenic acid is disadvantageous as it makes the oil susceptible to oxidation.

**Table 9.2 Fatty acid composition of various oils and fats**

| Fatty acid | Soya bean oil | Sunflower oil | Maize oil | Groundnut oil | Coconut oil | Palm oil | Lard oil | Beef tallow |
|---|---|---|---|---|---|---|---|---|
| | Percentage composition | | | | | | | |
| **Saturated** | | | | | | | | |
| Lauric acid | — | — | — | — | 47.0 | — | 0.5 | 0.5 |
| Myristic acid | 0.5 | 0.5 | — | — | — | — | 2.0 | 3.0 |
| Palmitic acid | 12.0 | 6.8 | 13.2 | 11.0 | 9.0 | 43.0 | 27.5 | 26.8 |
| Stearic acid | 4.0 | 3.5 | 2.5 | 3.0 | 2.0 | 5.0 | 13.3 | 17.5 |
| **Unsaturated** | | | | | | | | |
| Oleic acid | 23.3 | 24.7 | 32.3 | 53.0 | 7.0 | 38.0 | 42.8 | 40.5 |
| Linoleic acid | 52.3 | 63.2 | 49.3 | 27.0 | 2.0 | 10.0 | 7.3 | 2.5 |
| Linolenic acid | 7.2 | — | 1.0 | — | — | — | 0.8 | 0.8 |

Tocopherols, in particular alpha-tocopherol, act as natural antioxidants and are responsible for the keeping qualities of most vegetable oils. In spite of the relatively high content of tocopherol in soya oil (about 120 mg per 100 g oil), the shelf life of soya oil is short compared with such other oils as cotton-seed oil, maize oil and groundnut oil. This is attributed primarily to the relatively high content of linolenic acid, which auto-oxidizes, leading to the 'green', 'beany' taste of the oil, commonly referred to as 'flavour-reversion'. Thus the removal by liquid-liquid fractionation of most of the glycerides containing linolenic acid results in oil with better flavour stability

## *Antinutritional and organoleptic properties of soya bean*

The raw or undercooked soya bean has low food value because it contains a number of antinutritional factors. Some of the biologically active substances in soya beans that have been identified include trypsin inhibitors, haemagglutinins, urease enzymes, lipoxidase saponins and compounds with oestrogenic activity. It is generally known that undercooked soya bean oil meal has low nutritive value. The reason for this is believed to be the presence of a growth inhibitor, known as trypsin inhibitor. Trypsin is an enzyme in pancreatic juices which digests the protein that man and animals eat. The protein cannot be digested if the digestive process is inhibited as a result of the intake of undercooked soya beans. The lack of digestion of the protein results in increased secretion of pancreatic enzymes into the faeces. Excessive loss of essential amino-acids through the faeces is reported to be responsible for the inhibition of growth in young rats after eating soya beans (Rackis *et al.*, 1979).

Trypsin inhibitors are inactivated by moist heat treatment. Proper heat treatment also accomplishes another function, namely the reduction of urease activity in the soya bean. Urease is the main enzyme present in the soya bean, and it decomposes into ammonia and carbon dioxide. High urease activity causes the urease to decompose, perhaps releasing ammonia in sufficient concentration to make the product unpalatable or even hazardous.

Haemagglutinins in defatted soya bean meal consist of several glycoproteins capable of agglutinating red blood cells (Lis *et al.*, 1966; Turner and Liener, 1975). The agglutinating activity is readily destroyed by moist heat or by digestion with pepsin. Flavours and odours associated with soya beans are considered to be two of the major factors limiting their use in foods. Flavour compounds of various types, reported to be responsible for the objectionable flavour of soya bean products, have been isolated. These include carbonyl compounds, phenolic acids, volatile fatty acids and amines, alcohols and phosphatidylcholine. Phosphatidylcholine is a potential source of bitterness in soya bean products (Sessa *et al.*, 1974). The involvement of the hydroperoxides of linoleic and linolenic acids in the origin of some flavours (grassy, beany) of soya bean products has been investigated (Kalbrener *et al.*, 1974).

Table 9.4 summarizes some of the biologically active substances found in soya beans, their biological responses and the methods for their deactivation. For detailed discussions of nutritional, biological and physiological properties

**Table 9.4 Biologically active substances and flavour/odour compounds in raw soya-beans, their biological responses, and the methods for their destruction or reduction**

| Substance | Biological response | Method of destruction | Reference |
|---|---|---|---|
| Trypsin inhibitors | Increased synthesis and secretion of pancreatic enzymes, pancreatic hypertrophy and inhibition of growth | Moist heat treatment; drying; addition of proteolytics | Rackis *et al.* (1979) |
| Urease | Increased release of ammonia; intoxication | Cooking | |
| Hemagglutinins | Agglutination of red blood cells | Moist heat treatment; pepsin digestion | Lis *et al.* (1966) |
| Goitrogen | Enlargement of thyroid gland | Addition of iodine to diet; autoclaving | Turner and Liener (1975) |
| Oestrogens | Inhibition of growth; increased uterine weight | Extraction with aqueous ethanol; preparation of isolates | Kanijn *et al.* (1973) Carter *et al.* (1953) |
| Allergen | Allergy and asthma | Autoclaving | Rabner *et al.* (1955) |
| Raffinose and stachyose | Flatulence | Aqueous extraction as used in preparation of concentrates and isolates, soaking, germination or fermentation | Perlman (1965) Cristofaro *et al.* (1974) Rackis (1976) |
| Oxidized phosphatidylcholine | Bitterness | Moist heat treatment; soaking | Wolff (1982) |
| Phenolic acids | Astrigency | Cooking | Sugimoto and van Buren (1970), Mital and Steinkraus (1975) |
| Oxidized linoleic and linolenic acids | Beany flavour | Inactivation of lipoxygenase by wet grinding at 80°C or higher, dry heating of cracked beans followed by extrusion cooking; blanching of intact beans, soaking of beans in aqueous alcohol | |

of soya beans and soya bean proteins reviews by Liener (1972), Rackis (1972, 1974, 1976, 1979) and Wolff (1982) should be consulted.

## Processing of soya beans into various traditional products

Proper storage of soya beans is an important factor due to their high susceptibility to fungi. Their moisture content should not exceed 13 per cent in temperate regions, and 8 per cent in the tropics if spoilage is to be avoided.

The unprocessed soya bean does not play a significant role in human nutrition as, apart from being bitter, it is relatively difficult to cook well using conventional techniques. Soya beans can be made palatable, easily digestible and nutritious by processing into various products, the most popular of which are tempeh, tofu, soya milk, soya sauce and soya sprouts. The whole soya bean plant yields a palatable leafy vegetable and a high protein hay. The fully grown, half-ripe beans can be cooked relatively easily and eaten like any other cooked bean.

Soya bean food products can be classified into general types with respect to the processing technology used, namely, nonfermented and fermented.

### *Soya sprouts*

Soya beans are sprouted in the same way as mung beans (*Vigna radiata*). Sprouting takes 3–5 days at a temperature of 25°–30°C, and is carried out in the dark to produce white coloured, straight sprouts. After sprouting, the soya bean sprouts may reach a length of 3–5 cm.

According to Nichols *et al.* (1961) soya bean sprouts contain 10 mg vitamin C per 100 g sprout. In the original unsprouted bean vitamin C is not present. Fresh soya sprouts are boiled in water for 5 minutes prior to consumption as a vegetable.

### *Traditional method of preparing tempeh (soya steak)*

Tempeh is a white mould-covered cake produced by fungal fermentation of dehulled, soaked, and partially cooked soya beans. Tempeh has been prepared and consumed in Indonesia for centuries. There exists a broad range of tempeh-making processes in different localities and countries. The processing methods are, however, basically the same. Soya bean is processed into tempeh on a small-scale basis by soaking whole soya beans in water for about 18 hours. The soaked beans are then dehulled either by hand or by treading bare-footed on the beans in baskets. The dehulled beans are next cooked in water for about 5 minutes to soften the beans and to destroy contaminating micro-organisms. Following this heat treatment, the beans are spread in thin layers (1–2 cm) in a stainless, wooden, plastic or bamboo box to allow the water to drain and evaporate from the surface.

The air-cooled soya beans are next mixed with a starter or inoculum consisting of moulded soya bean material from a previous tempeh fermentation. The inoculated soya beans are then wrapped in banana leaves, provided

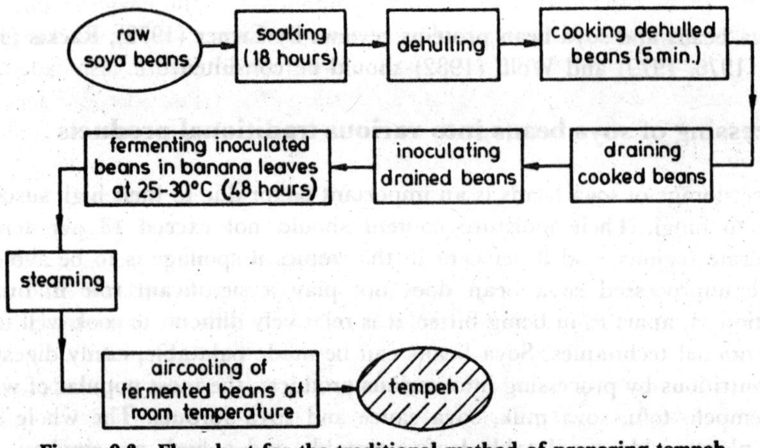


**Figure 9.3 Flowchart for the traditional method of preparing tempeh.**

with several little holes, and the material is allowed to ferment at a temperature between 25 and 30°C until the soya beans are completely mouldy, the resulting fermented soya beans being the product known as tempeh (Figure 9.3). The fermentation process is usually complete within 48 hours. It destroys the bad odour and flavour of the soya beans, as it apparently allows the micro-organism to produce enzymes that act on the proteins, carbohydrates and oil to make tempeh palatable and nutritious and to give a desirable flavour. The micro-organisms mainly responsible for tempeh fermentation are believed to be *Rhizopus oryzae* and *Aspergillus oryzae* which require oxygen for growth and the formation of enzymes. A study on the penetration of *Rhizopus digosporus* into soya beans in tempeh by Jurus and Sundberg (1976) revealed an extreme depth of mycelial infiltration which can partially explain the rapid physical and chemical changes which occur during tempeh fermentation.

The hyphae may mechanically push the bean cells apart, prior to, or in conjunction with enzymatic digestion. Thus, the beans become soft. Likewise, the penetration of enzymatic activity can also be enhanced, since the distance over which diffusion of enzymes must occur is greatly reduced. Under natural conditions in the tropics, tempeh production involves two distinct fermentations. The first, which occurs during soaking (hydration), is bacterial and results in acidification of the beans. The second is fungal and results in overgrowth of the bean cotyledons by the mould mycelia. Hydration is a process in which soya beans soaking in excess water absorb approximately their own weight of water, that is they double their original dry weight. During bacterial acid fermentation, the pH of the beans falls to a range of 4.5–5.3. This does not affect the mould growth, but it does prevent the development of undesirable bacteria that might spoil the tempeh.

### *Soya sauce*

Soya sauce is a light brown to black liquid with a meat-like, salty flavour prepared by fermenting soya beans, with or without the addition of wheat.

Soya sauce fermentation is basically a process of enzymatic hydrolysis of complex plant constituents to simpler compounds by the enzymes produced by micro-organisms. Usually, the fermentation is carried out in two successive stages. The first involves the preparation of a mould mixture, also called koji, under aerobic conditions with strains of *Aspergillus oryzae*. The koji serves as a source of enzymes and nutrients for the second fermentation, which is anaerobic and involves yeasts and bacteria (Young and Wood, 1974). Traditionally, soya sauce is made from whole soya beans; however, in recent years, defatted soya bean meal and flakes or mixtures of soya bean and wheat have been used for making soya sauce instead of soya beans alone (Figure 9.4). Using soya beans and wheat as the starting raw materials for soya sauce, the proportion of the latter to the former may vary, but the best soya sauce is generally believed to be made with a soya beans to wheat ratio of 50:50 by weight of raw materials. After the first fermentation, a 20 per cent salt solution is added to the koji for the next fermentation. Sodium chloride (NaCl) imparts a salty taste, acts as a preservative, and also has a selective action on the

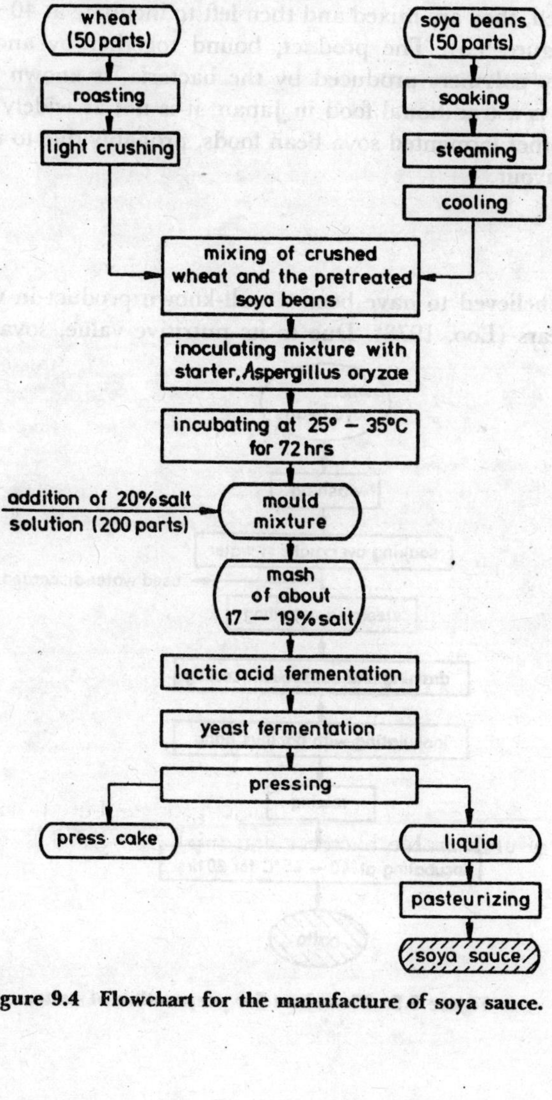


**Figure 9.4 Flowchart for the manufacture of soya sauce.**

micro-organisms that grow in the fermentation substrate. Salt plays an extraordinary role in soya sauce fermentation since it prevents dangerous spoilage by micro-organism and the growth of food poisoning bacteria, while permitting the development of flavour and aroma-forming yeasts and bacteria.

About 5000 litres of soya sauce can be obtained from 1 tonne of defatted soya bean meal, 1 tonne of wheat and 1 tonne of salt (Hesseltine and Wang, 1972).

### *Natto*

Like the production of other fermented soya products, the preparation of natto is essentially a fermentation process. In soya bean fermentations, moulds usually dominate, but natto fermentation is an exception, being predominantly carried out by bacteria. The bacterium, *Bacillus natto* (syn. *B. subtilis*), is the organism responsible for the fermentation. After pretreated soya beans have been fermented, they are mixed and then left to incubate at 40–45°C for up to 20 hours (Figure 9.5). The product, bound together by and covered with viscous, sticky polymers produced by the bacteria, is known as natto. Even though natto is a traditional food in Japan, it is not as widely consumed and accepted as other fermented soya bean foods, probably due to its less pleasant odour and flavour.

### *Soya milk*

Soya milk is believed to have been a well-known product in China for more than 2000 years (Loo, 1978). Due to its nutritive value, soya milk can be a

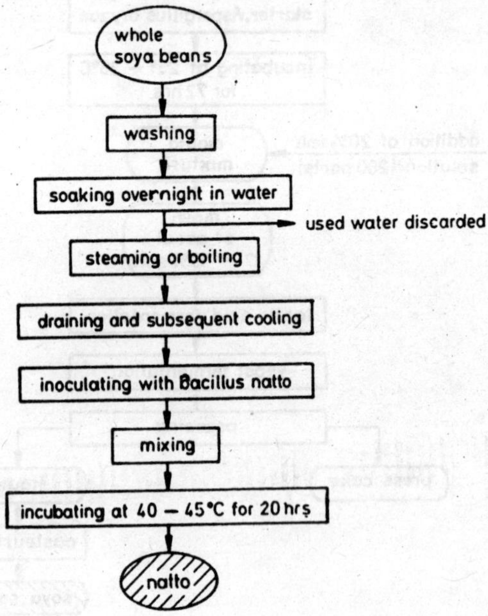


**Figure 9.5 Flowchart for preparation of natto.**

suitable substitute for cow's milk (Table 9.5) in areas where cattle rearing is a problem. Traditionally, soya milk is made by soaking the beans in water overnight, after which they are wet ground by means of a stone mill or mortar and pestle. The milky fluid produced is then separated from the solid residue by filtration through a cloth. Before consumption soya milk is boiled for 5–10 minutes to inactivate the trypsin inhibitor and the soya bean hemagglutinin. Preservation of soya milk for up to two years or more is done by heating the freshly filtered milk in steam for at least one hour in sealed containers.

## *Tofu*

Tofu is a popular soya product among the people of eastern Asia, particularly China and Japan. It is prepared by coagulating soya milk, for example with calcium sulphate ($CaSO_4$). The coagulated mass is filtered and then pressed to reduce the water content to about 60 per cent (Figure 9.6). The product, known as tofu, can either be further processed into tofu crisps, by drying at a temperature not exceeding 70°C, or fried in hot oil.

In the tropics, fresh tofu can be kept for only 1–2 days under water at room temperature without deterioration in the nutritive value. However, fresh tofu can be kept for as long as 6 months after sterilizing in a container for one hour.

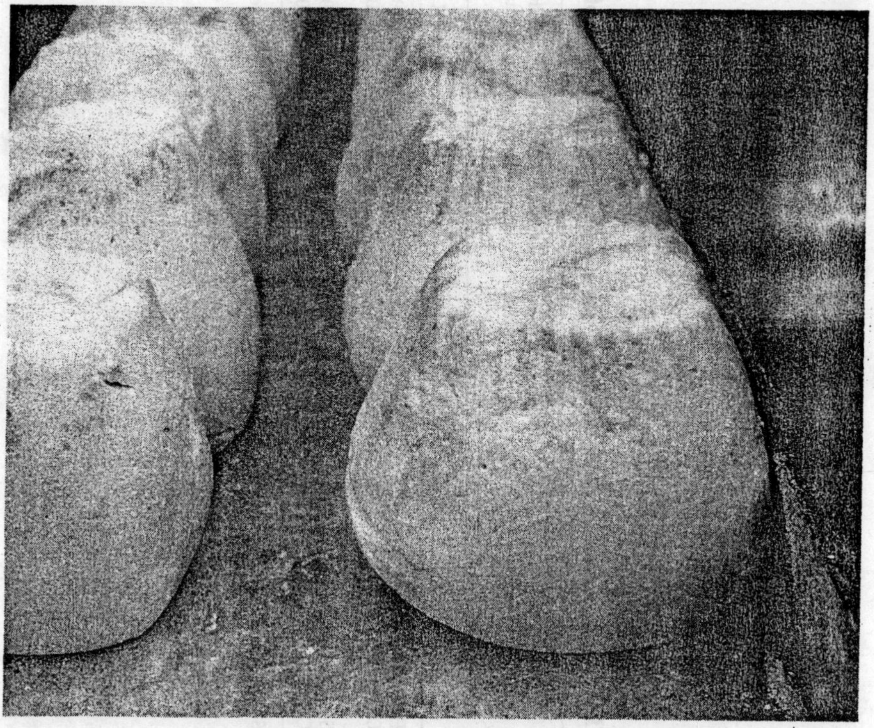

**Figure 9.6 Soya bean curd in rows, in Thailand.**

**Table 9.5 Protein and essential amino-acid composition of human milk, cow's milk and soya milk**

| | Nitrogen | Protein | Isoten-cine | Leucine | Lysine | Sulph-urous | Aromatic | Threo-nine | Trypto-phan | Valine | Essential amino-acids (total) | Amino-acids (total) |
|---|---|---|---|---|---|---|---|---|---|---|---|---|
| | (g/100 g of food) | | (mg/g of total nitrogen) | | | | | | | | | |
| Human milk | 0.19 | 1.2 | 254 | 548 | 428 | 185 | 421 | 280 | 105 | 284 | 2505 | 5530 |
| Cow's milk (pasteurized) | 0.55 | 3.5 | 399 | 782 | 450 | 201 | 820 | 278 | 91 | 463 | 3494 | 6698 |
| Soya milk | 0.56 | 3.2 | 305 | 497 | 348 | 191 | 551 | 229 | 85 | 294 | 3500 | 6011 |
| Cow's milk | 0.55 | 3.5 | 295 | 586 | 487 | 208 | 633 | 278 | 88 | 362 | 2947 | 6463 |

Source: Amino-acid content of foods and biological data on proteins. FAO Rome, 1970. FAO Nutritional Studies No. 24.

### *Soya yoghurt*

Washed beans are steamed for 30 minutes and then soaked for one hour in 0.2 per cent caustic soda (NaOH). The skins of the beans are then removed and the beans are soaked again in the same caustic soda solution for 4–5 hours. The treated soya beans are ground together with the caustic solution and filtered to yield a milky fluid. The fluid is neutralized with hydrochloric acid or citric acid to adjust the pH of the solution to about 7.0, after which sugar (about 75 g/500 ml solution) is added. Following this, the solution is boiled, while agitating continuously, for 5 minutes and left to cool to room temperature. The cooled solution is inoculated with the bacterium, *Lactobacillus bulgaricus*, and thoroughly mixed.

Finally, the whole mixture is kept in an oven (incubated) at a temperature of 37–43°C for 8–10 hours until the protein, which has been extracted from the soya beans and is dissolved in the mixture, coagulates. In a simplified form, the flowchart for processing soya milk, soya yoghurt and tofu, is illustrated in Figure 9.7.

## Composition of fermented and nonfermented soya bean foods

The composition of a soya bean food varies with the variety and maturity of soya beans used, as well as with the method employed for its production. The methods of preparing soya bean products vary from country to country. It is not uncommon, even within a country, for methods to vary, but the essential steps are similar.

Fermentation, being the most important processing step in the making of

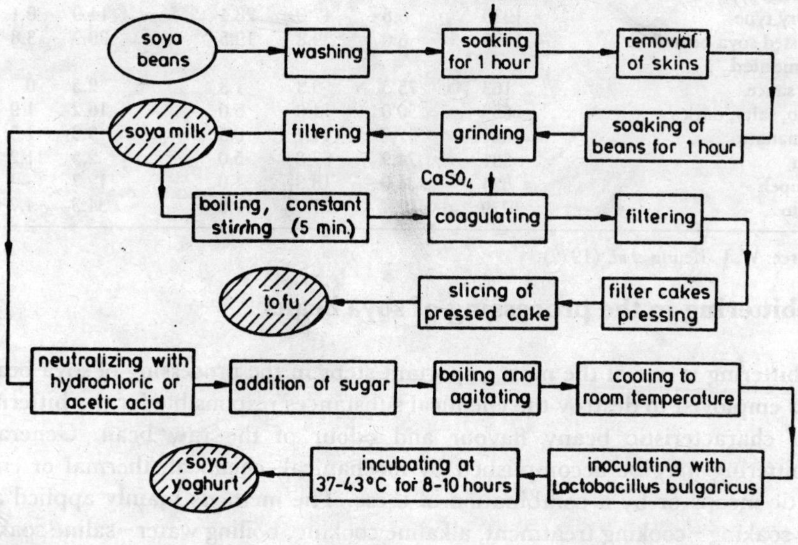


**Figure 9.7 Flowchart for processing of soya milk, soya yoghurt, and tofu.**

many soya foods, has received considerable attention in the last decade. As a result, pure culture fermentations have been developed, and the effects of fermentation on the composition of the soya bean have been studied by Jurus and Sundberg (1976), Yong and Wood (1974) and Leung *et al.* (1972), among others. Tables 9.6 and 9.7 show the composition of the most popular fermented and nonfermented soya bean foods.

The composition of the soya bean is not significantly changed by fermentation. For example, in tempeh free amino-acids increase, while total nitrogen and amino-acid composition remain relatively constant during fermentation (Table 9.8). It is assumed that perhaps the amount of mycelial protein in tempeh is not high enough to alter the protein content or the amino-acid composition of soya beans, nor does the mould depend on any specific amino-acid for growth. However, the mould has a greater capacity for synthesizing vitamins, such as niacin, which is increased after fermentation.

**Table 9.6 Composition of some typical soya bean foods (per 100 g of edible portion)**

| Food | Food energy (Joules) | Moisture (%) | Protein (g) | Fat (g) | Total carbo-hydrates (g) | Fibre (g) | Ash (g) |
|---|---|---|---|---|---|---|---|
| **Nonfermented** | | | | | | | |
| Mature soya bean seeds | 1675 | 10.2 | 35.1 | 17.7 | 32.0 | 4.2 | 5.0 |
| Fresh green soya beans, shelled | 582 | 68.2 | 13.0 | 5.7 | 11.4 | 1.9 | 1.7 |
| Sprouts | 259 | 81.5 | 7.7 | 1.8 | 8.0 | 0.7 | 1.0 |
| Milk | 155 | 91.4 | 2.8 | 1.5 | 3.6 | 0.1 | 0.7 |
| Curd (tofu) | 138 | 93.0 | 3.1 | 1.9 | 1.5 | — | 0.5 |
| | 192 | 89.7 | 4.9 | 2.8 | 1.5 | — | 1.1 |
| | 264 | 86.7 | 7.9 | 4.1 | 0.4 | 0.1 | 0.9 |
| | 473 | 76.0 | 13.3 | 6.5 | 3.1 | 0.2 | 1.1 |
| Protein-lipid film | | | | | | | |
| Moist type | 1193 | 43.2 | 29.9 | 18.8 | 5.9 | 0.1 | 2.2 |
| Dry type | 1930 | 6.8 | 47.0 | 28.4 | 14.9 | 0.1 | 2.9 |
| Roasted soya bean flour | 1796 | 6.0 | 29.8 | 19.5 | 39.9 | 3.8 | 4.8 |
| **Fermented** | | | | | | | |
| Soy sauce | 163 | 75.5 | 5.3 | 1.3 | 2.5 | 0 | 15.4 |
| Miso, salty, dark | 653 | 50.0 | 14.0 | 5.0 | 16.2 | 1.9 | 14.8 |
| Hamanatto | 661 | 52.7 | 14.7 | 8.3 | 9.3 | 1.5 | 15.0 |
| Sufu | 331 | 74.9 | 67.8 | 5.0 | 2.5 | 11.2 | 9.8 |
| Tempeh | 624 | 64.0 | 18.3 | 4.0 | 12.7 | — | 1.0 |
| Natto | 1779 | 7.4 | 31.7 | 17.9 | 34.3 | 4.7 | 4.0 |

Source: W.T. Leung *et al.* (1972).

## Debittering in the processing of soya beans

Debittering is one of the most important steps in the processing of soya beans. It is employed to destroy the chemical substances responsible for the bitterness and characteristic beany flavour and odour of the raw bean. Generally, debittering may be accomplished by mechanical, chemical, thermal or enzymatic means or by a combination of these. The methods mainly applied are: the soaking–cooking treatment, alkaline cooking, boiling water–saline soaking treatment, ammonium salt treatment and infra-red heating.

**Table 9.7 Mineral and vitamin composition of some typical soya bean foods (per 100 g of edible portion)**

| Food | Calcium (mg) | Phosphorus (mg) | Iron (mg) | Sodium (mg) | Potassium (mg) | β-carotene equivalent (μg) | Thiamin (mg) | Riboflavin (mg) | Niacin (mg) | Ascorbic acid (mg) |
|---|---|---|---|---|---|---|---|---|---|---|
| **Nonfermented** | | | | | | | | | | |
| Mature soya bean seeds | 226 | 546 | 8.5 | — | 1504 | 10 | 0.66 | 0.22 | 2.2 | 0 |
| Fresh, green soya beans, shelled | 78 | 158 | 3.8 | 15 | 607 | 360 | 0.40 | 0.17 | 1.5 | 27 |
| Sprouts | 52 | 58 | 1.1 | 30 | 279 | 25 | 0.19 | 0.15 | 0.8 | 10 |
| Milk | 18 | 36 | 1.2 | 15 | 58 | 30 | 0.05 | 0.02 | 0.3 | 0 |
| Curd (tofu) | 114 | 38 | 0.8 | 4.0 | 119 | — | 0.06 | 0.04 | 0.1 | — |
| | 150 | 56 | 1.1 | — | — | 0 | 0.02 | 0.02 | 0.3 | — |
| | 150 | 104 | 2.2 | 12 | 151 | — | 0.04 | 0.02 | 0.4 | 0 |
| | 136 | 145 | 2.8 | — | 154 | 20 | 0.06 | 0.03 | 0.2 | 0 |
| Protein-lipid film | | | | | | | | | | |
| Moist type | 508 | 375 | 3.8 | — | 169 | — | 0.07 | 0.10 | — | — |
| Dry type | 245 | 494 | 9.5 | 124 | 463 | — | 0.42 | 0.16 | 0.15 | — |
| Roasted soya bean flour | 189 | 540 | 7.5 | — | — | 0 | 0.45 | 0.15 | 2.0 | 0 |
| **Fermented** | | | | | | | | | | |
| Soy sauce | 59 | 100 | 4.9 | 5713 | 186 | — | — | — | — | — |
| Miso, salty, dark | 115 | 190 | 4.0 | 4600 | — | 0 | 0.03 | 0.10 | 1.5 | 0 |
| Hamanatto | 142 | 135 | 7.9 | 4842 | 580 | 30 | 0.06 | 0.27 | 0.6 | 9.0 |
| Sufu | 110 | 137 | 1.0 | 3700 | 55 | 40 | 0.01 | 0.05 | 0.5 | 0 |
| Tempeh | 129 | 154 | 10.0 | — | — | 30 | 0.17 | 0.07 | 3.2 | 0 |

**Table 9.8 Essential amino-acid content of some typical soya bean foods**

| Essential amino-acid | Fresh green soybeans | | Sprouts | | Milk | | Curd | | Soy sauce | | Miso | | Tempeh (dried) | | Natto | |
|---|---|---|---|---|---|---|---|---|---|---|---|---|---|---|---|---|
| | A | B | A | B | A | B | A | B | A | B | A | B | A | B | A | B |
| Isoleucine | 413 | 859 | 313 | 385 | 330 | 162 | 261 | 360 | 270 | 246 | 279 | 608 | 333 | 2837 | 322 | 931 |
| Leucine | 444 | 924 | 525 | 646 | 470 | 230 | 448 | 618 | 430 | 396 | 449 | 979 | 529 | 4507 | 470 | 1358 |
| Lysine | 531 | 1104 | 438 | 539 | 330 | 162 | 333 | 460 | 350 | 322 | 323 | 704 | 370 | 3152 | 352 | 1017 |
| Methionine | 50 | 104 | 63 | 77 | 86 | 42 | 78 | 108 | 49 | 45 | 74 | 161 | 71 | 605 | 66 | 191 |
| Cysteine | – | – | 25 | 31 | 46 | 23 | 78 | 108 | 57 | 52 | 69 | 150 | 100 | 852 | 80 | 231 |
| Phenylalanine | 313 | 651 | 250 | 308 | 330 | 162 | 321 | 443 | 210 | 193 | 272 | 593 | 305 | 2599 | 292 | 844 |
| Tyrosine | 175 | 364 | 206 | 253 | 220 | 108 | 169 | 233 | 59 | 54 | 226 | 493 | 170 | 1448 | 179 | 517 |
| Threonine | 119 | 248 | 363 | 446 | 310 | 152 | 170 | 235 | 190 | 175 | 210 | 458 | 245 | 2087 | 255 | 737 |
| Tryptophan | 63 | 131 | 144 | 177 | 85 | 42 | 96 | 133 | 30 | 28 | 54 | 118 | 77 | 656 | 75 | 217 |
| Valine | 350 | 728 | 350 | 431 | 360 | 176 | 264 | 364 | 290 | 267 | 270 | 589 | 332 | 2829 | 324 | 936 |

A = mg/g of total N
B = mg/100 g of food.
Source: Leung *et al.* (1972).

### *Soaking—cooking treatment*

This is the method most commonly applied traditionally. Prior to further processing, raw soya beans are soaked in water for several hours and then cooked for 15–30 minutes, depending on the quantity and degree of ripeness. The soaking removes the bitterness whilst the subsequent heat treatment inactivates the harmful trypsin inhibitors and the hemagglutinins in the beans. This method is simple and cheap. However, the heat treatment, if not controlled, can have adverse effects on certain amino-acids, especially lysine, cystine and tryptophan.

### *Alkaline cooking treatment*

Alkaline cooking is a method which enables the removal of undesirable compounds from the soya beans, while preserving the quality of the proteins and avoiding the destruction of the sulphur amino-acids. This treatment consists of cooking the soya in an aqueous suspension, followed by a drying operation. Cooking is carried out at a temperature of 80°C, with a pH value between 8 and 9 (Rambaud 1965, U.S. Patent 3220851). With this temperature and pH, the duration of cooking is less than 20 minutes for a concentration of soya beans of 20 per cent by weight.

The effect of cooking is completed by the addition of a proteolytic enzyme that may be of vegetable origin (such as papain or ficin), of animal origin (such as pancreatin) or of bacterial or fungal origin (obtained by biosynthesis). The addition of the enzymatic preparation (0.1–0.5 per cent) to the soya suspension solubilizes the proteins, fluidifies the mass and allows treatment at a higher concentration of up to 30 per cent or more. Without the enzyme the concentration is limited to 20–25 per cent of soya beans.

For example, at a concentration of 35 per cent of soya, the duration of

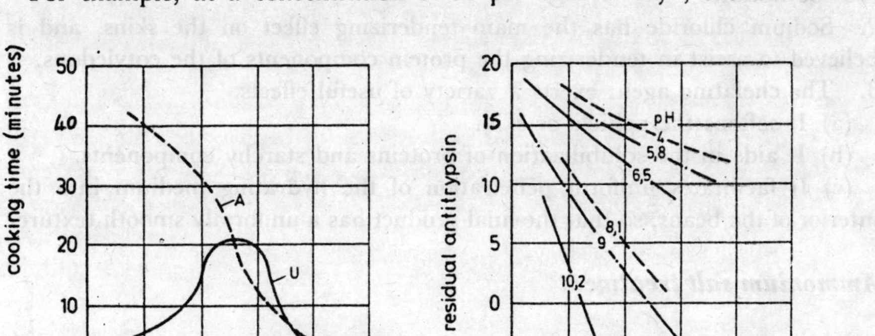


**Figure 9.8 (a) Cooking time in minutes at 80°C necessary to obtain a destruction of 90 per cent of the urease (U) and of the antitrypsin (A) as a function of the pH value of the suspension.**
**(b) Residual antitrypsin as a function of the cooking time for different pH values (Rambaud 1965, U.S. Patent 3220851).**

cooking, sufficient to destroy the undesirable constituents, is one hour at 80°C. After the destruction of antitrypsin and urease at pH 8–9, the pH must be adjusted to neutral for the drying operation since drying at too high a pH value results in degradation of the sulphur amino-acids. One means of achieving this objective – bringing the pH value to neutral – is by the use of ammonia as an alkalizing agent for the cooking. In the subsequent drying at a higher temperature, the ammonia evaporates and the suspension becomes neutralized. With this method of cooking and drying, the consumer is offered a product that is neither alkaline nor acid, and is free of antitrypsin and urease. The relationship between the cooking time and residues of urease and antitrypsin for different pH values of the suspension is shown in Figure 9.8.

### *Boiling water–saline soaking treatment*

This process includes conditioning of the beans by brief contact (1–2 minutes) with boiling water, followed by soaking for about 24 hours at room temperature in an aqueous solution containing sodium chloride, a chelating agent and an alkaline substance (Raymond, 1957, U.S. Patent 2795502). Usually the sodium chloride is present in the hydrating medium in the concentration of 1–3 per cent. The chelating agent added to the hydrating medium is in the concentration of 0.1–5 per cent.

Following hydration the beans are air dried at a temperature of 55–77°C. Alternatively, the hydrated beans may be preserved by freezing. Beans treated in this way have been shown to have excellent cooking qualities and are free of the characteristic bitterness of raw soya beans. The most important features of this process are:

1. The hydrating medium contains one or more tenderizing agents, which contribute significantly to the goal of attaining a product that is quick cooking (10–60 minutes).
2. Sodium chloride has the main tenderizing effect on the skins, and is believed to assist in tenderizing the protein components of the cotyledons.
3. The chelating agent exerts a variety of useful effects:
   (a) It softens the pellicle or skin.
   (b) It aids in the solubilization of proteins and starchy components.
   (c) It facilitates uniform penetration of the hydrating medium into the interior of the beans, so that the final product has a uniformly smooth texture.

### *Ammonium salt treatment*

The characteristic bitterness and the objectionable odour of raw soya beans are destroyed by using an ammonium salt (Raymond, 1957, U.S. Patent 2795502). Soya beans either whole or in pieces are first soaked in an aqueous solution (1–3 per cent) of ammonium carbonate or bicarbonate for about 10 hours at room temperature. During soaking, a certain degree of fermentation occurs due to the presence of natural enzymes in the beans. For increased fermentation, active yeasts are added to the ammonium solution to the extent of 1–5 per cent. Apart from increasing the fermentation activity, the added yeast changes the soya bean material into a more soluble and digestible form; starches, for instance, are more or less converted into sugars and the proteins into proteoses. Enzyme preparations from the fungus *Aspergillus* spp. may also be advantageously added to supplement the yeast and soya bean enzymes.

After the yeast has been added, the temperature is raised to 43–55°C to decompose the ammonium compound and expel any ammonia from the solution, as well as to complete the action of ammonia on the material. Following this step, the solution is drained and the skins removed from the beans if desired. The beans are then boiled for one and a half hours to yield a lighter coloured product. The material produced is found to be edible and palatable, and may be dried and ground into powder. If oil is the desired end-product the treated beans can be subjected to various oil-extracting methods.

In this processing method, the ammonium compound helps to saturate the beans with the solution. It also has a tenderizing, softening and dissolving effect on the beans and penetrates into the oil cells. By its chemical action and the evolution of gas, it also helps to break down the bean material and completely destroy the objectionable beany flavour and odour.

## *Mechanical peeling and debittering of soya beans*

In this process, soya beans are subjected to the effect of steam for a period of 2–5 minutes and then suddenly cooled with cold water. The beans are then passed with a stream of water into a closed peeling cylinder in which the water is subjected to a whirling motion by means of blades (Figure 9.9).

The hulls of the beans are split by the effect of the swirling water by which the beans are thrown against the interior walls of the vessel. The inner wall surfaces of the vessel are coarsened, being either rilled, grooved or wire-netted. By means of the hydrodynamic force of the streaming water, the split hulls are stripped from the beans and the embryos are separated from the cotyledons. The dehulled beans are then dried at a temperature range of 70–75°C, thus destroying the bitter constituents. The beans resulting from this process are not denaturalized, that is, they are neither taste cooked nor roasted (Hanson, 1974).

1 cylindrical vessel (inner wall surfaces are coarsened, rilled or grooved)
2 feeding funnel
3 exit conduit
4 rotating shaft
5 beater blades
6 inner coarse surface of vessel
7 tube for pretreated beans
8 pipe for water

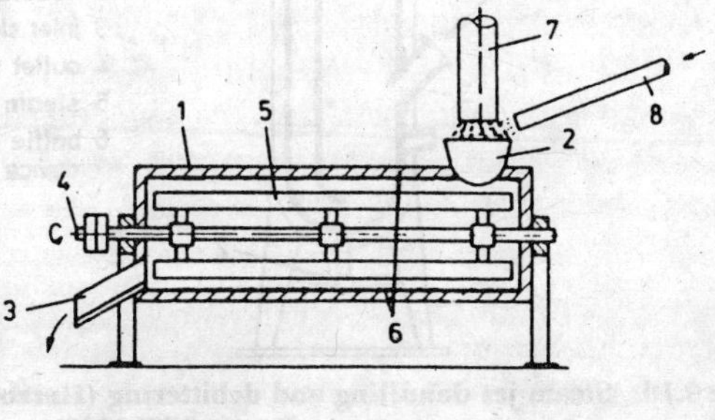


**Figure 9.9 Soya bean dehuller (I. Kovàsznay and E. Kovàsznay, U.S. Patent 3058829).**

### *Steam jet dehulling and debittering*

Raw soya beans which have been halved, quartered and preliminarily dehulled on corrugated plates, are fed through the inlet into the whirl chamber containing the baffle. Steam at a temperature of about 200°C is jetted through the nozzle assembly into the chamber (Figure 9.10).

As a result of the expansion of steam after its passage through the nozzles, the steam temperature falls to the working temperature of 135°C. The quantity in a single charge of material to be treated is preferably 3 kg, and the time interval for the treatment of material amounts to 20–60 seconds. The treated beans are subsequently ground or milled into full fat flours. These flours are found to be of high food value and digestibility, have a neutral flavour and are free of bitter substances and undesirable enzymes.

### *Large-scale tempeh production*

For medium and large-scale tempeh production, the traditional method would not be appropriate because of the following drawbacks:

1. The mass of soya beans that can be fermented is restricted to a small amount of moulded soya beans, since the required organisms are aerobic.
2. The dehulling process involves much hand work.
3. Fermentation times for the formation of tempeh are variable since the inoculum is highly variable with regard to the viability of the mould.
4. The lengthy soaking time required to remove the soya bean seed coats also restricts its mass production.

The preparation of tempeh by methods of fermenting soya beans rapidly in large amounts to yield a product of uniform size and shape has been described by Raymond (1957) and Hanson (1974).

Full fat grits or whole dehulled soya beans are inoculated with sporangiospores from *Rhizopus* and then put in perforated plastic bags. The ends of the bags are heat sealed and the beans are allowed to ferment in the package in which they

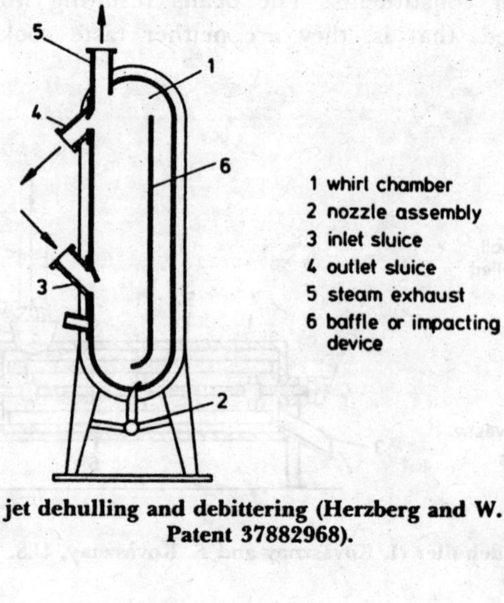


**Figure 9.10 Steam jet dehulling and debittering (Herzberg and W. Dollbaum 1974, U.S. Patent 37882968).**

are to be sold. The flexible perforated plastic bags (0.5 mm diameter perforations located not more than 13 mm apart) present an ideal fermentation environment for the fermentation of soya beans to tempeh.

### *Infra-red heating of raw soya beans*

The method of processing basically involves treating soya beans in their natural state to improve their utility as food or feed. As a first step the moisture content in soya beans is adjusted to 8–27 per cent by weight, preferably between 19 and 22 per cent. Beans with insufficient moisture tend to undergo protein modification during infra-red radiation.

The beans are then allowed to stand for a period of time so that the moisture content is uniformly distributed. The next step is to subject the moist beans to infra-red radiation. An infra-red radiation source which peaks at between 2 and 6 microns is satisfactory and one which peaks at 3 microns is preferred (Guidarelli and Lawrence, 1964).

The radiation process is carried out by passing the beans in a layer 2–4 beans thick on a conveyor into contact with the radiation. The extent and degree of radiation is regulated so that the final moisture content of the beans is not reduced below 7 per cent on a wet weight basis. This is important in order to preserve the nutritional value and edibility of the beans. Treated beans may undergo further processing operations, such as dehulling, grinding or flaking, to achieve the desired final product.

### *Infra-red treated ready-to-eat soya products*

Guidarelli (1968) modified the process of infra-red treatment using increased moisture content in the soya beans and utilizing infra-red radiation having a peak wave length outside the effective water absorption range. In general, the modified method includes swelling the beans, for example by soaking in water to increase the moisture content of the beans evenly, and then subjecting the swollen moist beans to infra-red heat treatment.

The moisture content of the beans may be increased by soaking to between 35 and 75 per cent on a wet weight basis. The infra-red heat treatment is carried out with radiation of peak wave lengths between 1 and 1.7 or between 7 and 30 microns. The soaking of the beans to increase the moisture content causes the beans to swell considerably. The infra-red heat treatment of the moist beans roasts the beans to destroy the bitter taste and the enzyme inhibitors without adversely affecting the soya bean protein.

In this process the typical spacing between the radiator and the beans is 150–450 mm. The dried beans are found to be crunchy, easily edible, palatable and readily acceptable as human food. Seasoning or flavouring may be incorporated during roasting.

## Simulated milk from soya beans

In the traditional method of processing soya milk a considerable amount of the useful content of the soya bean is lost or wasted through prolonged soaking (Table 9.9) and during the filtering process.

**Table 9.9 Influence of soaking period on the loss of dry matter content of soya beans**

| Quantity of soaked beans (g) | Soaking period (hours) | Weight of dry matter left in water (g) |
|---|---|---|
| 250 | 6 | 2.681 |
| 250 | 18 | 5.712 |
| 250 | 18 | 5.775 |
| 250 | 18 | 6.208 |
| 250 | 18 | 6.320 |
| 250 | 18 | 6.489 |
| 250 | 24 | 6.934 |
| 250 | 24 | 7.512 |
| 250 | 24 | 7.773 |
| 250 | 24 | 7.924 |
| 250 | 48 | 10.323 |

Source: Thio Goani Loo, 1978.

The process developed by Miles (1966) aims at improving the traditional practice by eliminating both the presoaking of the soya beans and the subsequent grinding process of the water-softened beans. Also, in this method, the possible bacterial growth and enzymatic action are arrested almost at the onset, that is, immediately after mixing the soya bean flakes with water. This is accomplished by cooling the slurry and then promptly subjecting it to high pressure homogenization. The process includes the following steps:

1. Soya beans with a moisture content of 10–14 per cent water by weight are 'dry' cracked to loosen their hulls. Dehulling is accomplished by winnowing. Immediately after this the dehulled beans are fed into flaking rolls where they are rolled under heavy pressure into thin flakes (0.07–0.2 mm).
2. Following this step, cold or hot water (in the proportion of 1 kg beans to 10 litres water) is added to the soya flakes to form a slurry and then agitated for almost half an hour.
3. A stabilizer of the phosphate type is added to the slurry, which is then pressure cooked at a temperature between 100 and 120°C for a maximum period of 10 minutes. After pressure cooking the slurry is subjected to high pressure homogenization. It is then clarified in a centrifugal separator such as a De Laval or Westphalia, with a flush cycle of 2–5 minutes. The liquid produced is put into a tank where oil, carbohydrates (including sugar), vitamins and minerals are added in accordance with the production requirements, either for whole milk or skimmed milk. The formulated milk is subjected to the conventional milk homogenizing process. It is then evaporated in a vacuum pan to yield the correct density for direct consumption, or to produce a condensed milk.
4. For longer storage, the milk is canned and sterilized at 110–125°C for 8–12 minutes, after which the cans are immediately cooled to between 32 and 38°C.

## Processing soya beans into oil and meal

To remove oil from oilseeds with a relatively low oil content, such as soya beans, cottonseed and rice bran, solvent extraction is the best process to employ. The extraction rate of solvent extraction can be as high as 99 per cent. Figure 9.11 illustrates the overall scheme for processing soya beans into oil and meal. The 'processing chain' begins with the cleaning of the beans, which are passed through a magnetic separator and a screening operation. In this way metal pieces are removed, and larger and smaller sized seeds are separated from the soya beans. Light foreign matter and dust are blown out or removed by aspiration.

After cleaning, the beans, containing about 17–19 per cent oil, are conveyed into crushing rolls. Cracking or crushing into pieces and loosening of the hulls are effected by pressure and friction between the surfaces of two or three pairs of rolls. The surfaces of the rolls may be smooth or fluted. For food and high-protein feeds, dehulling is usually carried out after crushing to reduce the fibre content. The crushed beans are then conditioned to 10–12 per cent moisture at about 80°C and then flaked. Conditioning and flaking serve to provide convenient feed for the extractor and to promote easy oil release by rupturing the cells of the beans. To a certain degree, conditioning also inactivates the undesirable enzyme systems in the beans.

The treated soya beans are fed into the extractor, where at temperatures of 50°C and above, for a period of two hours, oil is removed using hexane. Depending on availability, desired end-products and the outside temperature of the processing plant (explosion hazard), other solvents such as heptane, cyclohexane or trichloroethylene may be used.

From the solvent extractor the oil-hexane mixture, also called miscella, is brought to the distillation columns where hexane is separated from the crude soya oil. The hexane vapours are condensed and then returned to the

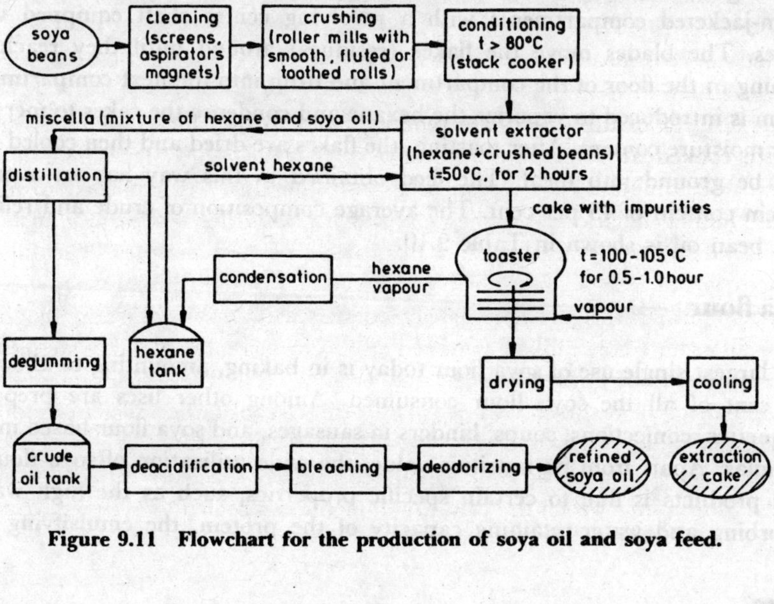


**Figure 9.11 Flowchart for the production of soya oil and soya feed.**

extractor. The soya oil from the distillation then undergoes a degumming process, in which the oil is agitated with hot water, and the hydrated and emulsified phosphatides are separated in a centrifuge.

The degummed soya bean oil is next deacidified or alkali-refined to remove most of the free fatty acids. Alkali-refining is carried out with caustic soda (sodium hydroxide), or soda ash (sodium carbonate), or a combination of both in tanks equipped with agitator-heating coils. Passing through the tanks the oil emulsified with the alkali solution is heated to 60–70°C and stirred until small clots are formed from the free fatty acids and the lipids. At this point, heating and stirring are discontinued and the so-called 'break' is allowed to settle to the bottom of the tank. The clear super-natant oil is withdrawn from the top. Centrifuges may also be used to separate the soapstock from the oil.

The alkali-refined oil is dried, and then conveyed to a bleaching kettle where Fuller's earth (hydrated aluminium silicate), activated carbon and other adsorptive agents are added. The mixture is stirred, heated, and filtered through filter presses to remove the spent bleaching agents. The bleached oil flows to the deodorizer where it is subjected to temperatures of 260°C or higher and pressures of 5–12 mm mercury. By the use of low pressures and high temperatures with the passage of steam through the oil, the volatile products condense on the sides of the deodorizer and are collected at the bottom.

If the oil is to be converted into shortening, margarine or similar products, it is hardened by hydrogenation prior to deodorizing. Hydrogenation is carried out by bubbling hydrogen gas through the hot oil in the presence of a nickel catalyst. The residue remaining after extraction of the oil contains approximately 1 per cent oil and over 30 per cent hexane. For use in feed the hexane-laden residue is passed through a desolventizer-toaster, which recovers the hexane and simultaneously toasts the residue to obtain a palatable product with high nutritive value. The desolventizer-toaster is a vertical series of steam-jacketed compartments with a revolving centre shaft equipped with blades. The blades move the flakes (residues) around until they reach an opening in the floor of the compartment and drop into the next compartment. Steam is introduced to vaporize the hexane and condense the cakes to increase their moisture content. After toasting, the flakes are dried and then cooled and may be ground into meal. The feed obtained in this way has an average protein content of 45 per cent. The average composition of crude and refined soya bean oil is shown in Table 9.10.

## Soya flour

The largest single use of soya flour today is in baking, amounting to about 40 per cent of all the soya flour consumed. Among other uses are prepared croquettes, confections, soups, binders in sausages, and soya flour-based infant formulas. Apart from its nutritive value, the wide utilization of soya flour in food products is due to certain specific properties, such as the high water-absorbing and water-retaining capacity of the protein, the emulsifying and

## Table 9.10 Average composition of crude and refined soya bean oil

| Composition | Crude oil | Refined oil |
|---|---|---|
| Triglycerides (%) | 95−97 | 99 |
| Phosphatides (%) | 1.5 −2.5 | 0.003−0.015 |
| Unsaponifiable matter (%) | 0.6 | 0.3 |
| Sterols | 0.33 | 0.13 |
| Tocopherols (%) | 0.15−0.21 | 0.11 −0.18 |
| Hydrocarbons (squalene) (%) | 0.014 | 0.01 |
| Free fatty acids (%) | 0.3 −0.7 | 0.05 |
| Trace metals: | — | — |
| Iron (ppm) | 1−3 | 0.1 −0.3 |
| Copper (ppm) | 0.03−0.05 | 0.02 −0.06 |

Source: American Soybean Association, 1981.

anti-oxidant properties of the lecithin, and the relatively low carbohydrate content of the flour. Soya flour does not, however, have the visco-elastic properties characteristic of wheat flour proteins.

Three types of soya flour, differing mainly in their fat and protein contents, may be produced: full-fat, low-fat and defatted flour. Full-fat flour is produced from whole soya beans. The beans are cleaned, and heat-treated (steaming) to eliminate the beany and bitter flavours characteristic of raw soya beans, to improve the biological food value of the protein and to inactive lipoxygenase and other enzymes. Next, the beans are dried to about 5−6 per cent moisture, passed through crushing rolls, dehulled and finally ground. Flour of the desired fineness is obtained by sieving. Full-fat soya flour contains about 40 per cent protein and 20 per cent fat.

Low-fat soya flour is made from the soya bean meal produced by the hydraulic or expeller processes that are sometimes used to express oil from soya beans instead of the solvent extraction process. In this case the soya meal is heat-treated, dried and ground into flour. Low-fat soya flour contains about 45 per cent protein and 6 per cent fat.

Defatted soya flour is obtained from meal derived from the solvent extraction process as outlined previously. The meal is made into flour in a manner similar to the method described above. Defatted soya flour contains about 50 per cent protein and 1−1.5 per cent fat.

## *References*

**American Soybean Association**. *Soya oil*. Hamburg 1981.

**Carter, M.W., Smart, W.W.G., Jr.** and **Matrone, G.** (1953). Estimation of estrogenic activity of genistein obtained from soybean meal, *Proc. Soc. Exp. Biol. Med.*, **84**: 506.

**Cristofaro, E. Mottu, F.**, and **Wuhrmann, J.J.** (1974). Involvement of the raffinose family of oligosaccharides in flatulence, in *Sugars in Nutrition* (Sipple, H.L. and McNutt, K.W., eds.), Academic Press, New York, 313.

**FAO** (1986). *FAO Mon. Bull. Statist.* Vol. 9 July/August 1986, Rome.

**FAO** (1970). *FAO Nutritional Studies*, No. 24, Rome.

**Guidarelli, E.J.** (1968). U.S. patent 3407073, October 22.

**Guidarelli, E.J.** and **Lawrence, J.F.** (1964) U.S. Patent 3141777.

**Hanson, L.P.** (1974). *Vegetable protein processing.* Noyes Data Corporation. Park Ridge, New Jersey. London/England.

**Herzberg, C.** and **Dollbaum, W.** (1974). U.S. Patent 37882968.

**Hesseltine, C.W.** and **Wang, H.L.** (1972). Fermented soybean products, in *Soybeans: Chemistry and Technology* (Smith, A.K. and Circle, S.J., eds.), AVI Publishing, Westport, Conn.

**Jurus, A.M.** and **Sundberg, W.J.** (1976). Penetration of *Rhizopus oligosporus* into soybeans in tempeh, *Appl. Environ. Microbiol.*, **32**: 284.

**Kalbrener, J.E., Warner, K.** and **Eldridge, A.C.** (1974). *Cereal Chem.* **51**: 406.

**Konijn, A.M., Gershon, B.** and **Guggenheim, K.** (1973). Further purification and mode of action of a goitrogenic material from soybean flour, *J. Nutr.*, **103**: 378.

**Kovásznay, I.** and **Kovásznay, E.** (1962). U.S. Patent 3058829.

**Leung, W.-T. Wu., Butrum, R.R.** and **Chang, F.H.** (1972). *Food composition table for use in East Asia,* U.S. Department of Health, Education and Welfare, and Food and Agriculture Organization.

**Liener, I.E.** (1972). In *Soybeans: Chemistry and Technology.* Proteins Vol (Smith, A.K. and Circle, S.J., eds.), AVI Publ., Westport, Conn. Chap. 7.

**Lis, H., Fridman, C., Sharon, N.** and **Katchalski, E.** (1966). Multiple hemagglutinins in soybean, *Arch. Biochem. Biophys.*, **117**: 301.

**Loo Thio Goan** (1978). *Small-scale and home processing of soya beans, with applications and recipes.* Communication 64a of the Department Agric. Research of the Royal Tropical Institute, Amsterdam.

**Miles, C.P.** (1966). U.S. Patent 3288614.

**Mital, B.K.** and **Steinkraus, K.H.** (1975). Utilization of oligosaccharides by lactic acid bacteria during fermentation of soy milk, *J. Food Sci.*, **40**: 114.

**Nichols, L., Sinclair, H.M.** and **Jeliffe, D.B.** (1961). *Tropical Nutrition and Dietetics.*

**Perlman, F.** (1965). Food allergy and vegetable proteins, *Food Technol.*, **20**: 1438.

**Pryde, E.H.** (1980) Composition of soybean oil, in *Handbook of Soy Oil Processing and Utilization,* American oil chemists society, Campaign, Ill., American Soybean Association.

**Rackis, J.J.** (1972). In *Soybeans: Chemistry and Technology.* Proteins, Vol. 1 (Smith, A.K. and Circle, S.J., eds.), AVI Publ., Westport, Conn. Chap. 7.

**Rackis, J.J. (1976).** Flatulence problems associated with soy products, in *World Soybean Research* (Hill, L.D., ed.), Interstate Printers and Publishers, Danville, Ill. 892.

**Rackis, J.J.** (1979). Flavor problems of vegetable food proteins, *J. Am. Oil Chem. Soc.*, **56**: 262.

**Rackis, J.J., McGhee, J.E., Gumbmann, M.R.** and **Booth, A.N.** (1979). Effects of soya proteins containing trypsin inhibitors in long term feeding studies in nats, *J. Am. Oil Chem. Soc.*, **56**: 162.

**Rambaud, M.** (1965). U.S. Patent 3220851), November 30.

**Raymond, C.A.** (1957) U.S. Patent 2795502, June 11.

**Ratner, B., Untracht. S., Crawford, L.V., Malone, H.J.** and **Retsina, M.**, (1955). Allergenicity of modified and processed foodstuffs, *Am. J. Dis. Child.*, **89**: 187.

**Sessa, D.J., Warner, K.** and **Honig, D.H.** (1974). *Journal Food Sci.* **39**: 69.

**Sugimoto, H.** and **Van Buren, J.P.**(1970). Removal of oligosaccharides from soy milk by an enzyme from *Aspergillus saitoi, J. Food Sci.*, **35**: 655.

**Tan Boe Han** (1958). Technology of soymilk and some derivatives, Ph.D. Thesis, Agricultural University, Wageningen, Netherlands.

**Turner, R.H.** and **Liener I.E.** (1975). The effect of the selective removal of hemagglutinins on the nutritive value of soybeans, *J. Agric. Food Chem.*, **23**: 284.

**Wang, H.L., Mustakas, G.C., Wolf, W.J., Wang, L.C., Hesseltine, C.W.** and **Bagley, E.B.** (1979). *Soybeans as human food – unprocessed and simply processed,* U.S. Department of Agriculture, Utilization Res. Rep. No. 5, January.

**Wolff, I.A.** (1982). *Handbook of Processing and Utilization in Agriculture.* Volume II: Part 1, Plant Products. CRC Press Inc., Boca Raton, Florida.

**Yong, F.M.** and **Wood, B.J.B.** (1974). Microbiology and biochemistry of soy sauce fermentation, *Adv. Appl. Microbiol.*, **17**: 157.

# 10

# Yam

## Classification and production

Yams are any of the economically useful plants belonging to the genus *Dioscorea*, or the tubers or rhizomes of these plants (Coursey, 1967). The genus *Dioscorea* contains about 600 species but the main varieties of food yams belong to the species *Dioscorea rotundata, Dioscorea cayenensis, Dioscorea alata, Dioscorea dumetorum* and *Dioscorea esculenta*.

The yam tuber is economically the most important part of the plant. The structure of the tuber is extremely variable, depending on the species. Most yam tubers however are more or less cylindrical in shape; weights of individual tubers may range from 200 g to 50 kg.

The tuber of *D. rotundata* (white yam) is more or less cylindrical in shape. The skin is smooth and brown, while the flesh is usually white and firm. Starch grains in tubers of *D. rotundata* are large and ovoid. The tubers of *D. esculenta* (lesser yams, Chinese yams) are small and characteristically borne in clusters by the plants, unlike most other yams, which normally produce only one or two large tubers per plant. Lesser yam tubers have thin, smooth, yellowish-brown skins. The tuber flesh is white and less fibrous than that of most other yams. It is sweeter and its starch is more finely grained than that of other yams.

*D. cayenensis* (yellow yam, yellow guinea yam) is in many respects similar to *D. rotundata*. The tuber flesh is usually yellow in colour due to the presence of carotenoids. The tuber is firmer and less extensively grooved than that of *D. rotundata* (Onwueme, 1978). *D. alata* is the most widely distributed species of yam. The shape of the tuber is usually cylindrical. The tuber flesh is white to purplish, and loose ('watery') in texture. The tuber of *D. dumetorum* (bitter yam, trifoliate yam) is large and coarse; each plant may produce one tuber or a cluster. The tubers are bitter and, in many cultivars, poisonous due to the relatively high content of the alkaloid dioscorine or its derivatives. The toxic substances are made harmless by soaking in water or salt water before boiling.

Yams are grown extensively in only three parts of the world: West Africa, the Caribbean Islands and Southeast Asia. The global production is 25.5 million tonnes, excluding the People's Republic of China (FAO, 1984). By far the largest acreage and greatest amount of yam production is in West Africa, where over 95 per cent of the total world production takes place. Within West Africa, yam production is confined to the region stretching from the Ivory Coast to Cameroon. The major yam producing countries in this region, in order of importance, are Nigeria, the Ivory Coast, Ghana and Togo. Nigeria alone accounts for 78 per cent of world production.

The West Indies is the second most important yam producing region in the world. Other countries where significant production of yam occurs are Brazil, Venezuela, Papua New Guinea, China and the Philippines.

## Composition

As with most other crops, there is considerable variation in the chemical composition of yams, not only between species, but also within a single species, according to the cultural, climatic and edaphic factors under which it was cultivated, its maturity at harvest, and the method and length of storage.

A general summary, showing the ranges of composition on a wet weight basis, for the different species of yam, is illustrated in Table 10.1. Data for potatoes and sweet potatoes are included for comparison. All the yam species presented in Table 10.1 are composed mainly of water and carbohydrates. The carbohydrate consists essentially of starch and negligible amounts of sugar (usually less than 1 per cent). In *D. esculenta*, which is sweetish, the sugar content may be as high as 2–4 per cent on a fresh weight basis. The main constituent of yam starch is amylopectin, which exists in the cells in the form of starch grains. The protein content of yams is rather low, ranging from 1–3 per cent of the fresh weight. The proteins are low in the sulphur-containing amino-acids, methionine and cystine. Traces of vitamins A and B are also present. Besides these, some yam species also contain small quantities of alkaloids, for instance dioscorine and steroid derivatives, such as diosgenin and sapogenin. Apart from being a good calorie source, yam is also a significant source of vitamin C, iron, calcium and nicotinic acid (Table 10.2).

**Table 10.1 Composition of various yam species and potatoes**

| Species | Moisture (%) | Carbohydrate (%) | Fat (%) | Crude protein (%) | Crude fibre (%) | Ash (%) |
|---|---|---|---|---|---|---|
| **Yams** | | | | | | |
| *Dioscorea alata* | 65–73 | 22–29 | 0.03–0.27 | 1.12–2.78 | 0.65–1.40 | 0.07–2.06 |
| *Dioscorea cayenensis* | 83 | 15 | 0.05 | 1.02 | 0.40 | 0.53 |
| *Dioscorea rotundata* | 53–73 | 23 | 0.12 | 1.09–1.99 | 0.35–0.79 | 0.63–2.56 |
| *Dioscorea opposita* | 70–80 | 16–29 | 0.06–1.10 | 1.11–3.10 | 0.33–1.00 | 0.59–1.10 |
| *Dioscorea esculenta* | 67–81 | 17–25 | 0.04–0.29 | 1.29–1.87 | 0.18–1.51 | 0.50–1.24 |
| *Dioscorea bulbifera* | 63–67 | 27–33 | 0.04 | 1.12–1.50 | 0.70–0.73 | 1.03–1.51 |
| *Dioscorea dumentorum* | 79 | 17 | 0.28 | 2.78 | 0.30 | 0.72 |
| *Dioscorea hispida* | 78 | 18 | 0.16 | 1.81 | 0.93 | 0.69 |
| *Dioscorea trifida* | — | 38 | 0.44 | 2.54 | — | — |
| **Potatoes** | | | | | | |
| Potato, *Solanum tuberosum* | 68–82 | 14–27 | 0.02–0.18 | 1.14–2.98 | 0.38–0.35 | 0.78–1.16 |
| Sweet potato, *Ipomoea batatas* | 58–81 | 17–43 | 0.18–1.66 | 0.45–4.37 | 0.60–4.54 | 0.66–1.38 |

Source: Coursey, 1967.

**Table 10.2 Nutritional value of 1 kg of yam**

| Constituent | Composition |
|---|---|
| Calories | 4186 Joule |
| Protein | 20 g |
| Calcium | 150 mg |
| Iron | 10 mg |
| Vitamin A | Traces |
| Thiamine | 1 mg |
| Riboflavin | 0.3 mg |
| Nicotinic acid | 4 mg |
| Vitamin C | 50 mg |

Source: Coursey, 1967.

## Storage of yams

A conservative estimate indicates that about 15 per cent of yams produced annually never reach the market, mainly because of post-harvest losses. Losses are due to the lack of appropriate storage facilities, a variety of diseases and pests, and sprouting. Even in storage, losses still occur. Coursey (1965) estimates that over one million tonnes of yam tubers are lost annually in storage in West Africa.

Sources of storage losses are rotting, pests, respiration and sprouting. Sprouting causes reduction in the food reserves by translocating carbohydrates from the tuber into the sprouts for metabolic processes. It also increases the respiration rate, thereby increasing the rate of dry matter loss, and accelerates moisture loss through the permeable surface of the sprouts. The yam, as a result, becomes progressively soft to the touch from the bottom upwards, resulting in rotting.

Sprouting of *D. rotundata* tubers can be delayed by non-lethal doses of gamma irradiation (Adesuyi and Mackenzie, 1973). Similar results were obtained by Rivera *et al.* (1974b), who exposed *D. alata* tubers to various levels of gamma irradiation and recommended 7.5 Krad as the optimum dose to inhibit sprouting. This technique has not yet been applied on a commercial scale.

Rotting causes the greatest amount of loss of dry matter in yam storage. It is due mostly to the effects of fungi and bacteria. Notably, *Serratia* has been implicated (Okafor, 1986). To prevent rotting during storage, wounding of the tubers during harvesting and handling is to be avoided. Curing of yam tubers at 25°C and low humidity (55–62 per cent) for five days before storage, prevents, to a certain degree, storage rot. Loss of stored yams due to attacks by pests (rodents and insects) can be substantial. Respiration is a serious source of storage loss in yams. Since harvested tubers are living, they are respiring, and the substrate for the respiration is the dry matter which is stored in the tubers.

Coursey and Walker (1960) estimate that yams stored for five months may lose over 10 per cent of their dry matter through respiration. One effective way to reduce respiratory loss is to keep the temperature low, but not below 16°C.

Yam is stored for two reasons: to preserve planting material for next year's crop and to preserve yam tubers destined to be used for food. Storage of fresh yam tubers for consumption should not last longer than six months.

## *Traditional methods of yam storage*

Storage of yams by the indigenous peoples of the tropics and subtropics is carried out mostly by traditional methods. The methods employed vary from country to country and from region to region; the variations are often related to climate but local natural resources and customs also influence the choice of storage method used.

*Burying (clamp) or pit storage*

This involves the digging of holes or pits about 1 m deep and 1 m in diameter in the ground and lining them with dried leaves and grasses. After selecting good quality yam tubers, they are placed in the pit, alternating each layer of tubers with dried grasses. They are finally covered with a layer of dried

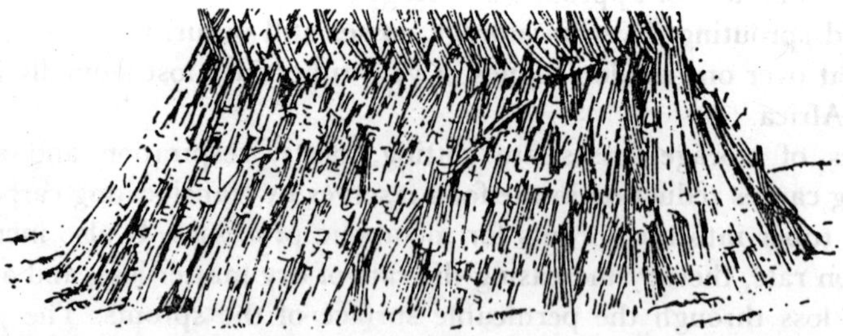

**Figure 10.1 Storage of yams in the field – sorghum stalk 'tent'.**

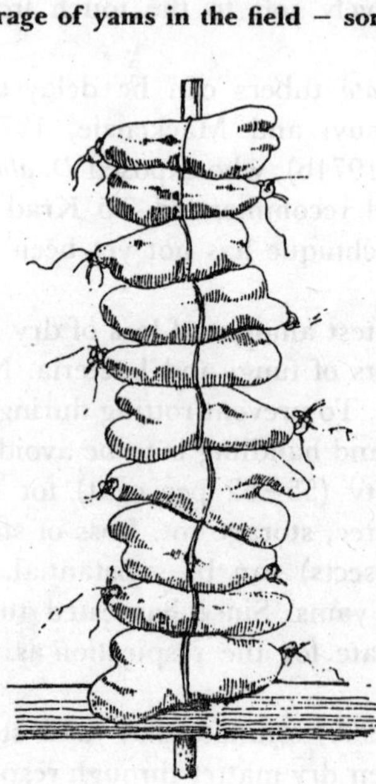

**Figure 10.2 Yams single-tied to a vertical wooden framework.**

grasses and a top layer of soil. This method of yam storage is not very effective due to the lack of sufficient ventilation, and the difficulty of inspecting the tubers.

*Field storage*

Leaving the yams in the field is the simplest method of storage. Tubers are stacked in piles and covered with grass, sorghum or millet stalks. A tent made of sorghum stalks is also used as a yam store (Figure 10.1).

*Barn storage*

In the humid forest areas of West Africa, yams are often kept in barns. Basically barns consist of a vertical wooden framework. To this framework the yam tubers are tied individually by means of tie-vines or strings (Figures 10.2 and 10.3), or they can be laid on simple openwork shelving inside the barn. In some cases the barns are covered with thatched roofs to protect the tubers from rain and the heat of the sun. Placing yams on the shelves requires less time and labour than tying. Shelving also has the advantage that insect-infested and rotting yams can be removed easily.

The disadvantage with shelving lies in the fact that it is easier for rats to hide amongst the yams. Losses due to decay in some cases can amount to 40–50 per cent (Olorunda and Adesuyi, 1973). Losses can be reduced by protecting the barn from rats by fixing a barrier made of iron sheeting around the barn.

*Compound storage*

Outside the forest zone, for example in the savanna regions of West Africa, yams can be stored either in the field, or in the farm compound. Yams stored in the farm compound are stacked in cribs like those used for maize. The cribs are raised well off the ground with rat guards fitted to the legs (Figure 10.4).

**Figure 10.3 Yam tubers tried to a vertical wooden framework in the shade for storage.**

In Papua New Guinea, the so-called Trobriand yam house is widely used (Figure 10.5). The typical Trobriand yam house consists of platforms or 'shelves' of wicker or of light poles laid together, supported on vertical wooden poles a metre or two above ground level. The vertical poles, which support the shelves, also carry a pitched roof of thatch, which protects the yam tubers from the sun and, to a certain extent, from rain. The tubers are placed on top of each other or beside each other on the open shelves. Since individual shelves would normally contain only a few dozen tubers, adequate ventilation is ensured. However, if the structure is not raised adequately above ground level, the tubers may not be sufficiently protected from rodent attack or from the danger of flooding. The Trobriand yam house is especially suitable for storage of the firmer or harder-fleshed yam varieties, which do not bruise easily.

*Cold storage of yam tubers*

The application of low temperature storage to yams is limited by the fact that yams are susceptible to low temperature injury at temperatures of 10–12°C or less (Coursey, 1968). Temperatures of 16–17°C have been successfully employed to prolong the storage time of yams (Gonzalez and Collazo de Rivera, 1972; Rivera *et al.* 1974; Passam, 1977). At 16°C the dormancy and hence the storage life of *D. alata* tubers may be extended by as much as four months, particularly if the tubers are cured prior to storage so as to reduce infection by wound pathogens (Gonzalez and Collzo de Rivera, 1972).

The storage of yam tubers at lowered temperatures has the advantage of reducing the major sources of storage loss (respiration, sprouting and rotting).

**Figure 10.4 Crib for yam storage.**

**Figure 10.5 Typical Trobriand yam house (Gideon, 1975).**

However, the widespread application of cold storage for yams is not yet economically feasible due to the relatively high cost of the technology involved.

## Uses of yam tubers in West Africa

The yam tuber, like other root crops, is essentially a starchy or carbohydrate food; its principal nutritional function being the supply of calories to the body. Today, the total contribution of yams towards feeding the people of West Africa is less than other staples, such as cassava, maize and rice, but the food value of yams is still an important one. The yam crop provides some 18 million tonnes of food for the people of the 'yam zone' in the tropical regions of West Africa.

The yam tuber is prepared for consumption in a variety of ways including boiling, frying and baking. The prepared yam is normally consumed with meat, fish or green vegetables. One of the most important traditional culinary preparations of yam in West Africa is fufu. The term 'fufu' is not confined to products made from yam, but is also used to describe similar doughs prepared from cassava, plantain or mixtures of both.

Apart from its nutritional value, yam plays an important role in social and religious festivals. In fact in the yam-growing areas of West Africa, yam is a vital, integral part of the cultural heritage for many people. For example, the new yam festival marks the earliest date on which new yams may be harvested or eaten. It ensures that the new crop is ushered in formally, and that its consumption does not occur until the community as a whole gives thanks to the gods and celebrates the event. Yams also occupy a place in many traditional marriage ceremonies and in special diets for fetish priests in sacrificial ceremonies.

## Yam processing

### *Boiled, fried or roasted yam*

Small tubers are boiled whole without peeling, for conservation of the vitamin content, or are roasted by placing unpeeled tubers in the ashes of a dying fire among stones previously heated by fire. Pieces of peeled yam can be fried in any convenient oil; palm oil is usually used in West Africa. The ascorbic acid content is reduced using these methods (Table 10.3).

**Table 10.3 Retention of ascorbic acid during boiling, frying or roasting of yams**

| Method of preparation | Ascorbic acid retention (%) |
|---|---|
| Boiling (without peeling) | 95 |
| Frying in palm oil | 93 |
| Roasting or baking | 85 |
| Boiling (after peeling) | 65 |

Source: Coursey, 1967.

### *Pounded yam*

Fufu, or foutou as it is known in parts of French speaking West Africa, is prepared by peeling the tubers, removing the inedible parts, cutting the remainder into pieces and boiling in water until soft. The boiled pieces are then strained, and pounded using a wooden mortar and pestle (Figure 10.6), until the cellular and, to a certain extent, the starch granular nature is destroyed, and a stiff glutinous dough is formed. The pounding process usually takes 15–60 minutes, depending on the quantity. After pounding, the fufu is moulded by hand into spherical shapes, from which pieces are detached by hand to be used to scoop up stew or soup. As an alternative to fufu, boiled yam is sometimes mashed and mixed with palm oil, meat or fish and spices, to form what can be described as yam pottage, which is known in Ghana as 'etoh'.

Generally *D. alata* makes poor pounded yam, while *D. rotundata* and *D. cayenensis* make better pounded yam. These differences in quality of the pounded yam result from differences in the properties of the starch grains in the various types of yam. Dioscorea rotundata is the most viscous of the yams. It also has a high gel strength. It is these properties that make *D. rotundata* the preferred species for the making of pounded yam or fufu, for which a stiff dough is required.

### *Yam flour*

Traditionally, yam flour is prepared by slicing yam tubers to a thickness of about 1 cm, peeling the pieces and then drying them in the sun until the moisture has been reduced to about 10 per cent. In some cases the slices are boiled or parboiled before sun drying, which softens the tissues considerably, and gives a more palatable product. After drying, the pieces are ground in

**Figure 10.6 Pounding yams with a wooden mortar and pestle in Nigeria.**

mortars or milled in corn mills to yield a coarse flour. Jarmal and Montford (1968) suggest the preparation of yam flour from pre-cooked yams, by oven drying in 2 cm layers at 50–70°C for 6–8 hours. The product, with less than 10 per cent moisture, was used to prepare fufu, which was reported to be indistinguishable from the fufu obtained from fresh tubers.

A slightly different method for the production of yam flour of acceptable quality is by oven drying the pre-cooked, grated yam at 50°C for 4 hours. The flour shows no sign of deterioration after storage at 30°C for 6 months. A comparison has been made of air drying 1–2 mm peeled yam slices at 42 ± 0.5°C for 4–6 hours, and freeze drying them at 180 mm Hg from an initial yam temperature of −31°C to a final temperature of 26°C after 6 hours. A loss of vitamin C from the original 5 mg per 100 g, was observed during both methods of drying. An outline of the processes involved in yam flour production is given in Figure 10.7.

A somewhat modified method is described by Kay (1973). Tubers are thoroughly washed, cut into slices (5 cm thick) and cooked for 20–25 minutes, after which they are peeled and mashed into a pulp, which is spread out to a depth of 2 cm and dried for 6–8 hours at 50–70°C to a moisture content of 10 per cent or less. The dried product is then finely ground and sieved to produce yam flour.

Large-scale industrial yam flour production has not been carried out to any appreciable extent due to problems such as consumer acceptability of the end product upon rehydration, availability of yam tubers and adaptability of the most cumbersome processing stage (peeling) to mechanization. However, the National Root Crops Production Company of Nigeria in co-operation with the University of Nigeria, has made great efforts to produce instant yam on an industrial scale. Instant yam is produced by peeling, cubing and sulphiting the cubed yams. These are steamed for a few minutes, then dried and ground mechanically into flour. The economic viability of the system depends largely on the peeling process which is done manually. This affects both the quality of the finished product and the amount of waste (Ihekoronye and Ngoddy, 1985).

*Yam flour in composite flours for bakery*

The partial substitution of wheat flour with yam flour (20 per cent) in breads has been shown to give satisfactory products. With other bakery products, such as pancakes and cupcakes, even higher percentages (50 and even 100 per

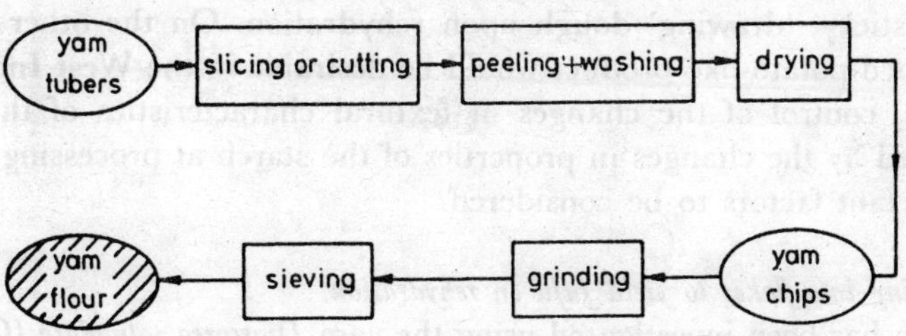


**Figure 10.7 Flowchart for the processing of yam flour.**

cent) are acceptable depending upon the yam variety (Dendy *et al.*, 1970). *Dioscorea alata* cultivars, Forestero and Farm Lisbon, were found to yield superior baking flours.

Dendy, Clarke and James (1970) studied various composite flours, including yam flour. They concluded that in general up to 20 per cent of a non-wheat material may be used; moreover the possibility of using higher levels may be viable if mechanical dough development instead of bulk fermentation is employed. Table 10.4 shows the results of baking tests using various composite flours based upon 15 per cent non-wheat flour. The substitution of wheat flour with yam flour in breads has not been carried out commercially because yam flour is probably still too expensive to compete with wheat or cassava flour.

**Table 10.4 Baking tests with composite flours**

| Flour mixture (15% HRS wheat flour replaced by non-wheat flour) | Moisture absorption on 14% m.b. (%) | Specific loaf volume (ml/100 g flour 14% m.b.) | Crumb compressibility 1/10 mm | | |
|---|---|---|---|---|---|
| | | | 1 day | 2 days | 3 days |
| HRS wheat flour only | 64.2 | 827 | 107 | 74 | 51 |
| HRS + cassava flour | 61.2 | 688 | 72 | 60 | 41 |
| HRS + yam flour (*D. rotundata*) | 60.3 | 740 | 77 | 56 | 53 |
| HRS + yam flour (*D. alata*) | 61.2 | 740 | 72 | 54 | 47 |
| HRS + yam flour (D. cayenensis) | 60.0 | 680 | 56 | 46 | 38 |
| HRS + plantain flour | 59.5 | 710 | 61 | 46 | 38 |
| HRS + konkonte | 60.5 | 705 | 62 | 48 | 42 |
| HRS + gari flour | 59.9 | 615 | 57 | 40 | 35 |
| HRS + sorghum flour | 61.5 | 680 | 75 | 44 | 37 |
| HRS + millet flour | 61.8 | 710 | 86 | 50 | 36 |

HRS = Hard red spring wheat.
Source: Rasper *et al.* (1974).

## Instant yam flakes

Fresh yam is processed into flakes by peeling, washing, cutting the tubers into slices, and cooking them for about 30 minutes. They are then mashed and dried to yield thin flakes. The type of flakes produced will, however, depend upon their intended use. In West Africa the end product must be capable of yielding a sticky, 'drawing' dough upon rehydration. On the other hand, a fluffy, mashed-potato-like product would be desirable in the West Indies. For this reason, control of the changes in textural characteristics of the flakes, accompanied by the changes in properties of the starch at processing, are the most important factors to be considered.

*Yam processing into flakes to yield fufu on rehydration*
Instant fufu has been investigated using the yam *Dioscorea rotundata* (Onayemi and Potter 1974). Pre-cooked, mashed yam was drum dried to yield flakes of

less than 10 per cent moisture content. Upon rehydration, a sticky dough similar to the conventionally prepared pounded yam and superior to the product from traditional yam flour, was obtained. Starch damage was not controlled in the test. The flakes, stored at 37.8°C, retained their acceptability for at least 90 days. Slight colour deterioration was, however, detected in samples with a higher moisture content. A correlation between the cooking time and mashing conditions of drum dried flakes, and the properties (such as stickiness and elasticity) of the fufu prepared from the flakes, was established by Ayernor (1975). Increased cooking time and more intensive mashing, although conferring the greatest starch damage and less rigidity to the resulting dough, were found to render the reconstituted product greater elasticity, a desirable property in pounded yam. Tables 10.5 and 10.6 show the influence of cooking time on the properties of the dried product and the reconstituted dough. A similar investigation – the dependance of stickiness on cooking time (degree of starch damage) – was carried out by Rodriguez-Sosa and Gonzalez (1972) (Table 10.7).

Starch damage is higher with increased cooking time for the yam *Dioscorea alata*, cultivar Florido. The gelatinization temperature of the mashed yam decreases with increased cooking; as a result a greater degree of cell rupture during subsequent drying is achieved as is shown by an increased Blue Value Index, which expresses the amount of free starch. Also, as cooking time is increased, the starch exhibits higher viscosity and greater stickiness on rehydration.

**Table 10.5 Influence of steam pressure (used to cook diced yam for 10 minutes) on flour powder characteristics**

| Steam pressure (psig) | Powder bulk density ($g/cm^3$) | Starch damage BVI | Swelling power ($cm^3/g$ solids) | Solubility ($g/cm^3$) |
|---|---|---|---|---|
| 0 | 0.31±0.05 | 70±2 | 11.7 | 2.4 |
| 5 | 0.30±0.07 | 100±3 | 11.7 | 2.6 |
| 10 | 0.28±0.03 | 110±1 | 13.2 | 3.1 |

BVI = Blue Value Index (amount of free starch).
Source: Ayernor 1976.

**Table 10.6 Rheological properties of yam dough 15 minutes after reconstitution**

| Steam * Pressure (psi) | Modulus of elasticity ($\times 10^5$ dyne/$cm^2$) | Ultimate (yield) deformation | Recoverable elasticity (%) | Energy ratio |
|---|---|---|---|---|
| 0 | 5.2 | 0.50 | 66 | 0.77 |
| 5 | 4.9 | 0.48 | 73 | 0.71 |
| 10 | 4.4 | 0.45 | '1 | 0.55 |

* 1 psi = 6.894 kPa
Source: Ayernor, 1976.

**Table 10.7 Pastiness characteristics of mashed yam, BVI and total solids content of yam flakes**

| Cooking period (min) | Mashed yam | | | Yam flakes | |
|---|---|---|---|---|---|
| | Gelatinization temperature (°C) | Viscosity (Brabender units) | Viscosity after 40 minutes cooking at 90°C (Brabender units) | BVI | Total solids (%) |
| 5 | 66.4 | 137 | 248 | 200 | 94.11 |
| 10 | 63.0 | 237 | 237 | 239 | 93.99 |
| 15 | 60.3 | 278 | 197 | 273 | 94.15 |
| 20 | 59.5 | 295 | 177 | 244 | 93.12 |

Source: Rodrigue Sosa and Gonzales.

*Yam flakes on rehydration*

Cell rupture – release of free starch and therefore stickiness – is an undesirable effect in the preparation of instant yam flakes meant for the West Indies regions. A method of dehydrating yam to instant flakes was investigated by Gooding (1972), in which excessive stickiness in the product was prevented.

Cooked, mashed yam was drum-dried and the degree of starch damage, as measured by the quantity of free starch – by determining the Blue Value of the mash – was monitored at all stages of the process. Cooking for up to 30 minutes was found to have little effect on the extent of cell rupture. Mashing intensity, on the other hand, had considerable influence on the stickiness. Gentle mashing by means of a fork gave almost no cell breakage whilst a mechanical mixer or a waring blender produced a very sticky mash with a large number of broken cells and a high Blue Value.

With temperatures up to 140°C in the drum drier, the cells remained intact, but at 150°C and above almost all cells were ruptured. The study recommended ricing of the cooked yam followed by gentle mashing in a mechanical mixer.

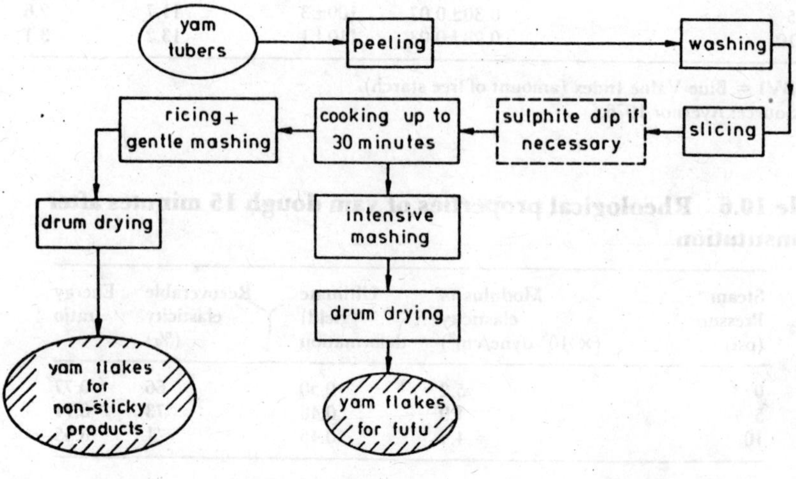


**Figure 10.8 Flowchart for the production of yam flakes.**

The addition of glyceryl monostearate at 0.5 per cent of the dry matter in the mash also reduced the unwanted stickiness. A general flow sheet for instant yam production is presented in Figure 10.8.

## Non-food uses of yam

Several species of yam contain small amounts of sapogenins and alkaloids. Particularly, the alkaloid dioscorine ($C_{13}H_{19}O_2N$) and the *Dioscorea* saponins and sapogenins occur in varying quantities in different species of yam. Saponin is usually hydrolyzed to give sapogenin (Figure 10.9). The major sapogenin present in yams is diosgenin.

All the important yam sapogenins are steroids in character and are closely related to cortisone. They are utilized as raw material for the manufacture of corticosteroid drugs in the pharmaceutical industry. Corticosteroids are used as anti-inflammatory agents, metabolic stimulants, and as general stress-reaction protectives.

Commercial utilization of yams for diosgenin production takes place particularly in Mexico, where *D. mexicana*, *D. floribunda* and *D. composita* are grown mainly for medicinal purposes. The alkaloids present in yams include dioscorine in *D. hispida*, and dihydrodioscorine in *D. dumetorum*. Dioscorine is toxic, causing general paralysis of the central nervous system in test animals, while saponins are fish poisons, haemolytic and surface active agents.

### *Preparation of yam sapogenins ($C_{27}H_{42}O_3$)*

Two main procedures are possible for the production of diosgenin or similar compounds:

1. Alcoholic extraction of the sapogenin, acid hydrolysis, and isolation of the sapogenin by filtration or solvent extraction.
2. Acid hydrolysis of the total plant material and solvent extraction of the sapogenin.

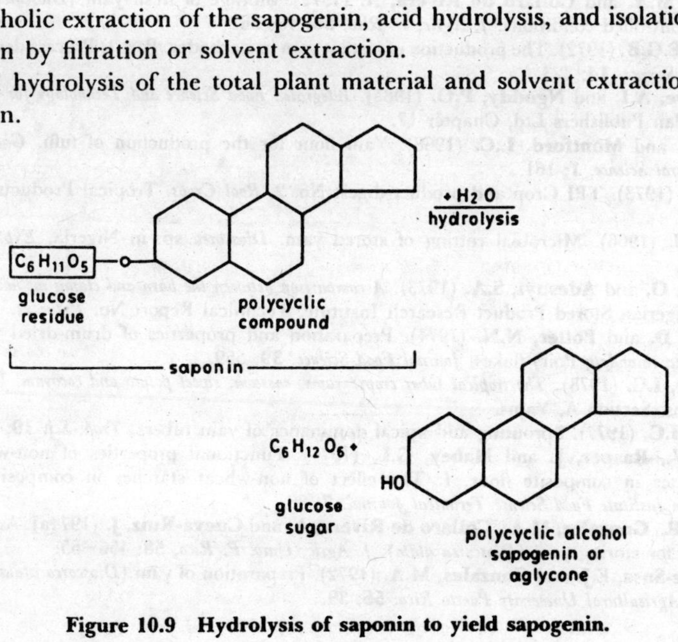


**Figure 10.9 Hydrolysis of saponin to yield sapogenin.**

In either case the tubers are chopped, air dried and then finely ground. In the first procedure initial extracting is by boiling methyl or ethyl alcohol; after filtration the alcoholic extract is concentrated to a syrup, which is hydrolyzed by heating with mineral acid (12−4 N) for 2−12 hours. Diosgenin eventually precipitates and is separated by filtration and washed with water; it may be purified by recrystallization from an organic solvent.

In the second method the dried ground material is refluxed or autoclaved with mineral acid (2−4 N) for 2−6 hours. After filtration the press cake is washed with water or dilute alkali until neutral and dried; it is then extracted with a hydrocarbon solvent (such as a petroleum heptane), from which the sapogenin separates on cooking. This approach avoids the formation of gums during alcoholic hydrolysis and yields a cleaner product.

## *References*


**Adesuyi, S.A.** and **Mackenzie, J.A.** (1973). The inhibition of sprouting in stored yams, *Dioscorea rotundata*, by gamma radiation and chemicals. Radiation preservation of food. *Int. Atom. Energy Ag. Bull.* 127−36.

**Ayernor, G.S.** (1976). Particulate properties and rheology of pregelled yam (*Dioscorea rotundata*) products. *Journal Food Science*, **14**: 180.

**Coursey, D.G.** (1965). The role of yam in West Africa food economics. *World Crops*. **17**: 74.

**Coursey, D.G.** (1967). *Yams*. Tropical Agriculture Series, London, Longmans.

**Coursey, D.G.** (1968). Low temperature injury in yams. *Food Technology*, **3**: 143−50.

**Coursey, D.G.** and **Walker, H.M.** (1960). *A study of the origins of weight loss in stored yams*. Rep. W. Afr. Stored Prod. Res. Unit. 61−64.

**Dendy, D.A.V., Clarke, P.A.** and **James, A.W.** (1970). The use of Blends of Wheat and Non-Wheat Flours in Bread Making. *Tropical Science*. **12**: 131.

**FAO** (1984). *FAO Production Yearbook*, Vol. 38, Rome.

**Gideon, R.** (1975). Preservation of yam, in *Papua New Guinea Agriculture and Resource Technology*, Vudal Agricultural College.

**Gonzalez, M.A.** and **Collaro de Rivera, A.** (1972). Storage of fresh yam (*Dioscorea alata* L.) under controlled conditions. *J. Agric. P. Rico*. **56**: 46−56.

**Gooding, E.G.B.** (1972). The production of instant yam in Barbados. Part 1: Process development. *Tropical Science*. **14**: 323.

**Ihekoronye, A.I.** and **Ngoddy, P.O.** (1985). *Integrated Food Science and Technology for the Tropics*, Macmillan Publishers Ltd, Chapter 17.

**Jarmai, S.** and **Montford, L.C.** (1968). Yam flour for the production of fufu. *Ghana Journal Agricultural Science*, **1**: 161.

**Kay, D.E.** (1973). TPI Crop and product digest No. 2, *Root Crops*. Tropical Products Institute, London.

**Okafor, N.** (1966). Microbial rotting of stored yam, *Dioscorea* sp. in Nigeria, *Expt. Agric.* **2**: 179−82.

**Olorunda, O.** and **Adesuyi, S.A.** (1973). *A comparison between the barn and clamp methods of storing yams*. Nigerian Stored Product Research Institute, Technical Report No. 757−59.

**Onayemi, D.** and **Potter, N.N.** (1974). Preparation and properties of drum-dried white yam (*Dioscorea rotundata* Poir) flakes. *Journal Food Science*. **39**: 559.

**Onwueme, I.C.** (1978). *The tropical tuber crops: yams, cassava, sweet potato and cocoyam*. John Wiley and Sons. Section A, Yams.

**Passam, H.C.** (1977). Sprouting and apical dominance of yam tubers. *Trop. Sci.* **19**: 29−39.

**Rasper, V., Rasper, J.** and **Mabey, G.L.** (1974). Functional properties of non-wheat flour substitutes in composite flour. I. The effect of non-wheat starches in composite doughs. *Canadian Institute Food Science Technical Journal*. **7**: 86.

**Rivera, J.R., Gonzalez, M.A., Collaro de Rivera, A.** and **Cueva-Ruiz, J.** (1974a). An improved method for storing yam (*Dioscorea alata*). *J. Agric. Univ. P. Rico*, **58**: 456−65.

**Rodriguez-Sosa, E.J.** and **Gonzales, M.A.** (1972). Preparation of yam (*Dioscorea alata* L.) flakes. *Journal Agricultural University Puerto Rico*: **56**: 39.

# Index